# Structure Preserving Energy Functions in Power Systems

## THEORY AND APPLICATIONS

# Structure Preserving Energy Functions in Power Systems

## THEORY AND APPLICATIONS

K. R. Padiyar

CRC Press
Taylor & Francis Group
Boca Raton London New York

CRC Press is an imprint of the
Taylor & Francis Group, an **informa** business

CRC Press
Taylor & Francis Group
6000 Broken Sound Parkway NW, Suite 300
Boca Raton, FL 33487-2742

First issued in paperback 2017

© 2013 by Taylor & Francis Group, LLC
CRC Press is an imprint of Taylor & Francis Group, an Informa business

No claim to original U.S. Government works

ISBN-13: 978-1-4398-7936-8 (hbk)
ISBN-13: 978-1-138-07771-3 (pbk)

---

**Library of Congress Cataloging-in-Publication Data**

---

Padiyar, K. R.
  Structure preserving energy functions in power systems : theory and applications / K.R. Padiyar.
       pages cm
  "A CRC title, part of the Taylor & Francis imprint, a member of the Taylor & Francis Group, the academic division of T&F Informa plc."
  Includes bibliographical references and index.
  ISBN 978-1-4398-7936-8 (hardcover : acid-free paper)
  1. Electric power systems--Control. 2. Electric power systems--Control--Mathematical models. 3. Electric power system stability. 4. Electric power systems--Security measures. 5. Synchronization. 6. Damping (Mechanics) I. Title.

TK1007.P33 2013
621.31'7--dc23                                                                                    2012045698

---

**Visit the Taylor & Francis Web site at**
**http://www.taylorandfrancis.com**

**and the CRC Press Web site at**
**http://www.crcpress.com**

# Contents

Preface ............................................................................................................. xiii
Acknowledgments ......................................................................................... xvii
Author ............................................................................................................. xix
Abbreviations and Acronyms ..................................................................... xxi

**1. Introduction** ................................................................................................. 1
   1.1    General ................................................................................................. 1
   1.2    Power System Stability ..................................................................... 2
   1.3    Power System Security ...................................................................... 3
   1.4    Monitoring and Enhancing System Security .................................. 6
   1.5    Emergency Control and System Protection .................................... 7
   1.6    Application of Energy Functions ..................................................... 8
   1.7    Scope of This Book ........................................................................... 14

**2. Review of Direct Methods for Transient Stability Evaluations
for Systems with Simplified Models** ........................................................ 17
   2.1    Introduction ..................................................................................... 17
   2.2    System Model ................................................................................... 18
         2.2.1   Synchronous Generators ................................................... 18
         2.2.2   Network Equations ............................................................ 21
         2.2.3   Load Model .......................................................................... 22
         2.2.4   Expressions for Electrical Power ..................................... 23
   2.3    Mathematical Preliminaries .......................................................... 25
         2.3.1   Equilibrium Points ............................................................ 26
         2.3.2   Stability of Equilibrium Point .......................................... 27
         2.3.3   Lyapunov Stability ............................................................ 27
         2.3.4   Theorem on Lyapunov Stability ...................................... 27
   2.4    Two-Machine System and Equal Area Criterion .......................... 30
         2.4.1   Equal Area Criterion ......................................................... 31
         2.4.2   Energy Function Analysis of an SMIB System ............... 32
   2.5    Lyapunov Functions for Direct Stability Evaluation .................... 34
         2.5.1   Construction of Lyapunov Function ................................ 38
   2.6    Energy Functions for Multimachine Power Systems .................... 39
         2.6.1   Characterization of Transient Stability .......................... 39
         2.6.2   Center of Inertia Formulations ........................................ 40
         2.6.3   Energy Function Using COI Formulation ....................... 43
   2.7    Estimation of Stability Domain ..................................................... 44
         2.7.1   Incorporating Transfer Conductances
                  in Energy Function .............................................................. 44

2.7.2    Determination of Critical Energy ...................................... 46
     2.7.2.1    Single-Machine System ............................... 46
     2.7.2.2    Multimachine System................................. 48
2.7.3    Potential Energy Boundary Surface...................... 48
2.7.4    Controlling UEP Method...................................... 50
2.7.5    BCU Method ....................................................... 51
2.8    Extended Equal Area Criterion................................... 53
    2.8.1    Formulation ....................................................... 53
    2.8.2    Approximation of Faulted Trajectory................ 54
    2.8.3    Identification of Critical Cluster ..................... 55

**3. Structure Preserving Energy Functions for Systems with
Nonlinear Load Models and Generator Flux Decay**........................... 57
3.1    Introduction ................................................................. 57
3.2    Structure Preserving Model....................................... 57
3.3    Inclusion of Voltage-Dependent Power Loads.............. 61
3.4    SPEF with Voltage-Dependent Load Models ................ 62
    3.4.1    Dynamic Equations of Generator........................ 62
    3.4.2    Load Model........................................................ 63
    3.4.3    Power Flow Equations........................................ 64
    3.4.4    Structure Preserving Energy Functions............. 64
    3.4.5    Computation of Stability Region ....................... 68
3.5    Case Studies on IEEE Test Systems ............................ 69
    3.5.1    Seventeen-Generator System.............................. 70
    3.5.2    Fifty-Generator System ..................................... 74
3.6    Solution of System Equations during a Transient .......... 76
3.7    Noniterative Solution of Networks with
Nonlinear Loads............................................................ 77
    3.7.1    System Equations............................................... 78
    3.7.2    Dynamic Equations of Generators ..................... 78
    3.7.3    Power Flow Equations during a Transient.......... 79
    3.7.4    Special Cases ..................................................... 81
    3.7.5    Solutions of the Quartic Equation ..................... 82
    3.7.6    Network Transformation for Decoupling
of Load Buses.................................................... 83
    3.7.7    Transformation of the Load Characteristics ...... 84
3.8    Inclusion of Transmission Losses in Energy Function.... 85
    3.8.1    Transformation of a Lossy Network ................... 85
    3.8.2    Structure Preserving Energy Function Incorporating
Transmission Line Resistances ........................... 87
3.9    SPEF for Systems with Generator Flux Decay .............. 90
    3.9.1    System Model ................................................... 90
      3.9.1.1    Generator Model................................. 91
      3.9.1.2    Load Model ....................................... 92
      3.9.1.3    Power Flow Equations ........................ 92

3.9.2    Structure Preserving Energy Function...............................93
3.9.3    Example.................................................................................96
3.10   Network Analogy for System Stability Analysis ......................97

**4. Structure Preserving Energy Functions for Systems
with Detailed Generator and Load Models** .......................................... 105
4.1    Introduction ................................................................................. 105
4.2    System Model................................................................................ 106
        4.2.1    Generator Model .............................................................. 106
        4.2.2    Excitation System Model................................................. 107
        4.2.3    Load Model....................................................................... 108
        4.2.4    Power Flow Equations..................................................... 108
4.3    Structure Preserving Energy Function with
        Detailed Generator Models ...................................................... 109
        4.3.1    Structure Preserving Energy Function............................ 109
        4.3.2    Simpler Expression for SPEF .......................................... 112
4.4    Numerical Examples .................................................................. 114
        4.4.1    SMIB System ..................................................................... 114
        4.4.2    Ten-Generator, 39-Bus New England Test System ......... 115
        4.4.3    Variation of Total Energy and Its Components .............. 121
4.5    Modeling of Dynamic Loads....................................................... 122
        4.5.1    Induction Motor Model.................................................... 124
        4.5.2    Voltage Instability in Induction Motors......................... 126
        4.5.3    Simpler Models of Induction Motors .............................. 128
        4.5.4    Energy Function Analysis of Synchronous
                   and Voltage Stability.......................................................... 128
                   4.5.4.1    Computation of Equilibrium Points ................. 130
                   4.5.4.2    Computation of Energy at UEP .......................... 132
        4.5.5    Dynamic Load Models in Multimachine Power
                   Systems............................................................................... 135
4.6    New Results on SPEF Based on Network Analogy ...................... 136
        4.6.1    Potential Energy Contributed by Considering the
                   Two-Axis Model of the Synchronous Generator............. 140
4.7    Unstable Modes and Parametric Resonance................................ 144
        4.7.1    Normal Forms ................................................................... 145
        4.7.2    Fast Fourier Transform of Potential Energy.................... 146
                   4.7.2.1    Results of the Case Study .................................. 146

**5. Structure Preserving Energy Functions for Systems
with HVDC and FACTS Controllers** ..................................................... 149
5.1    Introduction ................................................................................. 149
5.2    HVDC Power Transmission Links ............................................... 149
        5.2.1    HVDC Systems and Energy Functions ............................ 149
        5.2.2    HVDC System Model ......................................................... 150
                   5.2.2.1    Converter Model................................................ 150

               5.2.2.2   DC Network Equations ........................................ 152

               5.2.2.3   Converter Control Model ..................................... 152

       5.2.3   AC System Model........................................................... 155

               5.2.3.1   Generator Model........................................... 155

               5.2.3.2   Load Model ................................................... 156

               5.2.3.3   AC Network Equations.................................. 156

       5.2.4   Structure Preserving Energy Function....................... 156

       5.2.5   Example.......................................................................... 160

               5.2.5.1   Auxiliary Controller ..................................... 161

               5.2.5.2   Emergency Controller.................................... 162

               5.2.5.3   Case Study and Results ................................. 162

5.3   Static Var Compensator................................................................. 163

       5.3.1   Description..................................................................... 163

       5.3.2   Control Characteristics and Modeling of SVC
               Controller ...................................................................... 164

       5.3.3   Network Solution with SVC: Application of
               Compensation Theorem....................................... 166

               5.3.3.1   Calculation of $\phi_{\text{SVC}}$ in Control Region.............. 167

               5.3.3.2   Network Solution ....................................... 168

       5.3.4   Potential Energy Function for SVC ................................. 169

       5.3.5   Example.......................................................................... 171

       5.3.6   Case Study of New England Test System....................... 172

               5.3.6.1   Network Calculation with Multiple SVCs........ 173

               5.3.6.2   Structure Preserving Energy Function.............. 174

               5.3.6.3   Results and Discussion.................................... 175

5.4   Static Synchronous Compensator................................................. 175

       5.4.1   General .......................................................................... 175

       5.4.2   Modeling of a STATCOM .............................................. 176

       5.4.3   STATCOM Controller .................................................... 178

       5.4.4   Potential Energy Function for a STATCOM..................... 180

5.5   Series-Connected FACTS Controllers ............................................ 181

       5.5.1   Thyristor-Controlled Series Capacitor............................ 182

               5.5.1.1   Power Scheduling Control ............................. 182

               5.5.1.2   Power Swing Damping Control ......................... 183

               5.5.1.3   Transient Stability Control............................... 183

       5.5.2   Static Synchronous Series Compensator ......................... 184

5.6   Potential Energy in a Line with Series
       FACTS Controllers........................................................................ 185

       5.6.1   Thyristor-Controlled Series Capacitor............................ 186

       5.6.2   Static Synchronous Series Compensator ......................... 187

       5.6.3   Potential Energy in the Presence of CC and CA
               Controllers ................................................................... 188

               5.6.3.1   Potential Energy with CC Control.................... 188

               5.6.3.2   Potential Energy with CA Control.................... 189

5.7   Unified Power Flow Controller ..................................................... 189

     5.7.1   Description.................................................................... 189
     5.7.2   Energy Function with Unified Power
             Flow Controller ........................................................ 191

**6. Detection of Instability Based on Identification
   of Critical Cutsets**................................................................. 195
  6.1   Introduction.................................................................. 195
  6.2   Basic Concepts ............................................................. 196
  6.3   Prediction of the Critical Cutset.................................... 198
     6.3.1   Analysis............................................................... 198
     6.3.2   Case Study .......................................................... 203
     6.3.3   Discussion........................................................... 203
  6.4   Detection of Instability by Monitoring Critical Cutset.............. 203
     6.4.1   Criterion for Instability ....................................... 204
     6.4.2   Modification of the Instability Criterion ........... 206
  6.5   Algorithm for Identification of Critical Cutset ............ 207
  6.6   Prediction of Instability ............................................... 209
  6.7   Case Studies.................................................................. 210
     6.7.1   Ten-Generator New England Test System........... 210
     6.7.2   Seventeen-Generator IEEE Test System ............. 213
     6.7.3   Discussion........................................................... 214
  6.8   Study of a Practical System........................................... 215
     6.8.1   Discussion........................................................... 217
  6.9   Adaptive System Protection ......................................... 219
     6.9.1   Discussion........................................................... 225

**7. Sensitivity Analysis for Dynamic Security and Preventive
   Control Using Damping Controllers Based on FACTS** ...... 227
  7.1   Introduction.................................................................. 227
  7.2   Basic Concepts in Sensitivity Analysis ......................... 228
  7.3   Dynamic Security Assessment Based on
       Energy Margin .............................................................. 229
     7.3.1   Transient Energy Margin..................................... 229
     7.3.2   Computation of Energy Margin .......................... 229
           7.3.2.1   Evaluation of Path-Dependent Integrals.......... 231
           7.3.2.2   Computation of Energy Margin Based
                   on Critical Cutsets ................................. 231
  7.4   Energy Margin Sensitivity............................................. 232
     7.4.1   Application to Structure Preserving Model.............. 232
  7.5   Trajectory Sensitivity .................................................... 235
     7.5.1   Sensitivity to Initial Condition Variations .......... 235
     7.5.2   Discussion........................................................... 236
  7.6   Energy Function-Based Design of Damping Controllers ......... 236
     7.6.1   Series FACTS Controllers..................................... 237
           7.6.1.1   Linearized System Equations.............................. 238

          7.6.1.2  Synthesis of the Control Signal ......................... 242
          7.6.1.3  Case Study of a 10-Machine System ................. 244
     7.6.2  Shunt FACTS Controllers ...................................... 245
          7.6.2.1  Linear Network Model for
                   Reactive Current ....................................... 247
  7.7  Damping Controllers for UPFC ......................................... 251
     7.7.1  Discussion ........................................................ 254

**8. Application of FACTS Controllers for Emergency Control—I** ......... 255
  8.1  Introduction ................................................................... 255
  8.2  Basic Concepts ............................................................... 256
  8.3  Switched Series Compensation ......................................... 257
     8.3.1  Time-Optimal Control ........................................ 257
  8.4  Control Strategy for a Two-Machine System ..................... 260
  8.5  Comparative Study of TCSC and SSSC ............................. 264
     8.5.1  Control Strategy ................................................ 265
  8.6  Discrete Control of STATCOM ........................................ 269
  8.7  Discrete Control of UPFC ............................................... 272
     8.7.1  Application of Control Strategy to SMIB System
              with UPFC ...................................................... 278
     8.7.2  Discussion ........................................................ 278
  8.8  Improvement of Transient Stability by Static Phase-Shifting
       Transformer .................................................................. 280
  8.9  Emergency Control Measures ........................................... 282
     8.9.1  Controlled System Separation and
              Load Shedding ................................................. 283
     8.9.2  Generator Tripping ............................................ 284

**9. Application of FACTS Controllers for Emergency Control—II** ........ 285
  9.1  Introduction ................................................................... 285
  9.2  Discrete Control Strategy ................................................ 285
  9.3  Case Study I: Application of TCSC .................................... 289
     9.3.1  Fault at Bus 26: Without Line Tripping ................. 289
     9.3.2  Fault at Bus 26: Cleared by Line Tripping ............. 292
  9.4  Case Study II: Application of UPFC .................................. 292
     9.4.1  Single UPFC ..................................................... 293
     9.4.2  Multiple UPFC .................................................. 298
     9.4.3  Practical Implementation .................................... 299
  9.5  Discussion and Directions for Further Research ................. 302

References .................................................................................................... 305

Appendix A: Synchronous Generator Model .............................................. 315

Appendix B: Boundary of Stability Region: Theoretical Results............ 327

Appendix C: Network Solution for Transient Stability Analysis ........... 331

Appendix D: Data for 10-Generator System ................................................. 341

Index ............................................................................................................... 345

# *Preface*

Electrification was considered the greatest engineering achievement of the twentieth century. However, at the turn of the century, the electric power industry is undergoing momentous changes in the organizational structure that drastically affects the power system operation. The economic and environmental constraints restrict the choices available to the system planners to meet the demand requirements. The load growth is also accompanied by high expectations on the performance and quality of power supply. The role of renewable energy (wind and solar) is expected to grow while reducing the contribution of nuclear- and fossil fuel-based generation.

The great blackout in 1965 in the United States led to the development of computer-based modern energy management (or control) centers with emphasis on system security. State estimation, short term load forecasting, on-line power flow, security monitoring and assessment are functions combined with economic dispatch and automatic generation control (AGC). The objective was to implement preventive control to minimize the system transition to emergency state. However, these developments did not incorporate system dynamics and thus did not prevent the occurrence of major power blackouts in 1996 and 2003 in North America. Europe also witnessed similar blackouts in 2003.

The research on direct methods of transient stability evaluation based on Lyapunov functions commenced in the 1960s. One could argue that the equal area criterion was the first application of direct method for two- or one-machine systems. By the early 1980s, the development of energy functions (Lyapunov-like) and application of PEBS and controlling UEP methods for the computation of the stability region (surrounding the stable equilibrium point) for a specified major disturbance (such as a three-phase fault followed by a line tripping) created an interest among utility engineers to apply these techniques to determine stability limits. Some utilities in the United States, Canada, and Japan have applied the available techniques to estimate the limits on generation and interface flows in interconnected power systems. However, major limitations were the use of simplified system models and reduced network representation. Although some attempts were made to address these issues, there was not much progress. The state of the art in direct methods of stability analysis (till the 1990s) is reported in books by Pai (1989), Fouad and Vittal (1992), and Pavella and Murthy (1994).

The concept of structure preserving model (SPM) retaining the network structure in defining energy functions was proposed by Bergen and Hill (1981). This was a major departure in the application of direct methods. The energy functions based on SPM were called "topological" or "structure preserving" energy functions (SPEF) to distinguish them from transient energy function defined on reduced network model (RNM). The SPEF expanded the scope of the application of energy functions to systems with detailed generator and load models, and inclusion of network controllers such as HVDC and FACTS controllers. Although it would appear that the complexity is increased by retaining the network structure, sparsity-oriented computational structure is more efficient and has been employed in developing computer programs for power flow and stability analysis (by simulation) for very large systems.

The present trend is to develop technological solutions for improving security in power system operation. A major development is the wide area measurement system (WAMS) that enables real-time measurements of phasors, based on GPS technology. Actually, one can generalize the concept to include simultaneous measurements of several quantities at different locations (such as bus frequencies and line flows) in a large system that can be processed locally and/or at a central location. This approach can be extended to systems that can be divided into coherent areas (which can be defined as areas having coherent generators and coherent buses with similar variations in frequency) interconnected by tie lines. There is an urgent need for reliable and effective analytical techniques to implement real-time processing of the measurements to detect emergencies and initiate automated control actions using available network controllers to achieve "self-healing" grids. Transmission planning must incorporate these ideas to optimize system security. With increasing renewable energy generation, it is anticipated that efficient and cost-effective energy storage technologies will be introduced that will add a new dimension to the system control.

This book is aimed at presenting some analytical techniques, based on SPEF, for (a) on-line detection of loss of synchronism and suggesting adaptive system protection, (b) design of effective linear damping controllers using FACTS, for damping small oscillations during normal operation to prevent transition to emergency states, and (c) emergency control, based on FACTS, to improve first swing stability and also provide rapid damping of nonlinear oscillations that threaten system security during major disturbances. The detection and control algorithms are derived from theoretical considerations and illustrated through several examples and case studies on test systems.

This book has nine chapters and four appendices. The first chapter introduces the basic issues affecting secure system operation. The second chapter reviews the earlier literature on direct methods using RNM. The third chapter introduces SPM and energy functions defined for simpler generator models and nonlinear load models. It also presents techniques for efficient computations when the loads are unconnected and modeled by second-degree polynomials. It is also possible to transform the network in a coherent area such that the transformed network has unconnected load buses. The chapter also presents work on handling of line resistances (in formulating SPEF). A major contribution is the development of a network analogy for lossless power systems. The analogous network contains nonlinear inductors and linear capacitors (representing machine inertias). The fourth chapter extends the analysis to systems with detailed generator models and dynamic loads. The network analogy is utilized to derive new results that are used in later chapters for on-line detection and devising control strategies. The fifth chapter presents SPEF, incorporating HVDC and FACTS controllers.

Chapter 6 presents a technique for on-line detection of loss of synchronism, from identification of the critical cutsets. This is based on the proposition that there is a unique cutset of series elements (transmission lines and transformers) that separates the group of advanced generators from the rest when a major disturbance occurs. The theory is tried on three test systems, including a large practical system. The chapter also discusses controlled system separation based on adaptive system protection.

Chapter 7 reviews the sensitivity analysis and its application for dynamic security assessment and the determination of the location of FACTS controllers (both series and shunt). The concept of preventive control by linear damping controllers based on FACTS is proposed. The locations and control law are obtained from energy concepts using SPM. Chapters 8 and 9 present a new algorithm for stabilizing the system (using FACTS controllers such as TCSC, SSSC, and UPFC) following a large disturbance by first ensuring stability in the first swing, followed by damping the nonlinear oscillations to rapidly steer the system to the postdisturbance equilibrium state whose stability can be improved by the linear damping controllers discussed in Chapter 7. The emergency control is discrete in the sense that it is initiated only when a large disturbance is detected and discontinued when the system reaches close to the equilibrium state. To summarize, a new paradigm of dynamic preventive and emergency control concepts is elucidated in Chapters 7 through 9 with examples and case studies.

MATLAB® is a registered trademark of The MathWorks, Inc. For product information, please contact:

The MathWorks, Inc.
3 Apple Hill Drive
Natick, MA 01760-2098 USA
Tel: 508 647 7000
Fax: 508-647-7001
E-mail: info@mathworks.com
Web: www.mathworks.com

# Acknowledgments

It is a pleasure to acknowledge the contributions of several individuals over the years, which helped me in pursuing research on the theory and applications of SPEF and writing this book. First, I wish to acknowledge professor M.A. Pai, who introduced me to the topic of direct methods for stability evaluation when he was a visiting professor at the University of Waterloo in the early 1970s. Subsequently, he encouraged me to pursue research on SPEF in the 1980s. I also wish to thank professors Pravin Varaiya and Felix Wu for fruitful discussions during my visits to the University of California, Berkeley during the summers of 1982 and 1983. The research reported in this book is based on the work done by my PhD students—Drs. H.S.Y. Sastry, K.K. Ghosh (at IIT Kanpur), Vijayan Immanuel, K. Uma Rao, Anil Kulkarni, and S. Krishna (at I.I.Sc.). The results presented in Chapter 6 are taken from Dr. Krishna's thesis, and the results on TCSC control in Chapters 8 and 9 are taken from the thesis of Dr. Uma Rao. The results on STATCOM and UPFC in Chapters 8 and 9 are taken from Dr. Krishna's thesis. Dr. H.V. Saikumar provided results on the decoupled damping controllers for a UPFC reported in Chapter 7. Saichand and Anshu Verma, ME students at I.I.Sc., have helped in the preparation of some of the figures in the book. Mrs Sudha Aithal has typed the manuscript with painstaking care and also helped in preparing some figures.

I would like to thank Dr. Gagandeep Singh, Senior Editor and Ms Jennifer Ahringer, Project Coordinator for their cooperation and support in the publication of this book.

I am grateful to the Indian Institute of Science for providing facilities to pursue my academic activities as an emeritus professor.

Finally, I wish to thank my wife Usha for her patience, understanding, and support during the long period of work on this book.

I dedicate this book to the power engineers who worked to do the impossible—make electrical energy affordable to every citizen and yet improve reliability of supply in spite of the growing demand.

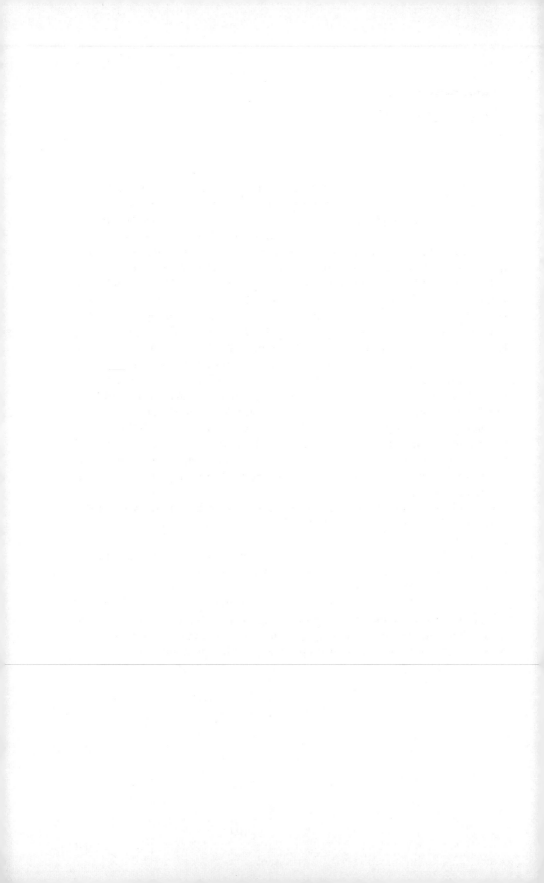

# Author

**Professor K. R. Padiyar** is with the Indian Institute of Science, Bangalore, India since 1987, where he is presently an emeritus professor in the department of electrical engineering. Previously, he was with the Indian Institute of Technology, Kanpur, India from 1976 to 1987, where he became a professor in 1980. He obtained his BE degree from University of Pune, Pune, India in 1962, ME degree from the Indian Institute of Science in 1964, and PhD degree from the University of Waterloo, Canada in 1972.

His research interests are in the areas of power system dynamics and control, HVDC power transmission, and FACTS controllers. He has authored over 200 papers and five books, including *HVDC Power Transmission Systems, Power System Dynamics, Analysis of Subsynchronous Resonance in Power Systems,* and *FACTS Controllers in Power Transmission and Distribution.* He is a fellow of the Indian National Academy of Engineering and life senior member of IEEE. He was awarded the Department of Power Prize twice by the Institution of Engineers (India). He is the recipient of the 1999 Professor Rustom Choksi Award for Excellence in Research in Engineering. He was ABB chair professor during 2001–2003.

# Abbreviations and Acronyms

| | |
|---|---|
| AC | Alternating current |
| AVR | Automatic voltage regulator |
| BCU | Boundary (of stability region based) controlling unstable (equilibrium point) |
| BTB | Back-to-back (HVDC link) |
| CA | Constant angle (control) |
| CC | Constant current (control) |
| CCT | Critical clearing time |
| CEA | Constant extinction angle (control) |
| CUEP | Controlling UEP |
| CVD | Constant voltage drop (control) |
| DAE | Differential Algebraic Equations |
| DC | Direct current |
| EAC | Equal area criterion |
| EEAC | Extended equal area criterion |
| EM | Energy margin |
| EP | Equilibrium point |
| EPC | Equidistant pulse control |
| FACTS | Flexible AC transmission system |
| GTO | Gate turn-off (thyristor) |
| HB | Hopf bifurcation |
| HVDC | High-voltage direct current |
| IGBT | Insulated gate bipolar transistor |
| IPC | Individual phase control |
| IPFC | Interline power flow controller |
| IRCFN | Incremental reactive current flow network |
| KE | Kinetic energy |
| MOI | Mode of instability |
| MTDC | Multiterminal direct current |
| NLF | Normalized location factor |
| PE | Potential energy |
| PEBS | Potential energy boundary surface |
| PMU | Phasor measurement unit |
| PSDC | Power swing damping control |
| PSS | Power system stabilizer |
| PST | Phase shifting transformer |
| PWM | Pulse width modulation |
| RCFN | Reactive current flow network |
| rms | root mean square |
| RNM | Reduced network model |

| | |
|---|---|
| SCR | Short circuit ratio |
| SEP | Stable equilibrium point |
| SIME | Single machine equivalent |
| SM | Swing mode |
| SMC | Supplementary modulation controller |
| SMIB | Single machine infinite bus |
| SNB | Saddle node bifurcation |
| SPEF | Structure preserving energy function |
| SPM | Structure preserving model |
| SPST | Static phase shifting transformer |
| SSR | Subsynchronous resonance |
| SSSC | Static synchronous series compensator |
| STATCOM | Static (synchronous) compensator |
| SVC | Static var compensator |
| TC | Transfer conductance |
| TCSC | Thyristor-controlled series capacitor |
| TEF | Transient energy function |
| UEP | Unstable equilibrium point |
| UPFC | Unified power flow controller |
| VDCOL | Voltage-dependent current order limiter |
| VSC | Voltage source converter |
| WAMS | Wide area measurement system |

# 1

## Introduction

### 1.1 General

Electric power systems are more than 125 years old and their growth has been phenomenal due to the extensive use of electrical energy in industrial, commercial, and residential establishments. Although the initial application was for street lighting, the versatility of electrical energy is spurring new applications, including rail and road transportation with electric or hybrid vehicles. The development of information technology has resulted in new avenues for the utilization of quality power.

The growth in electric energy supply industry is also due to the successful implementation of providing reliable power at minimum, affordable costs. However, these two criteria are not always compatible. Reliability implies minimizing the duration and frequency of power interruptions caused by failure of equipment or other factors that lead to breakdowns. The reduction of costs is achieved by designing efficient electrical equipment (generators, transformers, motors, capacitors, etc.) and reliability standards are met by providing the required redundancy. Minimization of costs and improvement of reliability simultaneously in power delivery systems were made feasible by adopting alternating current (AC) grids made up of high-voltage (HV) or extra-high-voltage (EHV) transmission lines with overhead conductors suspended by insulator strings on transmission towers. The economics of scale dictated the adoption of large generators of ratings up to 1300 MW. AC transmission lines do not require control, which minimizes complexity that can affect reliability. However, lack of control in AC lines can result in overloading of some lines whenever there is a contingency (tripping of a line or generator due to a fault). Assuming there is enough reserve in generation and transmission facilities, rescheduling of generation can ensure redistribution of power flows in the AC transmission network to relieve overloads and also maintain bus voltages in acceptable range. However, this will affect economic dispatch and increase the production costs. A major constraint that significantly affects power system operation is the lack of large energy storage facilities (except for pumped hydro schemes that are limited). The energy storage technologies are still being developed to make them efficient and cost effective.

## 1.2 Power System Stability

Another major constraint that affects reliable system operation is the requirement of synchronous operation of all the synchronous generators in the system. This implies that the average frequencies of all the generators have to be identical for satisfactory operation. Any major disturbance such as a fault on a major line and cleared by tripping the line can affect synchronous operation even if the postfault system has a stable equilibrium. One or more generators can go out of step or lose synchronism with respect to the remaining generators. It becomes necessary to trip generators that go out of step; otherwise, there could be damage to the generators apart from the fact that oscillating power outputs from out-of-step generators pose a major disturbance to the rest of the system.

The transient stability problem (loss of synchronous operation when the system is subjected to major disturbances) was a major problem affecting the reliability of remote generating stations connected to load centers. The problem was tackled by

a. Reducing the clearing time of the circuit breakers and high-speed reclosing
b. Introducing switching stations in long lines to limit the increase in the reactance of the postfault system (as viewed by the generating station)
c. Increasing the speed of response of the excitation system
d. Application of braking resistors (in hydro station) and fast valving (in thermal stations)

High-speed excitation systems using high-gain automatic voltage regulators (AVR) and static exciters also help in improving steady-state stability of power systems; particularly with modern turbo-generators having high synchronous reactances. Steady-state stability as defined in books by Crary (1947) and Kimbark (1948) refers to the stability of the system under slow and small changes (in the load) in the absence of dynamic controllers. Thus, the steady-state stability limit of a generator refers to the maximum power that can be provided by the generator for a specified operating condition. However, mathematically, it is accurate to define "small signal stability" as the stability under small disturbances (irrespective of the modeling), which is equivalent to the stability of the equilibrium point.

In the absence of AVR, the loss of small signal stability occurs by saddle-node bifurcation (a real eigenvalue of Jacobian matrix becoming positive). However, with high-speed excitation systems, the loss of small signal stability occurs by Hopf bifurcation (a complex pair of eigenvalues of Jacobian moving to the right half of the "s" plane). This fact was discovered in the mid-1960s when it was observed that there were spontaneous oscillations in the power flows in lines that increased in magnitude for no apparent reason. The problem is serious as undamped oscillations can result in loss of synchronous operation. The solution to the problem was to introduce power system stabilizers (PSS) that apply limited modulation of the terminal voltage to damp the oscillations caused by the generator rotor swings (Anderson and Fouad 1977, Kundur 1994, Padiyar 1996, Sauer and Pai 1998).

In the late 1970s, it was observed that under certain conditions, a disturbance can cause uncontrolled decrease in voltage, leading to voltage collapse. This phenomenon also leads to power blackouts, although generators remain in synchronism. This has been termed as voltage stability problem. This can also be divided into small-disturbance voltage stability, transient voltage stability, and long-term voltage stability. The primary factor that is responsible for voltage collapse is the mechanism of load restoration (inherent or controlled). Thus, induction machines, high-voltage direct current (HVDC) converters, and on-load tap changers (OLTC) contribute to the voltage instability problem (Kundur 1994, Padiyar 2002, Van Cutsem and Vournas 1998).

The system frequency is regulated under normal conditions by load frequency control (LFC). However, under disturbed conditions that lead to uncontrolled tripping of tie lines, islands can be formed with excess generation or excess load. The reduction of frequency below 5% of the rated frequency can result in damage to steam turbines apart from affecting customer equipment. Uncontrolled variation of frequency is termed as frequency instability (IEEE/CIGRE Joint Task Force 2004).

## 1.3 Power System Security

Although power system planners try to design the system to handle various contingencies that may arise during system operation, there could be several unforeseen contingencies that can lead to catastrophic failures resulting in major power blackouts spread over large areas affecting millions of customers. It may take hours or days to restore full power to customers. A major blackout in north-east United States occurred in November 1965. About 30 million people were affected in eight states and part of Canada. The initiating event was the faulty operations of a relay on a transmission line carrying 300 MW from a power plant on the Niagara river. This tripped the line, transferring the load carried by the line to the parallel lines that all tripped out. This resulted in a surge of power flow into the east-west transmission lines in New York state that resulted in tripping of the interconnecting line and seven generating units feeding the north-east grid. Loss of these plants placed a heavy drain on the generators in New York City. The result was a

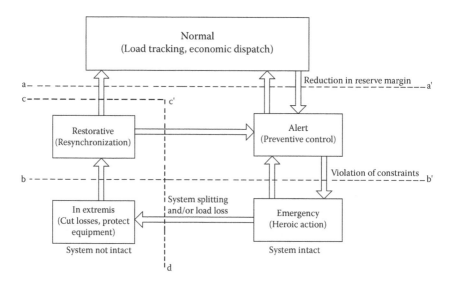

**FIGURE 1.1**
Power system operating states.

complete collapse of the system. It took over 12 h to restore power to New York City (Glover and Sarma 2002).

This major blackout showed that the reliable design is not adequate in maintaining security during system operation. System security implies taking proactive steps to monitor security and taking effective steps to prevent catastrophic failures when a major contingency occurs. The system operating states can be divided into five broad categories as shown in Figure 1.1 (DyLiacco 1979, Fink and Carlsen 1978). The definitions of these categories are based on answering the following questions:

1. Can the system meet the load demand? Mathematically, this is equivalent to verifying whether the following equality constraint is satisfied:

$$S_{Li} = S_{Li}^d, \quad i = 1, 2, \ldots, N \tag{1.1}$$

where $S_{Li}$ is the complex (load) power supplied at bus $i$ and $S_{Li}^d$ is the complex (demand) at bus $i$ of the $N$ bus system. Note that the two quantities are not equal as shortage of generation and/or transmission congestion can restrict the power supplied and limit it to a value less than the unrestricted demand.

2. Are all the equipment (generators, transformers, and transmission lines) operating within their feasible region (when they can operate continuously without any damage caused by heating or overvoltages)?

For example, the operating region of a generator in the P–Q plane is determined by constraints of limits on the armature current, field current, prime mover output, and stability considerations. Similarly, the insulation and power transfer characteristics force the transmission line voltages to remain in a feasible range. The power flow in an AC line is limited by the temperature rise of conductors, voltage regulation, and stability constraints depending on the line length (St. Clair 1953). Mathematically, the following constraint equation expresses the requirement of an equipment (or component) $k$ operating in the feasible region:

$$X_k \in A_k, \quad k = 1,2,\ldots,N_c \tag{1.2}$$

where $X_k$ is the state of the component $k$ and $A_k$ is its feasible region. $N_c$ is the total number of components.

3. Assuming the answers to the above two questions is affirmative, does the system (at the current state of operation) have sufficient margin such that any major credible contingency (loss of a heavily loaded line or a generator) will not result in overloading of any of the equipment? If the answer is affirmative, the system is said to satisfy security constraints and the operating state is classified as normal. If the answer is negative, the operating state is said to be in the "alert" mode.

When the system is in the alert state and a credible contingency occurs, the system can enter into an emergency state that is characterized by one or more equipment getting overloaded and operating outside its feasible region. This state of affairs cannot continue as the protection system will act automatically (with or without some time delay). If there is a fault (short circuit) in a transmission line, the protection system acts within 3–5 cycles. However, overloading of a line may be permitted for a few minutes. In any case, the protection system acts decisively to trip the component.

When the constraint (1.2) is violated, even if primary protection fails to act, back-up protective relays operate to disconnect the component. As the protection starts to operate, the system enters "In Extremis" state and this results in the equality constraints (1.1) being violated in addition to the constraint (1.2). Depending on the severity of the disturbance, load shedding will be limited to a localized area or widespread. The disturbances can result in the tripping of interconnecting lines leading to uncontrolled system separation and formation of islands where there could be deficit or excess of generation. The imbalance in the generation and load will result in tripping of generation and/or loads that exacerbate the problem. In general, it can be said that widespread blackouts are caused by cascading outages of transmission lines and generators.

Although the condition (1.2) will be satisfied in the "In Extremis" state, if sufficient time is allowed for the protection system to act and all the islands can reach equilibrium state when the operating generators are able to meet the curtailed load, this state of affairs cannot be allowed as it becomes imperative for the system operator to cut the losses and start restoring the load at the earliest. The system enters into the "Restorative" state when there is no violation of the condition (1.2) and the operator resynchronizes the tripped generators and transmission lines. At the end of restorative actions, the system enters either the "normal" or "alert" state.

The dashed lines in Figure 1.1 divide the system states into two groups. The line aa' divides the system states into (1) normal and (2) the rest (not normal) groups. The line bb' separates the system states into the groups: (1) where constraints (condition 1.2) are satisfied and (2) where condition (1.2) is not satisfied. The line cc'd separates the system states into groups: (1) where the system is intact and (2) where it is not intact.

It is obvious that apart from the normal and alert states, the remaining three states are temporary (or transient). It should be noted that the loss of transient stability following a major disturbance can also result in the transition from alert state into the emergency state. When there is a danger of loss of stability, there is hardly any time for the system operator to act to avert the situation. In general, the operator has little control over the system when it enters the emergency state. That is why the improvement of system security has been predicated on the preventive control by redeploying generation and transmission resources such that the system does not remain in the alert state. The preventive control requires checking for security constraints by conducting on-line security analysis by considering a list of credible contingencies. Static security analysis involves performing fast on-line power flow computations based on the operating data and knowledge of short-term load forecasts.

## 1.4 Monitoring and Enhancing System Security

Modern energy control (or management) centers have computer-based supervisory control and data acquisition systems (SCADA) for monitoring security. The measurements of power and reactive power flows, voltage magnitudes, and breaker status are gathered at remote terminal units (RTU) and the data are telemetered to control centers. Although the raw data can be used directly for preliminary checking of security by observing violations of the inequality constraints (that define the feasible operating regions of various equipment) if any, it is necessary to filter out bad data and also compute power flows in lines where the measurements are not available. This is possible by state estimation (SE) where states are defined by voltage magnitudes and phase angles at all the buses. For an $N$ bus system, there are $(2N - 1)$ states as one phase angle can be assumed arbitrarily (as zero). To ensure complete and accurate SE, there is a need for redundancy in the measurements, and the placements of meters should be such that the network is observable. On the basis of SE, it is possible to compute power and reactive power at all the load buses and in all lines. Unfortunately, SE does not consider the system dynamics and is not updated sufficiently fast to accommodate sudden and major disturbances.

As mentioned earlier, the security analysis function in energy control centers refers to static security analysis where the major assumption is that any contingency under consideration does not result in loss of stability. Thus, dynamic security analysis is not performed. This is a seriously flawed approach and is justified only on the grounds that an accurate dynamic security analysis is not feasible unless on-line transient stability type of simulation can be performed. While attempts have been made to apply parallel processing techniques to speed up computations, this approach has not been quite successful. Transient energy function (TEF) methods have been tried out in Canada, the United States, and Japan for both dynamic security assessment (DSA) and preventive control (El-Kady et al. 1986, Nishida 1984, Pavella et al. 2000, Rahimi et al. 1993).

## 1.5 Emergency Control and System Protection

Instead of relying purely on preventive control, it makes lot of sense to detect emergency states fast enough and deploy emergency control or remedial action schemes to steer the system to an alert or normal state. This approach is based on the utilization of technological advances in communications, computers, and control. A major development is the technology of synchronized phasor measurements (Horowitz and Phadke 1995) that enables direct and fast measurement of voltage phase angles (in addition to their magnitudes) that were earlier estimated by SE. This can help in real-time transient stability prediction (Liu and Thorp 1995). The essential feature of the phasor measurement unit (PMU) technology is the measurement of positive sequence voltages and currents in a power system in real time with precise time synchronization. This is achieved with a global positioning satellite (GPS) system. GPS is a U.S. government-sponsored program that provides worldwide position and time broadcasts free of charge. The current wide area communication technologies (such as SONET) are capable of delivering messages from one area of a power system to multiple nodes on the system in less than 6 ms (Adamiak et al. 2006, Begovic et al. 2005).

The emergency state is associated with the operation of protection systems based on the relays that monitor the voltages and currents at different locations in the system. Misoperation of protection systems is often a contributing factor for the cascading outages that lead to system collapse. Among possible failures in the protection systems, the most troublesome are the "hidden" failures (HF), which are defined as "a permanent defect that will cause a relay system to incorrectly and inappropriately remove circuit element(s) as a direct consequence of another switching event" (De La Ree et al. 2005).

It should be possible to eliminate hidden failures by modifying the protection system and introducing new concepts and technologies. It must be realized that equipment protection must be distinguished from system protection. While equipment protection must be designed to protect an equipment such as a transmission line from high fault currents that can damage the conductors, the out-of-step protection is essentially a system protection against transient (synchronous) instability. The complexities of a large power system imply the necessity for adaptive protection (Centeno et al. 1997).

The other technological innovations that can be applied for emergency control are based on power electronic converter controls used in flexible AC transmission system (FACTS) and HVDC controllers. The speed of control actions enables their application not only for the improvement of first swing stability but also for damping of initial large oscillations (Krishna and Padiyar 2005, Padiyar and Uma Rao 1997).

---

## 1.6 Application of Energy Functions

The well-known "equal area criterion" (EAC) applied for the stability evaluation of a two-machine system with simplified models is the first application of energy functions. The application to multimachine power systems with detailed generator and load models has been attempted by Tsolas et al. (1985), Padiyar and Sastry (1987), Padiyar and Ghosh (1989a) based on structure preserving models (SPM) (Bergen and Hill 1981). The work carried out in the 1980s has been summarized in monographs by Pai (1989), Fouad and Vittal (1992), and Pavella and Murthy (1994).

The basic idea behind the application of energy functions is to develop direct methods of stability evaluation without solving the system equations. A power system is described by a set of nonlinear differential and algebraic equations (DAE)

$$\dot{x} = f(x, y, p) \tag{1.3}$$

$$0 = g(x, y, p) \tag{1.4}$$

where $x \in R^n$, $y \in R^m$, and $p$ is a vector of specified parameters that include network admittances and load parameters (Arapostathis et al. 1982, Ilic and Zaborszky 2000, Zou et al. 2003).

The DAE system can be interpreted as an implicit dynamic system defined on the algebraic manifold defined by

$$L = \{(x,y) : g(x,y,p) = 0\} \tag{1.5}$$

On the algebraic manifold $L$, singular surfaces defined by

$$S = \left\{(x,y) \in L, \ \det\left[\frac{\partial g}{\partial y}\right] = 0\right\} \tag{1.6}$$

may exist and decompose the manifold $L$ into several disjoint components. A component where the Jacobian matrix $[\partial g/\partial y]$ has all eigenvalues with negative real parts is a stable component. The Jacobian is also nonsingular in each of the disjoint components and hence, by implicit function theorem, there exists a function $h(x)$ such that the local behavior of the system trajectory around $(x, y)$ can be described by

$$y = h(x) \tag{1.7}$$

By substituting Equation 1.7 in Equation 1.3, we get

$$\dot{x} = f[x, h(x)] = F(x) \tag{1.8}$$

In the above equation, we are not considering explicitly the dependence on the parameter vector, which is assumed to be constant. We assume that $x_s$ is the stable equilibrium point (SEP) for the system described by Equation 1.8. Note that $x_s$ satisfies

$$0 = F(x_s) \tag{1.9}$$

and the Jacobian matrix $[\partial F/\partial y]$ evaluated at $x = x_s$ has all eigenvalues with negative real parts.

Let us assume that the power system is subjected to a set of disturbances and the postdisturbance system is described by Equation 1.8. The initial condition at $t = t_0$ (when the system enters the postdisturbance system) is $x = x_0$. The question that we pose is whether the system trajectory approaches $x_s$ as $t \to \infty$. In such a case, we say the system is stable, otherwise (when the trajectory does not reach $x_s$), the system is unstable. While we discuss the mathematical issues in Chapter 2, we can state that the system is stable if the initial condition for the postdisturbance system lies in the region of attraction of the SEP. In applying the energy function theory, we assume an energy function (Hirsch et al. 2004, Zou et al. 2003) defined by

$$W = W(x,y) \tag{1.10}$$

such that the system is stable if

$$W(x_0, y_0) \leq W_{cr} \tag{1.11}$$

where $W_{cr}$ is the critical energy.

Note that the determination of $x_0$ and $y_0$ requires the solution of the system equations during the disturbances, starting from the initial conditions corresponding to the predisturbance state. It is generally assumed that the system is in stable equilibrium before the disturbances occur. This assumption may not always be true. The typical disturbance is the occurrence of a bus or line fault followed by the clearing of the fault by the protection system, which may result in tripping of line(s).

The issues in applying the energy function theory are

1. Defining the energy function
2. Accurate computation of critical energy
3. Computing initial values of $x$ and $y$ for the postdisturbance system

This book focuses on the development of the structure preserving energy function (SPEF) that takes into account the structure or the topology of the transmission network. It is possible to define energy for each element or component of the system. The advantages of SPEF over earlier approaches are

1. Inclusion of accurate models of loads as functions of voltage and frequency. Even dynamic load models can be considered.
2. The retention of network sparsity that enables efficient computations in stability analysis.
3. Elimination of the transfer conductances introduced by the reduction of the system by eliminating load buses and retaining only the internal nodes of generators.
4. Application of network theory for the study of system stability based on energy functions.
5. The representation of network controllers such as HVDC and FACTS controllers.
6. Application of SPEF for on-line detection of loss of synchronism and synthesizing remedial actions.

Energy functions can also be applied for designing damping controllers to improve the small signal stability of power systems.

To illustrate the application of energy concepts in conservative systems, we will consider two examples involving L–C circuits.

## Example 1

Consider a linear charged capacitor connected to a linear inductor through a switch as shown in Figure 1.2. The switch is initially open and is closed at $t = 0$. The equations for $t > 0$ are

$$C\frac{dv}{dt} = -i, \quad L\frac{di}{dt} = v \tag{1.12}$$

The initial conditions are $v(0) = v_0$ and $i(0) = 0$. The initial energy stored in the capacitor is given by

$$W_C(t = 0) = \frac{1}{2}Cv_0^2 \tag{1.13}$$

Even without solving Equation 1.12, we can observe that the current through the inductor increases from zero and reaches a maximum when the initial energy stored in the capacitor is transferred to the magnetic (field) energy in the inductor. Since the expression for the energy stored in the linear inductor is $(1/2)Li^2$, we get

$$i_{max} = \sqrt{\frac{C}{L}}v_0 \tag{1.14}$$

The current continues to flow in the same direction and starts charging the (discharged) capacitor in the reverse direction until current $i$ reduces to zero and the capacitor voltage reaches $-v_0$. If there is no dissipation of energy in the circuit, the exchange of energy between the capacitor and inductor continues forever. The variations of $v$ and $i$ are periodic. The voltage oscillates between $+v_0$ and $-v_0$ and the current oscillates between $+i_{max}$ and $-i_{max}$. At any given time $t$, the total energy in the circuit remains constant and is given by

$$W(v,i) = W_C(v) + W_L(i) = \frac{1}{2}Cv^2 + \frac{1}{2}Li^2 = \frac{1}{2}Cv_0^2 = \text{constant} \tag{1.15}$$

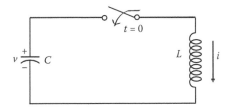

**FIGURE 1.2**
A linear L–C circuit.

**Example 2**

Consider a circuit consisting of a linear capacitor connected to a nonlinear inductor and excited by a direct current (DC) current source $I$ (see Figure 1.3). The initial current in the inductor is $I$ and the initial voltage across the capacitor is $v_0$. The capacitor is linear while the inductor is nonlinear defined by the equation

$$i = I_m \sin \lambda \tag{1.16}$$

where $\lambda$ is the flux linkage of the inductor. The equations for this circuit are given by

$$C\frac{dv}{dt} = I - i, \quad \frac{d\lambda}{dt} = v \tag{1.17}$$

The energy associated with the capacitor is $(1/2)Cv^2$ while the change in energy of the inductor is given by

$$\Delta W_L = \int_{\lambda_0}^{\lambda} i d\lambda \tag{1.18}$$

Substituting from Equation 1.16, we get

$$\Delta W_L = I_m \left[ \cos \lambda_0 - \cos \lambda \right] \tag{1.19}$$

The variation of the current with $\lambda$ is sinusoidal as shown in Figure 1.4. The initial value of $\lambda = \lambda_0$ is given by

$$\lambda_0 = \sin^{-1}\left( \frac{I}{I_m} \right) \tag{1.20}$$

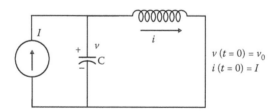

**FIGURE 1.3**
A linear capacitor connected to a nonlinear inductor.

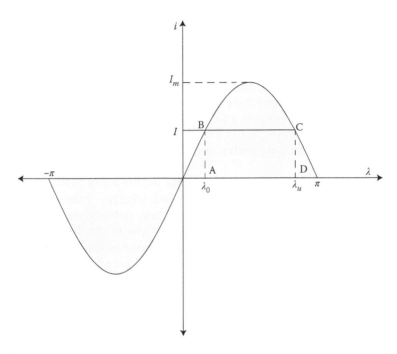

**FIGURE 1.4**
Current versus flux linkages in the nonlinear inductor.

The maximum increase in energy of the inductor is given by the area ABCD in Figure 1.4 (which shows current in the inductor as a function of flux linkages). This is given by Equation 1.19 when we substitute $\lambda = \lambda_u = \pi - \lambda_0$.

The increase in the energy of the inductor is contributed by (a) the constant DC current source and (b) the capacitor. The constant DC current source contributes the energy $\Delta W_s$ given by

$$\Delta W_s = I(\lambda_u - \lambda_0) \tag{1.21}$$

Note that the maximum energy transfer from the capacitor to the inductor is only $(\Delta W_L - \Delta W_s)$. If this energy is greater than the initial energy stored in the capacitor $\left((1/2)Cv_0^2\right)$, the capacitor discharges completely to the nonlinear inductor and the maximum flux linkage $(\lambda_m)$ is less than or equal to $\lambda_u$. The oscillation of the energy between the capacitor and the inductor continues indefinitely unless damped by dissipation in resistors in the circuit (which have been neglected in this analysis). If $(1/2)Cv_0^2 > (\Delta W_L - \Delta W_s)$, then the inductor cannot fully absorb the initial energy in the capacitor and $\lambda$ continues to increase implying that the capacitor voltage continues to increase.

It will be shown in Chapter 3 that the above example is similar to the determination of transient stability in a lossless single machine infinite bus (SMIB) system. The SMIB system can be modeled by a linear capacitor connected to a nonlinear inductor. The loss of synchronous stability in the SMIB system is analogous to the failure of the capacitor to completely transfer the stored energy to the nonlinear inductor (which represents the series reactance between the internal bus of the generator and the infinite bus). The current in the inductor is analogous to the power flow in the line and the voltage across the capacitor is analogous to the deviation in the rotor speed of the generator following a disturbance. Thus, the stored energy in the linear capacitor is analogous to the kinetic energy (KE) of the generator rotor moving with respect to a synchronously rotating reference frame.

It will be shown in Chapters 3 and 4 that the circuit or network analogy helps in deriving the total energy for detailed, structure preserving system models. Apart from direct evaluation of system stability, the energy functions can also be used for on-line detection of loss of synchronism and the mode of instability accurately. In addition, it becomes feasible to determine accurately the optimum locations and the control laws for the network controllers (FACTS controllers) for enhancing system stability.

## 1.7 Scope of This Book

The direct methods of transient stability analysis are primarily meant for off-line studies that can guide the system operator in maintaining system security, when there are adequate margins. However, they have not helped to develop control measures that can prevent major blackouts in 2003 (Andersson et al. 2005). Economic and environmental constraints dictate a paradigm shift that involves the adoption of technological solutions (wide area measurements, digital communication and controls, HVDC, and FACTS) for implementation of on-line emergency control to achieve the objective of "self-healing grids." The energy functions defined on SPM can help in designing real-time monitoring and control of dynamic security. Even for off-line studies, the use of SPEF permits the inclusion of detailed system models and provides improved accuracy of the results. Also, SPEF provides a tool for the study of both synchronous (angle) and voltage stability.

While Chapters 3 and 4 introduce SPEF and their applications for direct stability evaluation of systems with detailed load and generator models, Chapter 5 presents the analysis with HVDC and FACTS controllers. Chapter 6 describes a methodology for on-line detection of loss of synchronism by identifying the unique critical cutset that initially separates the system into two parts. Chapter 7 presents an approach based on the network analogy, for the synthesis of network (based on FACTS) damping controllers, which are optimal for damping small oscillations. Chapters 8 and 9 present an algorithm for emergency control that provides first swing stability and non-linear damping of large oscillations to steer the system close to the postfault equilibrium.

The next chapter reviews the literature on direct methods of transient stability evaluation based on simplified models.

# 2

## Review of Direct Methods for Transient Stability Evaluations for Systems with Simplified Models

## 2.1 Introduction

In this chapter, we will review the development of direct methods based on energy functions for evaluation of transient stability of power systems with simplified models. In particular, the generators are represented by classical models and the loads by constant impedances. The simplified system models do not consider the presence of controllers either in generating stations or in the transmission network (HVDC and FACTS controllers).

The application of energy functions to power systems has a long history. The earliest work was reported by Magnusson in 1947. An energy integral criterion was proposed by Aylett in 1958. As Lyapunov functions are considered as generalization of energy functions, developments of Lyapunov's method for direct stability analysis were reported by Gless (1966), El-Abiad and Nagappan (1966), and Willems and Willems (1970). There were several developments since then, which are summarized in the monographs by Pai (1981, 1989) and Fouad and Vittal (1992). There are also state-of-the-art papers by Ribbens-Pavella and Evans (1985), Varaiya et al. (1985), and Fouad and Vittal (1988).

While most of the developments are based on the reduced network models (RNM) (where only generator internal nodes are retained), the SPM was proposed by Bergen and Hill (1981). This has led to the developments of inclusion of detailed load and generator models as described in Narasimhamurthi and Musavi (1984), Tsolas et al. (1985), Bergen et al. (1986), Padiyar and Sastry (1987), and Padiyar and Ghosh (1989a). We will present the SPEF in the next chapter.

It can be shown that the well-known EAC for two-machine or one-machine, infinite-bus (OMIB) systems (Crary 1947, Kimbark 1948) is equivalent to the application of energy functions. (Note that a two-machine system can be reduced to an OMIB or SMIB.) In this approach, a major disturbance such as a three-phase fault followed by clearing results in increase of the transient energy. If the total energy at the time of clearing is less than the critical energy, the system is assumed to be transiently stable in the first swing. The critical energy is the potential energy (PE) at the unstable equilibrium point (UEP), corresponding to the postfault system. In extending this approach to a multimachine power system, it is observed that there can be up to $2(2^{n-1} - 1)$ UEPs for a system with $n$ machines. The critical energy was assumed to be the lowest energy computed at all UEPs. This method is quite cumbersome and involves significant computational complexity. The UEP corresponding to the lowest energy is said to be the closest UEP (to the SEP of the postfault system). This method also resulted in very conservative results.

A major departure from this approach was proposed by Athay et al. (1979) who proposed the concept of controlling unstable equilibrium point (CUEP) that determines the critical energy. The CUEP is defined as the UEP that lies on the boundary of system separation. These concepts were put on a firm mathematical foundation in papers by Chiang et al. (1987, 1988). The potential energy boundary surface (PEBS) method for computing the critical energy was first proposed by Kakimoto et al. (1978) and was extended by Athay et al. (1979).

The extension of EAC (called as extended equal area criterion) to multimachine power systems was proposed by Xue et al. (1989, 1992). This method is based on the conjecture that the loss of synchronism of a multimachine system, whenever it occurs, is triggered by the machines' irrevocable separation into two groups (Pavella and Murthy 1994). Subsequently, the extended equal area criterion (EEAC) has undergone modifications and has been transformed into single-machine equivalent (SIME), which is a "hybrid, temporal-direct method: temporal, since it relies on multimachine system evolution with time; direct, like the EEAC, from which it originates" (Pavella et al. 2000).

The direct methods based on Lyapunov or energy functions have been typically applied for direct stability evaluation for a specified fault clearing time or computation of critical clearing time (CCT). Using energy margin (EM) as an index for the proximity to transient instability (loss of synchronous operation), it is possible to apply energy functions for DSA and devise preventive control strategies based on sensitivities of EM with respect to control variables (such as generator power outputs, switched series capacitors, and power flow in critical tie lines).

## 2.2 System Model

### 2.2.1 Synchronous Generators

The stator equations in Park's reference frame (Padiyar 2002), expressed in per unit, are

$$-\frac{1}{\omega_B}\frac{d\psi_d}{dt} - \frac{\omega}{\omega_B}\psi_q - R_a i_d = V_d \tag{2.1}$$

$$-\frac{1}{\omega_B}\frac{d\psi_q}{dt} + \frac{\omega}{\omega_B}\psi_d - R_a i_q = V_q \qquad (2.2)$$

It is assumed that the zero-sequence currents in the stator are absent. If stator transients are to be ignored, it is equivalent to ignoring $p\psi_d$ and $p\psi_q$ terms in the above equations (note that $p$ is the differential operator). Since stability studies do not involve consideration of frequencies above 5 Hz, it is adequate to ignore network transients by modeling them by algebraic equations. Hence, it is consistent to ignore $p\psi_d$ and $p\psi_q$ terms.

The electrical torque on the generator rotor is expressed as

$$T_e = \psi_d i_q - \psi_q i_d \qquad (2.3)$$

If the armature resistance ($R_a$) is neglected, then it can be observed from Equations 2.1 and 2.2 that

$$T_e = \frac{(V_d i_d + V_q i_q)}{\bar{\omega}} \qquad (2.4)$$

where $\bar{\omega}$ is the per-unit speed (of the rotor). Note that Equation 2.4 is obtained by neglecting $p\psi_d$ and $p\psi_q$ terms and armature resistance. The rotor mechanical equations in per unit can be expressed as

$$M\frac{d^2\delta}{dt^2} + D'\frac{d\delta}{dt} = T_m - T_e \qquad (2.5)$$

where $M = 2H/\omega_B$, $T_m$ is the mechanical torque, and $D' = D/\omega_B$. $D$ is the per-unit damping. $\delta$ is the rotor angle measured in radians, with respect to a synchronously rotating reference frame (rotating at average system frequency of $\omega_0$ radians per second). Thus, $\delta$ is defined as

$$\delta = \theta_r - \omega_0 t, \quad \frac{d\theta_r}{dt} = \omega \qquad (2.6)$$

where $\theta_r$ is the rotor angle with respect to a stationary axis. Note that the average system frequency is not necessarily equal to $\omega_B$ (rated frequency), although we often assume that $\omega_0 = \omega_B$ in the normal state. Also, it is a common assumption to neglect variations in $\omega$, even during a transient, and consider $T_m \simeq P_m$ and $T_e \simeq P_e$ (in per-unit quantities).

### Expressions for the Stator Flux Linkages

The stator flux linkages $\psi_d$ and $\psi_q$ are functions of the stator currents (in $d$–$q$ variables) and currents in the rotor coils. The most detailed model of a generator considers the field winding on the *d*-axis and three damper windings (two in the *q*-axis and one in the *d*-axis). Since rotor currents cannot be considered as state variables when the stator and network transients are neglected (note that $i_d$ and $i_q$ are nonstate variables, when they are described by algebraic equations), it is convenient to express $\psi_d$ and $\psi_q$ in terms of rotor flux linkages and armature currents $i_d$ and $i_q$.

When we consider the so-called classical model for a synchronous generator, the following assumptions are made:

1. The damper windings are neglected
2. The flux decay in the field winding is neglected

With these assumptions, we can express $\psi_d$ and $\psi_q$ as

$$\psi_d = x'_d i_d + E'_q, \quad \psi_q = x_q i_q \tag{2.7}$$

where

$$x'_d = x_d - \frac{x_{df}^2}{x_f}, \quad E'_q = \frac{x_{df} \psi_f}{x_f} \tag{2.8}$$

The equation for the field winding is

$$\frac{dE'_q}{dt} = \frac{1}{T'_{d0}} [-E'_q + (x_d - x'_d) i_d + E_{fd}] \tag{2.9}$$

where

$$T'_{d0} = \frac{x_f}{\omega_B R_f}, \quad E_{fd} = \frac{x_{df}}{R_f} V_f \tag{2.10}$$

Substituting Equation 2.7 in Equations 2.1 and 2.2, we get

$$E'_q + x'_d i_d - R_a i_q = V_q \tag{2.11}$$

$$-x_q i_q - R_a i_d = V_d \tag{2.12}$$

Substituting Equation 2.7 in Equation 2.3, we get

$$T_e = E'_q i_q + (x'_d - x_q) i_d i_q \tag{2.13}$$

Note that if $T'_{d0}$ (open circuit field time constant) is considered very high (in comparison with the time required to test the first swing stability), we can ignore Equation 2.9 and assume $E'_q$ = constant. Thus, the classical model of a synchronous generator is described by a second-order swing Equation 2.5.

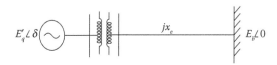

**FIGURE 2.1**
A generator connected to an infinite bus.

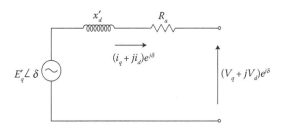

**FIGURE 2.2**
Equivalent circuit for the generator stator.

If we neglect armature resistance and consider the generator connected to an infinite bus (see Figure 2.1) through a lossless network with the series impedance $jx_e$, we can show that

$$T_e = \frac{E_q' E_b \sin\delta}{(x_d' + x_e)} + \frac{E_b^2(x_d' - x_q)\sin 2\delta}{2(x_d' + x_e)(x_q + x_e)} \tag{2.14}$$

Note that the electrical torque has two components, one dependent on the field voltage and the other independent of the field voltage. In other words, $T_e \neq 0$ even when the field current is zero. This is due to transient saliency. To simplify the analysis, a third assumption is introduced:

3. Transient saliency is neglected by assuming

$$x_q = x_d' \tag{2.15}$$

With this assumption, we can represent the stator Equations 2.11 and 2.12 by an equivalent circuit shown in Figure 2.2. For simplifying the notation, we can drop the subscript $q$ in $E_q'$.

The detailed model of a synchronous machine is described in Appendix A.

## 2.2.2 Network Equations

The system network is made of transformers, transmission lines, shunt reactors/capacitors, and series capacitors, if any. A transformer is usually represented by a series impedance of leakage reactance and winding resistances.

The magnetizing impedance is neglected. A long transmission line is represented by a $\pi$ equivalent circuit corresponding to the system frequency.

Since network transients are neglected, we can express the network equations using the admittance matrix as

$$YV = I \tag{2.16}$$

The size of the admittance matrix $Y$ is $n \times n$ for a network with $n$ nodes. $V$ and $I$ are both column vectors of dimension $n$. Without any loss of generality, we can assume that the first $m$ nodes are the terminal nodes of $m$ generators. From Figure 2.2, we can express the injected current by the $j$th generator as

$$I_j = (E'_j - V_j)y_{gj}, \quad j = 1, 2, \ldots, m \tag{2.17}$$

where $y_{gj} = 1/(R_{aj} + x'_{dj})$.

## 2.2.3 Load Model

The load dynamics is generally neglected unless the motor loads have a major impact on stability. However, the dependence of load on voltage magnitude is considered. In general, the load power at bus $i$ is expressed as

$$\frac{P_{Li}}{P_{Li0}} = a_0 + a_1 \left( \frac{V_{Li}}{V_{Li0}} \right) + a_2 \left( \frac{V_{Li}}{V_{Li0}} \right)^2, \quad i = 1, 2, \ldots, n \tag{2.18}$$

where the subscript 0 indicates values corresponding to operating SEP. The coefficients $a_0$, $a_1$, and $a_2$ satisfy the relationship

$$a_0 + a_1 + a_2 = 1$$

We can express the reactive power load at bus $i$ in a similar fashion as

$$\frac{Q_{Li}}{Q_{Li0}} = b_0 + b_1 \left( \frac{V_{Li}}{V_{Li0}} \right) + b_2 \left( \frac{V_{Li}}{V_{Li0}} \right)^2 \tag{2.19}$$

where $b_0 + b_1 + b_2 = 1$.

The first component of the load is termed as the constant power load, the second component is termed as the constant current load, and the third component as the constant impedance load.

There are other ways of representing static loads (IEEE Task Force 1993). In general, the active power load can be represented as

$$\frac{P_L}{P_{L0}} = c_1 \left( \frac{V_L}{V_{L0}} \right)^{m_{p1}} (1 + k_p \Delta f) + (1 - c_1) \left( \frac{V_L}{V_{L0}} \right)^{m_{p2}} \tag{2.20}$$

where $c_1$ is the frequency-dependent fraction of the active power load, $m_{p1}$ and $m_{p2}$ are the voltage exponents for (i) the frequency-dependent and (ii) frequency-independent components of the load, respectively, $\Delta f$ is the per-unit frequency deviation (from nominal), and $k_p$ is the frequency sensitivity coefficient for the active power load.

Bergen and Hill (1981) represent the load as the sum of two components given by

$$P_L = P_{L0} + D\Delta\omega \qquad (2.21)$$

where $\Delta\omega = d\delta/dt$.

The dynamics of induction motor load is usually represented by a third-order model when stator transients are neglected (Brereton et al. 1957).

A generic (aggregated) load model proposed by Hill (1993) is given by

$$T_p \frac{dP}{dt} + P = P_s(V) + k_p(V)\frac{dV}{dt} \qquad (2.22)$$

A similar equation applies for the reactive power. Here, the load behavior is determined by two load functions and a time constant $T_p$. $P_s$ is called the static load function and is applicable in steady state. $k_p$ is called the dynamic load function.

### 2.2.4 Expressions for Electrical Power

Neglecting mechanical damping, we can express the swing equations for $m$ generators in a power system as

$$M_i \frac{d^2\delta_i}{dt^2} = P_{mi} - P_{ei}, \quad i = 1, 2, \ldots, m \qquad (2.23)$$

On the basis of the stator equivalent circuit shown in Figure 2.2, applicable to all machines and assuming constant impedance type of load, we can obtain expressions for the electrical power outputs of the generators.

Consider the network shown in Figure 2.3 with $n$ buses with first $m$ buses corresponding to the terminal buses of $m$ generators.

The network Equation 2.16 can be expressed in the expanded form as

$$\begin{bmatrix} Y_{GG} & Y_{GL} \\ Y_{LG} & Y_{LL} \end{bmatrix} \begin{bmatrix} V_G \\ V_L \end{bmatrix} = \begin{bmatrix} I_G \\ I_L \end{bmatrix} \qquad (2.24)$$

where

$$I_G = [y_G] (E' - V_G) \qquad (2.25)$$

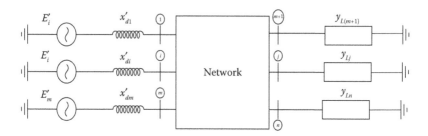

**FIGURE 2.3**
System network connected to $m$ generators.

$$[y_G] = \text{Diag}[y_{g1} \dots y_{gi} \dots y_{gm}], \quad E' = [E'_1 \cdots E'_i \cdots E'_m]^t$$

$$I_L = -[y_L]V_L \tag{2.26}$$

$$[y_L] = \text{Diag}[y_{L(m+1)} \dots y_{Ln}]$$

Substituting Equations 2.25 and 2.26 in Equation 2.24, we get

$$\begin{bmatrix} Y'_{GG} & Y_{GL} \\ Y_{LG} & Y'_{LL} \end{bmatrix} \begin{bmatrix} V_G \\ V_L \end{bmatrix} = \begin{bmatrix} I'_G \\ 0 \end{bmatrix} \tag{2.27}$$

where

$$[Y'_{GG}] = [Y_{GG}] + [y_G], \quad [Y'_{LL}] = [Y_{LL}] + [y_L], \quad I'_G = [y_G]E'$$

Eliminating the load buses and solving for $V_G$ (terminal voltage vector), we get

$$V_G = [Z_{GG}]I'_G \tag{2.28}$$

where

$$[Z_{GG}] = \left\{ [Y'_{GG}] - [Y_{GL}][Y'_{LL}]^{-1}[Y_{LG}] \right\}^{-1}$$

We can finally solve for $I_G$ from Equation 2.25 and obtain

$$I_G = [Y^R]E' \tag{2.29}$$

where

$$[Y^R] = [y_G] - [y_G] [Z_{GG}] [y_G]$$

is the reduced bus admittance matrix of the overall system network retaining only the internal buses of the generators.

The expression for $P_{ei}$ is obtained from

$$P_{ei} = \text{Re}[I_{Gi}^* E_i'] = E_i'^2 G_{ii} + \sum_{\substack{j=1 \\ j \neq i}}^{m} E_i' E_j' \left[ G_{ij} \cos(\delta_i - \delta_j) + B_{ij} \sin(\delta_i - \delta_j) \right] \quad (2.30)$$

**Remarks**

1. $G_{ij}(i \neq j)$ terms are called transfer conductances, which are often neglected to obtain a Lyapunov function.
2. The first term in the right-hand side (RHS) of Equation 2.30 can be absorbed in the mechanical power, which is normally assumed to be constant in the short time frame in which transient stability (in the first swing) is tested.
3. From Equations 2.1 and 2.2, it is obvious that the internal electromotive forces (EMFs) of generators vary with the rotor speeds during a transient. Since this variation is neglected in deriving the power expressions, Equation 2.30 can also be viewed as the expression for the electrical torque $T_{ei}$. Note that there is no difference between per-unit torque and power when the per-unit frequency is constant (in steady state) and assumed to be 1.0.
4. The above-mentioned fact will be utilized in deriving a network analogy, which helps in interpreting the energy functions and looking at different ways in which they can be computed. This will be discussed in Chapter 3.
5. The derivation of Equation 2.29 is based on the assumption that there are no loads at the generator terminals. However, this assumption can be relaxed by modifying the matrix $Y_{GG}$ to account for the shunt admittances due to load.

## 2.3 Mathematical Preliminaries

The dynamics of power system even with simplified models of generators and constant impedance-type loads is nonlinear. It is described by the following second-order vector differential equation:

$$M\ddot{x} + D\dot{x} + f(x,u) = 0 \quad (2.31)$$

where $x \in R^m$, $M$ is a diagonal matrix with positive elements, $D$ is also a diagonal matrix with positive elements, and $u$ is a vector of constant parameters (the network and load parameters).

Equation 2.31 can be rewritten as a set of first-order differential equations

$$\dot{X} = F(X, u) \tag{2.32}$$

where $X^t = (x \quad \dot{x})$. Note that both $X$ and $F$ are vectors of the same dimension (say $N$). Since $u$ is a constant vector, the system is said to be autonomous. If the elements of $u$ are explicit functions of time $t$, then the system is said to be nonautonomous. If the initial conditions are specified, that is

$$X(t_0) = X_0$$

then the solution of Equation 2.32 can be expressed as $\phi_t(X_0)$ to show explicitly the dependence on initial conditions. (Note that since $u$ is a constant vector, it is not essential to show explicitly the dependence of the solution on $u$.)

$F$ is called a vector field and $\phi_t(X_0)$ is called the trajectory through $X_0$. $\phi_t(X)$, $(X \in R^N)$ is called the flow. The function $F$ is said to be continuously differentiable or $C^1$ if all the partial derivatives of the elements of $F$ exist and are continuous. The matrix of partial derivatives is termed as Jacobian and denoted by $DF_X$. The elements of $DF_X$ are defined by

$$DF_X(i, j) = \frac{\partial F_i}{\partial X_j}, \quad i = 1, 2, \ldots, N; j = 1, 2, \ldots, N$$

The solution of Equation 2.32 has the following properties:

1. The solution exists for all values of $t$ and is unique for a specified $X_0$ at $t = t_0$. This implies that $\phi_t(X) = \phi_t(Y)$ if and only if $X = Y$.
2. $\phi_{(t_1+t_2)} = \phi_{t_1} \cdot \phi_{t_2}$ This implies that a trajectory is uniquely specified by its initial condition and distinct trajectories do not intersect.
3. The derivative of a trajectory with respect to the initial condition exists and is nonsingular. For fixed $t_0$ and $t$, $\phi_t(X_0)$ is continuous with respect to $X_0$.

## 2.3.1 Equilibrium Points

An equilibrium point (EP) $X_e$ is a constant solution of Equation 2.32 such that

$$X_e = \phi_t(X_e) \tag{2.33}$$

This implies that $X_e$ satisfies the equation

$$0 = F(X_e, u) \tag{2.34}$$

This shows that $X_e$ depends on $u$.

In general, there are several EPs that are obtained as real solutions of Equation 2.34 (Hirsch et al. 2004).

### 2.3.2 Stability of Equilibrium Point

An EP $X_e$ is said to be asymptotically stable if all nearby trajectories approach $X_e$ as $t \to \infty$. Otherwise, it is said to be unstable. We say that an EP is hyperbolic if the Jacobian matrix $DF_X$ evaluated at that EP has no eigenvalues with zero real parts. A hyperbolic EP is stable if all eigenvalues have negative real parts; otherwise it is a UEP. If all eigenvalues of the Jacobian have positive real parts, the UEP is called a source (all trajectories starting from the source leave the EP).

If the Jacobian at a UEP has exactly one positive real eigenvalue, the UEP is classified as type 1. Similarly, a type $k$ UEP has exactly $k$ eigenvalues with positive real parts.

If an EP is not hyperbolic, then the determination of the stability of the EP is not straightforward. However, there is an alternate method proposed by the Russian mathematician Lyapunov, which can be used to check whether the EP is asymptotically stable.

### 2.3.3 Lyapunov Stability

Consider the nonlinear autonomous system defined by

$$\dot{X} = F(X)$$

Let $X_e$ be an EP of the system. If every solution $X(t)$ with $X(0) = X_0$ in the open neighborhood $O_1$ of $X_e$ is defined and remains in the open neighborhood $O$ of $X_e$ for all $t > 0$, where $O_1 \subset O$, the EP is said to be stable. If in addition to being stable, we have $\lim_{t \to \infty} X(t) = X_e$, we say that the EP is asymptotically stable.

An alternative definition of Lyapunov stability can be stated as follows:

$X_e$ is stable in the sense of Lyapunov if for each arbitrary real number $\varepsilon$, there exists a positive real number $\delta(\varepsilon)$ such that $\|X_e - X_0\| < \delta$, implies $\|X(t) - X_e\| < \varepsilon$ for $\forall\, t \geq 0$.

### 2.3.4 Theorem on Lyapunov Stability

Let $X_0$ be an EP for $\dot{X} = F(X)$. Let $V(X)$ be a differentiable function defined on an open set containing $X_e$. If

    a. $V(X_e) = 0$ and $V(X) > 0$, if $X \neq X_e$
    b. $dV/dt \leq 0$ in $O - X_e$

        then $X_e$ is stable. Furthermore, if $V$ also satisfies

    c. $\dot{V} < 0$ in $O - X_e$

        then $X_e$ is asymptotically stable

A function $V(X)$ satisfying (a) and (b) is called a Lyapunov function. If (c) also holds, we call $V$ a strict Lyapunov function (Hirsch et al. 2004).

The advantages of Lyapunov's method are

i. There is no need to solve the differential equations for checking stability

ii. In addition to the determination of stability of the EP, it is also possible to estimate the size of the "basin of attraction" of an asymptotical SEP

The "basin of attraction" is defined as the set of all initial conditions whose solution tends to the EP. The major disadvantage is that there is no well-defined algorithm for finding Lyapunov functions or obtaining exactly the basin of attraction. However, as we will see later for power systems, it is possible to define the boundary of basin of attraction of the SEP under certain assumptions (Chiang et al. 1987).

Hirsch et al. (2004) give some examples of nonlinear systems for which the stability of the EP can be checked with the help of suitably defined energy functions. We give here an example of a nonlinear pendulum for illustration.

**Example**

We consider a nonlinear pendulum with some damping. The equations can be normalized and expressed as

$$\frac{d^2\theta}{dt^2} = -b\frac{d\theta}{dt} - \sin\theta$$

Let $x_1 = \theta$. The state equations can be expressed as

$$\dot{x}_1 = x_2, \quad \dot{x}_2 = -bx_2 - \sin x_1$$

We have two EPs (mod$2\pi$); the downward rest position at $x_1 = 0$, $x_2 = 0$ and the upward position at $x_1 = \pi$, $x_2 = 0$. The first is an SEP while the second is a UEP. For $b > 0$, the origin is asymptotically stable EP as the eigenvalues have negative real parts. If we define the following Lyapunov function

$$V(x) = \frac{1}{2}x_2^2 + 1 - \cos x_1, \quad \frac{dV}{dt} = -bx_2^2 \leq 0$$

The above equation shows that although the EP is asymptotically stable, the Lyapunov function is not strict. The following theorem gives a criterion for asymptotic stability and the size of the basin of attraction even when the Lyapunov function is not strict.

**Theorem (Lasalle's Invariance Principle)**

Let $X_e$ be an EP for $\dot{X} = F(X)$ and let $V(X)$ be a Lyapunov function for $X_e$ defined over the open set $O$ containing $X_e$. Let $P \subset O$ be a neighborhood of $X_e$ that is closed and bounded. Suppose $P$ is positively invariant and that there is no entire solution in $P - X_e$ on which $V$ is constant. Then, $X_e$ is asymptotically stable and $P$ is contained in the basin of attraction.

Note that a set $P$ is called invariant if for each $X \in P$, $\varphi_t(X)$ is defined and in $P$ for all $t \in R$. The set $P$ is positively invariant if for each $X \in P$, $\varphi_t(X)$ is defined and in $P$ for all $t \geq 0$.

For the example of the nonlinear pendulum, we note that the origin is asymptotically stable for $b > 0$. The invariant set $P$ is obtained as $V \leq c$ and $|x_1| < \pi, 0 < c < 2$.

**Remarks**

1. The Lyapunov function defined for the nonlinear pendulum is also called the energy function. If $b = 0$, then $\dot{V} = 0$. This implies that $V = $ constant along all solutions. Hence, $V = c$ gives the solution curves. The phase portrait for $b = 0$ is shown in Figure 2.4. The solutions encircling the origin have the property that $-\pi \leq x_1(t) \leq \pi$ for all $t$. $V = 2$ corresponds to the separatrix that is the boundary of the region of stability surrounding the origin (an SEP).

2. The equations of a nonlinear pendulum are similar to the swing equations of a single-machine (connected to infinite bus) system with $P_m = 0$.

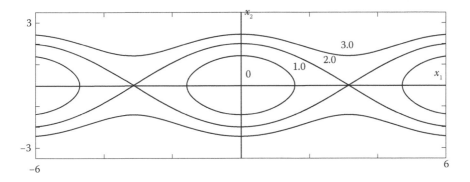

**FIGURE 2.4**
Phase portrait for a nonlinear pendulum (neglecting damping).

## 2.4 Two-Machine System and Equal Area Criterion

Consider a two-machine system shown in Figure 2.5. Assuming that both machines are represented by classical models of constant voltage sources behind transient reactances (see Figure 2.2), each machine can be modeled by a second-order swing equation (neglecting damping). Thus, the system model is described by

$$M_1 \frac{d^2 \delta_1}{dt^2} = P_{m1} - P_{e1} \tag{2.35}$$

$$M_2 \frac{d^2 \delta_2}{dt^2} = P_{m2} - P_{e2} \tag{2.36}$$

where

$$P_{e1} = G_{11}E_1^2 + G_{12}E_1E_2 \cos(\delta_1 - \delta_2) + B_{12}E_1E_2 \sin(\delta_1 - \delta_2)$$

$$P_{e2} = G_{22}E_2^2 + G_{12}E_1E_2 \cos(\delta_2 - \delta_1) + B_{12}E_1E_2 \sin(\delta_2 - \delta_1)$$

Note that we are using symbols $E_1$ and $E_2$ for internal EMFs in the two machines to simplify the notation. Note also that $G_{11} > 0$, $G_{22} > 0$, $B_{12} > 0$, and $G_{12} < 0$.

Multiplying Equations 2.35 and 2.36 by $M_2/(M_1 + M_2)$ and $M_1/(M_1 + M_2)$, respectively, and subtracting Equation 2.36 from Equation 2.35, we get

$$M_{eq} \frac{d^2 \delta}{dt^2} = P_m^{eq} - P_e^{eq} \tag{2.37}$$

where $M_{eq} = M_1 M_2/(M_1 + M_2)$, $\delta = \delta_1 - \delta_2$, and $P_m^{eq} = (M_2 P_{m1} - M_1 P_{m2})/(M_1 + M_2)$, $P_e^{eq} = (M_2 P_{e1} - M_1 P_{e2})/(M_1 + M_2)$.

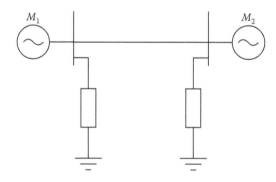

**FIGURE 2.5**
A two-machine system.

Substituting the expressions for $P_{e1}$ and $P_{e2}$ and simplifying, we can express Equation 2.37 as

$$M_{eq}\frac{d^2\delta}{dt^2} = P_i - GE_1E_2\cos\delta - BE_1E_2\sin\delta \qquad (2.38)$$

where

$$P_i = P_m^{eq} - (G_{11}E_1^2 - \alpha G_{22}E_2^2)/(1 + \alpha),$$

$$G = G_{12}\frac{1-\alpha}{1+\alpha}, \quad B = B_{12}\frac{1-\alpha}{1+\alpha}, \quad \text{and} \quad \alpha = \frac{M_1}{M_2}.$$

This shows that a two-machine system can be reduced to an SMIB system.

**Special Case**

When $M_1 \ll M_2$ we can approximate $\alpha \simeq 0$ and obtain

$$M_{eq} = M_1, \quad P_m^{eq} = P_{m1}, \quad P_i = P_{m1} - G_{11}E_1^2 \quad \text{and} \quad G = G_{12}, \quad B = B_{12}$$

In this case, we can assume bus 2 in Figure 2.5 as an infinite bus with constant voltage magnitude ($E_2$) and phase angle as zero.

### 2.4.1 Equal Area Criterion

This is a simple and direct method that does not require the solution of the swing equation of an SMIB system following disturbance(s). The assumptions used in applying this criterion are

1. Constant mechanical power
2. No damping
3. Classical machine model

The basis for this method is that if the system is stable (in the first swing), the rotor angle (after the disturbance) reaches a maximum value (assuming that the rotor initially accelerates) and then oscillates around the final steady-state value. It is assumed that the SEP exists for the postdisturbance system. The transient stability is checked by monitoring the deviation of the rotor speed ($d\delta/dt$) and ensuring that it becomes zero following the disturbance.

Let the swing equation be given by

$$M\frac{d^2\delta}{dt^2} = P_a = P_m - P_e$$

Multiplying both sides by $(d\delta/dt)$ and integrating with respect to time $t$, we get

$$M \int_{t_0}^{t} \frac{d\delta}{dt} \cdot \frac{d^2\delta}{dt^2} dt = \int_{t_0}^{t} (P_m - P_e) \frac{d\delta}{dt} dt$$

This leads to

$$\frac{1}{2} M \left( \frac{d\delta}{dt} \right)^2 = \int_{\delta_0}^{\delta} (P_m - P_e) d\delta \qquad (2.39)$$

It is assumed that at $t = t_0$, the system is at rest (equilibrium) and speed deviation $(d\delta/dt)$ is zero. The RHS of Equation 2.39 can be interpreted as the area between the curves $P_m$ versus $\delta$ and $P_e$ versus $\delta$ (the power angle curve). If the system is to be stable, then

$$\frac{d\delta}{dt}\bigg|_{\delta=\delta_{\max}} = 0$$

This implies that the area denoted by

$$A = \int_{\delta_0}^{\delta_{\max}} (P_m - P_e) d\delta = 0 \qquad (2.40)$$

must have a positive portion $A_1$ for which $P_m > P_e$ and a negative portion $A_2$ for which $P_m < P_e$. The magnitudes of the areas $A_1$ and $A_2$ are equal such that

$$A = A_1 - A_2 = 0 \qquad (2.41)$$

Hence, the nomenclature of equal area criterion.

**Remarks**

1. Neglecting damping, the EAC is also applicable for a two-machine system as it can be converted into SIME (as shown earlier).
2. The EAC is a special case of the direct method for stability evaluation using energy functions (Pai 1989, Padiyar 2002).

### 2.4.2 Energy Function Analysis of an SMIB System

Equation 2.39 can be rewritten as

$$W = W_1(\dot{\delta}) + W_2(\delta) = \text{constant} = k_1 \qquad (2.42)$$

where $W_1$ is called the KE given by

$$W_1(\dot{\delta}) = \frac{1}{2}M\dot{\delta}^2$$

$W_2$ is called the PE defined as

$$W_2(\delta) = \int_{\delta_0}^{\delta}(P_e - P_m)d\delta = \int_{\delta_0}^{\delta}P_e\,d\delta - P_m(\delta - \delta_0)$$

We have assumed that the system is initially at an equilibrium state with $\delta = \delta_0$ and $\dot{\delta} = 0$ at $t = t_0$. The disturbance(s) start at $t = t_0$, and at $t = t_{cl}$, the system enters the postdisturbance (or postfault) state. The total energy $W(\delta, \dot{\delta})$ at $t = t_{cl}$ is given by

$$W(t_{cl}) = W_1(\dot{\delta}_c) + W_2(\delta_c)$$

Note that $W_1(\dot{\delta}_c)$ is given by

$$W_1(\dot{\delta}_c) = \frac{1}{2}M(\dot{\delta}_c)^2 = \int_{\delta_0}^{\delta_c}(P_m - P_e)d\delta = A_1 \tag{2.43}$$

where $A_1$ is the area shown in Figure 2.6. This figure shows the power angle curve for a lossless system for the prefault and postfault system (assumed to be identical). The power output during the fault is assumed to be zero.

The system is stable (from EAC) if the area $A_2 \geq A_1$, where $A_2$ is defined by

$$A_2 = \int_{\delta_c}^{\delta_u}(P_e - P_m)d\delta \tag{2.44}$$

where $\delta_u = \pi - \delta_0$ is the rotor angle at the UEP for the postfault system. This stability criterion can be expressed as

$$W_1(t = t_{cl}) + W_2(t = t_{cl}) \leq W_{cr}, \quad W_{cr} = \int_{\delta_0}^{\delta_u}(P_e - P_m)d\delta \tag{2.45}$$

It is convenient to express the PE with respect to the postfault SEP ($\delta_s$). In general, $\delta_s \neq \delta_0$.

## Remarks

1. The PE for the SIME (Equation 2.38) for the two-machine system is given by

$$W_2(\delta) = -P_i(\delta - \delta_s) + GE_1E_2(\sin\delta - \sin\delta_s) + BE_1E_2(\cos\delta_s - \cos\delta)$$

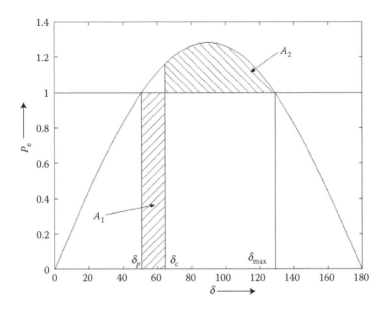

**FIGURE 2.6**
Power angle curves for a SMIB system.

This shows that the transfer conductance $G_{12}$ in a two-machine system contributes a path-independent component to the energy function. However, this is not true when there are more than two machines.

2. For constant $P_i$, the energy function $W(\delta,\dot{\delta})$ is a Lyapunov function with $\dot{W} = 0$. This shows that the EP($\delta_s$, 0) is stable. If mechanical damping is considered and assumed to be uniform $(D_1/M_1 = D_2/M_2)$, then the EP is asymptotically stable.

3. If the mechanical power $P_m$ is variable and can be computed as a function of $\delta$, the EAC still applies. However, the energy function is not a Lyapunov function.

4. EEAC has been reported by Xue et al. (1989, 1992) and Pavella and Murthy (1994).

## 2.5 Lyapunov Functions for Direct Stability Evaluation

The early literature on direct methods for transient stability evaluation concentrated on applying Lyapunov function theory. The major impediment is the lack of formal procedures to construct Lyapunov functions for a nonlinear system. Also, Lyapunov stability theorems only give sufficient conditions for stability and not necessary condition. Even when a Lyapunov function exists for a nonlinear system, it may give only conservative results for estimating the domain of stability. However, based on the work of many researchers, a broad categorization of methods for the construction of Lyapunov functions has been reported (Pai 1981). These are

1. Methods based on first integrals
2. Methods based on quadratic forms
3. Methods based on Popov's criterion
4. Methods based on the solution of partial differential equations (such as Zubov's method and variable gradient method)

The first three methods will be briefly reviewed here.

1. *Methods based on first integrals*   The Lyapunov function is a linear combination of the first integrals of the system equations for a conservative system where the total energy is constant along a system trajectory. A necessary and sufficient condition for a system, $\dot{x} = f(x), x \in R^n$ and $f(0) = 0$ (origin is an EP), to have a first integral is given by

$$\sum_{i=1}^{n} \frac{\partial f_i}{\partial x_i} = 0 \tag{2.46}$$

There are no general methods to construct the first integral except for second-order systems.

2. *Methods of quadratic forms (Krasovskii's method)*   This is an extension of the Lyapunov function applicable to a linear system

$$\dot{x} = Ax \tag{2.47}$$

The quadratic form, where $P$ is a symmetric positive definite matrix, is a Lyapunov function for the linear system described by Equation 2.47. The origin $(x = 0)$ is an EP that is asymptotically stable if

$$\dot{V} = \frac{d}{dt}(x^t Px) = -(x^t Qx) < 0 \tag{2.48}$$

This is equivalent to the condition that the symmetric matrix $Q$ is positive definite. $Q$ is defined by

$$A^T P + PA = -Q \tag{2.49}$$

To construct a Lyapunov function for a linear system defined by Equation 2.47, it is convenient to select $Q$ matrix and solve the matrix Equation 2.49. This is equivalent to a set of $(n(n + 1))/2$ linear equations for the elements of $P$. The matrix equation has a unique solution for $P$ if and only if $\lambda_i + \lambda_j \neq 0$ for all $i,j = 1,2,\ldots,n$, where $\lambda_i, \lambda_j$ are the eigenvalues of the matrix $A$.

*Krasovskii's Criterion*

If there exists a constant positive definite matrix $P$ such that

$$Q(x) = PJ(x) + J(x)^T P \tag{2.50}$$

is negative definite, then the origin of the system $\dot{x} = f(x), f(0) = 0$ is asymptotically stable in the large (ASIL).

ASIL implies that the EP is globally asymptotically stable. Any trajectory starting at any initial point in the entire state space reaches the SEP as $t \to \infty$. It is to be noted that the linear system satisfying Equation 2.50 is also ASIL.

3. *Methods based on the Popov criterion* Some nonlinear systems can be represented by linear systems with nonlinearity in the feedback path (see Figure 2.7). The nonlinearity lies in the first and third quadrant such that $\sigma f(\sigma) > 0$. Luré constructed a Lyapunov function for such systems, which can be expressed as a sum of quadratic function (of state variables) and integral of the nonlinearity (Pai 1981). Subsequently, Popov (1962) proposed a frequency domain criterion for the global asymptotic stability of a system with a single nonlinearity. Kalman (1963) and Yakubovitch (1962) established the connection between the Popov's criterion and the Lyapunov function proposed by Luré by proving that the satisfaction of Popov's criterion is a necessary and sufficient condition for the existence of the Lyapunov function. A generalized Popov criterion for multiple nonlinearities was derived by Moore and Anderson (1968). Willems (1969), Willems and Willems (1970), Pai and Murthy (1974), and

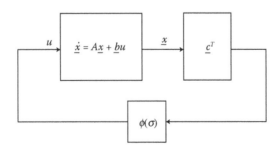

**FIGURE 2.7**
System with nonlinearity in the feedback path.

Henner (1974) applied Moore's criterion to power systems. Willems and Willems (1970) applied the Lyapunov function to the computation of transient stability regions for multimachine power systems with classical machine models and neglecting transfer conductances. Further developments consisted of extending the method to multiple argument nonlinearities that enable the treatment of flux decay and voltage regulators (Kakimoto et al. 1980, Pai and Rai 1974).

Willems and Willems (1970) consider a system of $n$ generators modeled by their swing equations of the form

$$M_i \frac{d^2 \delta_i}{dt^2} + D_i \frac{d\delta_i}{dt} + P_{ei} - P_{mi} = 0, \quad i = 1, 2, \ldots, n \tag{2.51}$$

Neglecting transfer conductances, $P_{ei}$ can be expressed as

$$P_{ei} = E_i^2 G_i + \sum_{\substack{i=1 \\ j \neq i}}^{n} E_i E_j B_{ij} \sin(\delta_i - \delta_j), \quad i = 1, 2, \ldots, n \tag{2.52}$$

The above implies that

$$\sum_{i=1}^{n} P_{mi} = \sum_{i=1}^{n} E_i^2 G_i \tag{2.53}$$

is a necessary condition for the existence of an equilibrium. The dynamic equations describing the motion of the system about the EP can be written in the state form as

$$\frac{dx}{dt} = Ax - Bf(Cx) \tag{2.54}$$

where $x = \begin{bmatrix} y \\ z \end{bmatrix}$ is a $2n$-dimensional vector.

$$y_i = \omega_i = \frac{d\delta_i}{dt}, \quad z_i = \delta_i - \delta_{i0}, \quad i = 1, 2, \ldots, n$$

$\delta_{i0}$ is the equilibrium angle of the machine $i$.

The vector valued function $f(d)$ has $m$ elements and is of the diagonal type, which implies that the $i$th component $\sigma_i$ of the vector $d$ determines the $i$th component of $f_i$ such that

$$f_i(\sigma_i) = \sin(\sigma_i + \sigma_{i0}) - \sin \sigma_{i0}, \quad i = 1, 2, \ldots, m, \quad m = \frac{n(n-1)}{2}$$

where $\sigma_k = z_i - z_j$, $i = 1, 2, \ldots, n$, $j = 1, 2, \ldots, n$, and $i \neq j$.

### 2.5.1 Construction of Lyapunov Function

The Lyapunov function is formulated based on the following theorem:

### Theorem

The null solution of the differential Equation 2.54 is asymptotically stable in the large if

1. $0 < \sigma_i f_i(\sigma_i)$ for all $i = 1,2,\ldots,m$ and $\sigma_i \neq 0$
2. There exists a diagonal $(m \times m)$ matrix. $Q = \text{diag} (q_i); q_i \geq 0$

such that $Z(s) = (I_m + Qs)C(sI_{2n} - A)^{-1}B$ is a positive real matrix, that is, $Z(j\omega) + Z^T(-j\omega)$ is a nonnegative definite Hermitian matrix for all real $\omega > 0$.

The Lyapunov function is given by

$$V(x) = x^t P x + \sum_{i=1}^{m} 2q_i \int_0^{C_i x} f_i(\sigma_i) d\sigma_i \qquad (2.55)$$

where $C_i$ is a $2n$-dimensional vector, that is, the $i$th row of the matrix $C$ and $q_i$ is the $i$th entry of the diagonal matrix $Q$.

$P$ is a real $(2n \times 2n)$ symmetric, positive definite matrix, which is the solution of the three algebraic matrix equations (Willems and Willems 1970).

The positive definite solution for $P$ can be obtained by

a. Spectral factorization methods
b. Direct solution of nonlinear algebraic equations
c. Finding the steady state of a nonlinear matrix differential equation

It is to be noted that nonlinearities appearing in the power system model do not satisfy condition (1) of the above theorem. Actually, the inequality $\sigma_i f(\sigma_i) \geq 0$ is only satisfied for a range of values of $\sigma$. Hence, the system is not asymptotically stable in the large. However, the Lyapunov function can be used to estimate the domain or basin of attraction (surrounding the SEP). Defining $S$ as the region where the conditions of the theorem are satisfied, it can be said that $\dot{V}(x)$ is nonpositive in the region $S$. If $\partial S$ is the boundary of $S$, let $V_{cr}$ denote the minimum of $V(x)$ over all $x$ in $\partial S$. The equation $V(x) = V_{cr}$ defines a bounded surface inside $S$ and contains the origin. The region enclosed by this surface belongs to the basin of attraction. The estimate of the basin of attraction can be improved by the following methods outlined in Willems (1969).

## 2.6 Energy Functions for Multimachine Power Systems

We consider here the classical model of generators—the stator is represented by a constant voltage source $E$ in series with a transient reactance $x'$. The generators are mathematically modeled by the swing equations given by Equation 2.51. As discussed in Section 2.2.4, we have the expression of electrical power output ($P_{ei}$) of machine $i$ given by

$$P_{ei} = \sum_{\substack{j=1 \\ j \neq i}}^{n} [C_{ij} \sin(\delta_i - \delta_j) + D_{ij} \cos(\delta_i - \delta_j)] \tag{2.56}$$

where $C_{ij} = E_i E_j B_{ij}$ and $D_{ij} = E_i E_j G_{ij}$.

Note that we use a reduced network retaining only the internal nodes of generators by eliminating all other modes. This is possible if the loads are assumed to be linear, that is, they can be modeled as constant impedances.

### 2.6.1 Characterization of Transient Stability

The question that needs to be answered is when the multimachine system is transiently stable following major disturbance(s) such as three-phase faults followed by clearing of the fault that is often accompanied by tripping of the faulted line. (Note that clearing of a bus fault need not result in tripping of line(s)). When connecting a large load, the tripping of a line due to overload-can also be viewed as large disturbance. Sometimes, the tripped line can be reclosed after allowing adequate time for the extinction of the arc and recovery of the air insulation in overhead lines.

The system is said to be in synchronous equilibrium if relative velocities and accelerations between any two generators is zero. Mathematically, we can state the conditions as

$$\frac{d\delta_{ij}}{dt} = 0, \quad \frac{d^2\delta_{ij}}{dt} = 0, \quad i = 1, 2, \ldots, (n-1), \quad j = 2, 3, \ldots, n \tag{2.57}$$

There are $(n(n-1))/2$ such conditions, where $n$ is the number of generators.

We can state that a power system is synchronously stable if the synchronous equilibrium is reached in finite time. The time is not specified; however, the literature refers to the first swing and multiswing stability. While swinging of an SMIB system is well defined, defining the first swing in a multimachine system based on the swing curves is not easy.

In checking transient stability, it is assumed that the postfault system has an SEP. If the EP is unstable and is of type 1 (with one positive real eigenvalue of the system Jacobian matrix evaluated at the EP), the system trajectory becomes unbounded. On the other hand, if the UEP is of type 2 (with two complex eigenvalues with positive real parts), the trajectory following a large disturbance is initially bounded. However, the machine rotor oscillations increase in amplitude with time and when the amplitude reaches a certain level, the system loses synchronous stability. The instability in such cases is primarily due to the presence of fast-acting excitation control with high-gain AVR (DeMello and Concordia 1969). When generators are described by classical models, we can generally ignore the possibility of the negatively damped oscillations.

Aylett (1958) considers $(n(n-1))/2$ swing equations for describing relative accelerations between any two machines in an $n$-machine system. These equations are given by

$$\frac{M_i M_j}{M_T}(\ddot{\delta}_i - \ddot{\delta}_j) = \frac{M_j}{M_T}P_{mi} - \frac{M_i}{M_T}P_{mj} - \frac{M_j}{M_T}P_{ei} + \frac{M_i}{M_T}P_{ej} \qquad (2.58)$$

The energy function for the system is given by

$$W = \sum_{i=1}^{n-1}\sum_{j=i+1}^{n}\left[\begin{array}{c} \dfrac{1}{2M_T}M_i M_j(\omega_i - \omega_j)^2 - \dfrac{1}{M_T}(P_i M_j - P_j M_i)(\delta_{ij} - \delta_{ij}^s) \\ \\ -C_{ij}(\cos\delta_{ij} - \cos\delta_{ij}^s) \end{array}\right] \qquad (2.59)$$

where $P_i = P_{mi} - E_i^2 G_{ii}$, $\omega_i = (d\delta_i/dt)$, $M_T = \sum_{i=1}^{n}M_i$, and $\delta_{ij}^s$ is the value of $(\delta_i - \delta_j)$ at postfault SEP.

In deriving the energy function, it is assumed that the transfer conductances $(G_{ij}, i \neq j)$ are neglected. The presence of the transfer conductances introduces energy components that are path dependent and this fact implies that the energy function is not a Lyapunov function unless the transfer conductances are neglected. Note that for a two-machine system, the transfer conductance does not pose any problem as the two-machine system can be reduced to an SMIB system.

It is inconvenient to consider $n(n-1)/2$ swing equations for intermachine oscillations in a large system. There is a simpler approach that is described next.

## 2.6.2 Center of Inertia Formulations

In classical mechanics (Goldstein 1959), the concept of a center of mass is used. We can define a rotational analog of the center of mass. This is termed as center of inertia (COI) by Stanton (1972) and center of angle (COA) by Tavora and Smith (1972a). We will use the former terminology as more appropriate. The COI is described by the following equation (neglecting damping):

$$M_T \ddot{\delta}_0 = \sum_{i=1}^{n} P_{mi} - \sum_{i=1}^{n} P_{ei} = P_{COI} \tag{2.60}$$

where $\delta_0$ is called the COA defined as

$$\delta_0 = \frac{1}{M_T} \sum_{i=1}^{n} M_i \delta_i \tag{2.61}$$

The COA can be used as a convenient reference for the measurement of rotor angles of machines. Thus

$$\theta_i = \delta_i - \delta_0, \quad i = 1,2,\ldots,n \tag{2.62}$$

where $\theta_i$ is called the off-center angle of machine $i$.

Using Equation 2.61, we can show that

$$\sum_{i=1}^{n} M_i \theta_i = 0, \quad \sum_{i=1}^{n} M_i \dot{\theta}_i = 0, \quad \sum_{i=1}^{n} M_i \ddot{\theta}_i = 0 \tag{2.63}$$

The above equations imply that $\theta_n$ is a dependent variable and can be expressed as

$$\theta_n = -\frac{1}{M_n} \sum_{i=1}^{n-1} M_i \theta_i \tag{2.64}$$

The above equation shows that it is adequate to consider swing equations only for $(n-1)$ machines. These are

$$M_i \ddot{\theta}_i = P_{mi} - P_{ei} - \frac{M_i}{M_T} P_{COI}, \quad i = 1,2,\ldots,(n-1) \tag{2.65}$$

An alternate formulation using internode coordinates is defined as (Tavora and Smith 1972a)

$$\alpha_i = \theta_i - \theta_n = \delta_i - \delta_n, \quad i = 1,2,\ldots,(n-1) \tag{2.66}$$

For a lossless system, we can express $P_{ei}$ as

$$P_{ei} = C_{in} \sin(\alpha_i) + \sum_{\substack{j=1 \\ j \neq i}}^{n-1} C_{ij} \sin(\alpha_i - \alpha_j) \tag{2.67}$$

The system is said to be in synchronous equilibrium if

$$\dot{\theta}_i = 0 \quad \text{and} \quad \ddot{\theta}_i = 0, \quad i = 1,2,\ldots,(n-1) \tag{2.68}$$

The system is said to be in frequency equilibrium if

$$\ddot{\delta}_0 = 0 \tag{2.69}$$

## Remarks

1. When the electrical network is assumed to be lossless and the loads are modeled as constant power type, $P_{COI}$ is independent of the network variables. This assumption helps to simplify the system modeling by considering only the $n$ (second-order) equations for $\theta_1, \theta_2, \ldots, \theta_{n-1}$ and $\delta_0$.

2. The system is transiently stable even if all the generators accelerate (the system is not in frequency equilibrium).

   The conditions (2.68) for synchronous equilibrium require that the system has adequate damping. However, in most of the studies—whether using direct methods or system simulation, it is customary to check whether $\theta_i$ $(i = 1,2,\ldots,(n-1))$ remains bounded following a major disturbance.

3. The rotor swings are usually damped by the mechanical damping—assumed to be linear. There are three major sources of damping: (i) prime mover, (ii) generator, and (iii) system (Crary 1947). The prime mover damping results from increase in shaft torque with decrease in speed. The per-unit damping torque coefficient is about 1.0. The generator damping is due to currents induced in the rotor windings when there is change in the rotor speed with respect to the speed of rotation of the armature magneto-motive Force (MMF) wave. The torque developed due to the induced currents is similar to that of an induction motor. The third source of damping is due to the system load characteristics.

   The speed governor control associated with the turbine(s) driving the generator can also contribute damping. Similarly, power system stabilizers (PSS) associated with the excitation controller, power swing damping controller (PSDC), or supplementary modulation controller (SMC) associated with FACTS controllers (Padiyar 2007) can also help in damping rotor swings if properly designed.

4. If damping torque is to be modeled, Equations 2.60 and 2.65 will be coupled unless the damping is uniform, that is, $D_i/M_i = \lambda$ for all $i$. For this case, Equations 2.60 and 2.65 are changed to

$$\ddot{\delta}_0 = -\lambda\dot{\delta}_0 + \frac{P_{COI}}{M_T} \tag{2.70}$$

$$\ddot{\theta}_i + \lambda\dot{\theta}_i = \frac{1}{M_i}(P_{mi} - P_{ei}) - \frac{P_{COI}}{M_T} \tag{2.71}$$

### 2.6.3 Energy Function Using COI Formulation

Neglecting transfer conductances, it is possible to obtain a simpler expression for the energy function, using COI formulations. The total energy $W$ can be expressed as

$$W = W_1(\bar{\omega}) + W_2(\theta) \tag{2.72}$$

where

$$W = W_1(\bar{\omega}) = \frac{1}{2}\sum_{i=1}^{n} M_i\bar{\omega}_i^2, \quad \bar{\omega}_i = \omega_i - \omega_0, \quad \omega_0 = \dot{\delta}_0$$

$$W_2(\theta) = -\sum_{i=1}^{n} P_i(\theta_i - \theta_i^s) - \sum_{i=1}^{n-1}\sum_{j=i+1}^{n}\left[C_{ij}(\cos\theta_{ij} - \cos\theta_{ij}^s)\right]$$

The superscript "s" indicates the value at SEP of the postfault system. It can be shown that the expression (2.72) is identical to the expression (2.59) (Athay et al. 1979).

**Remarks**

1. It can be shown that

$$\frac{1}{2}\sum_{i=1}^{n} M_i\bar{\omega}_i^2 = \frac{1}{2}\sum_{i=1}^{n} M_i\omega_i^2 - \frac{1}{2}M_T\omega_0^2 \tag{2.73}$$

2. $W_1$ is called the KE and $W_2$ is called the PE
3. The KE ($W_1$) using the internode coordinates, $\alpha_i$ ($i = 1,2,\dots,(n-1)$) can be expressed as

$$W_1 = \frac{1}{2}\sum_{i=1}^{n-1} M_i\dot{\alpha}_i^2 - \frac{1}{2}\frac{1}{M_T}\left(\sum_{i=1}^{n-1} M_i\dot{\alpha}_i\right)^2 \tag{2.74}$$

4. The KE responsible for the system separation does not include the KE associated with the COI (see Equation 2.73). As a matter of fact, the KE that affects transient stability is dependent on the mode of instability (MOI) that depends on the type and location of the disturbance or fault considered. The MOI refers to the classification of generators into two groups that separate from each other. The KE responsible for the system separation depends on the MOI. This will be discussed later.

## 2.7 Estimation of Stability Domain

The major efforts in applying direct methods using Lyapunov functions have been in the direction of estimating the region of attraction (stability domain) surrounding the SEP of the postfault system. By choosing the appropriate Lyapunov function, one tries to obtain the largest connected region of stability. However, the power systems modeled by swing equations readily permit the construction of energy functions using analogy based on classical mechanics. Neglecting damping, we get $\dot{W} = 0$ as the system is conservative. The boundary of the region of stability is defined by $W = W_{cr}$, where $W_{cr}$ is the critical energy that needs to be computed. If we assume that the system reaches the postfault state at $t = t_{cl}$, the system is said to be stable if $W(t = t_{cl}) \le W_{cr}$. We can determine stability without solving the system equations beyond $t = t_{cl}$. Also, we can get an estimate of the stability margin (using energy margin). The knowledge of the sensitivities of the EM with respect to the control parameters enables the design of preventive control in addition to DSA.

### 2.7.1 Incorporating Transfer Conductances in Energy Function

Since loads are modeled as constant impedances to reduce the system to an $n$ bus system with $n$ generators, the transfer conductances are not small and need to be taken into account in the energy function. The additional term to be added to the PE $W_2(\theta)$ in Equation 2.72 is obtained as

$$W_{23}(\theta) = -\sum_{i=1}^{n-1} \sum_{j=i+1}^{n} \int_{\theta_i^s + \theta_j^s}^{\theta_i + \theta_j} D_{ij} \cos\theta_{ij}\, d(\theta_i + \theta_j) \tag{2.75}$$

This is a path-dependent term and has to be computed by numerical integration if the system trajectory is known or can be computed. However, when $W_{cr}$ is to be obtained as the energy computed at a relevant UEP, the path from the SEP to UEP is not known. Hence, approximation is made by assuming a linear trajectory of the system between $\theta_i^s$ to $\theta_i^u$. Thus, we assume

$$\theta_i = \theta_i^s + p(\theta_i^u - \theta_i^s), \quad p \in [0,1], \ (i = 1,2,\dots,n) \qquad (2.76)$$

From Equation 2.76, we get

$$d\theta_i = (\theta_i^u - \theta_i^s)dp, \quad i = 1,2,\dots,n$$

$$d(\theta_i + \theta_j) = (\theta_i^u - \theta_i^s + \theta_j^u - \theta_j^s)dp$$

$$d(\theta_i - \theta_j) = (\theta_i^u - \theta_i^s - \theta_j^u + \theta_j^s)dp$$

Hence, we can express

$$d(\theta_i + \theta_j) = \frac{\theta_i^u + \theta_j^u - \theta_i^s - \theta_j^s}{\theta_i^u - \theta_j^u - (\theta_i^s - \theta_j^s)}d(\theta_i - \theta_j) \qquad (2.77)$$

by eliminating $dp$.

Substituting Equation 2.77 in Equation 2.75, we get

$$W_{23}^{cr} = -\sum_{i=1}^{n-1}\sum_{j=i+1}^{n}\int_{\theta_i^s+\theta_j^s}^{\theta_i^u+\theta_j^u} D_{ij}\cos\theta_{ij}\,d(\theta_i + \theta_j)$$

$$= -\sum_{i=1}^{n-1}\sum_{j=i+1}^{n}D_{ij}\frac{\theta_i^u + \theta_j^u - \theta_i^s - \theta_j^s}{\theta_i^u - \theta_j^u - (\theta_i^s - \theta_j^s)}[\sin\theta_{ij}^u - \sin\theta_{ij}^s] \qquad (2.78)$$

Note that $W_{23}^{cl}(\theta^{cl})$ at $t = t_{cl}$ can also be computed in a similar manner and the approximate expression is obtained by substituting $\theta_i^{cl}$ and $\theta_j^{cl}$ in Equation 2.78 in place of $\theta_i^u$ and $\theta_j^u$.

## Remarks

1. The approximation introduced by the assumption of a linear trajectory is rather crude and introduces errors (Athay et al. 1979). At the same time, neglecting the transfer conductances in the construction of energy functions is also not desirable (Kakimoto et al. 1978) as the total energy in the postfault system is not constant.
2. If the numerical integration is performed along the actual system trajectory (from the initial EP) during the fault, the results are likely to be more accurate.

3. According to Chiang (1989), there is no general Lyapunov-like energy function for power systems with losses when the number of generators exceeds two. Thus, we have seen that for a two-generator system, a Lyapunov function exists with transfer conductance. (Note that an infinite bus is to be viewed as an idealization of a large generator with high inertia and negligible transient reactance.) If there is a Lyapunov-like energy function for a nonlinear system, $\dot{x} = f(x)$, then it does not have a limit cycle. In contrast, it is shown by Abed and Varaiya (1984) that a three-machine system with transfer conductances can exhibit limit cycles.

4. It is feasible to construct local Lyapunov functions for systems with transfer conductances. A typical (local) Lyapunov function is $x^t P x$.

Bretas and Alberto (2003) use an extension of the invariance principle to construct Lyapunov functions for systems with small transfer conductances. They have applied their method to the 10-generator, 39-bus New England test system.

### 2.7.2 Determination of Critical Energy

#### 2.7.2.1 Single-Machine System

We have observed from the stability analysis of an SMIB system (from EAC) that the critical energy is the PE evaluated at the UEP. If the power angle curve is sinusoidal, it is routinely assumed that the rotor angle corresponding to UEP is $(\pi - \delta_s)$ in radians, where $\delta_s$ is the angle corresponding to the (postfault) SEP. However, an SMIB system has two UEPs in the range of $-2\pi \leq \delta \leq \pi$ (note that there are also two SEPs).

The UEPs are (a) $\pi - \delta_s$ and (b) $-(\pi + \delta_s)$. We can pose the following questions:

1. Which UEP determines transient stability?

2. Is it possible that the system settles down to the second SEP $(\delta_s - 2\pi)$?

Both these questions can be answered from the EAC that is related to the energy function. If $\delta_{cl} > \delta_s$, this implies that the generator rotor accelerated during the fault and gained KE. The increase in the PE ($\Delta W_2$) as $\delta$ increases beyond $\delta_{cl}$ is limited and reaches a maximum at $\delta = \pi - \delta_s$. The KE starts reducing as $\delta$ increases beyond $\delta_{cl}$. If the KE becomes zero ($\dot{\delta} = 0$) when $\delta < \delta_{u1}$ ($\delta_{u1} = \pi - \delta_s$), then the system is transiently stable. Note that $\delta$ starts reducing after it reaches the maximum value ($\delta_{max} < \delta_{u1}$). When $\delta$ reaches $\delta_s$ in the backswing, ($\dot{\delta} < 0$), and the angle continues to reduce until the acceleration gained when $\delta < \delta_s$ ensures $\dot{\delta}$ reaches zero. In the absence of the damping, $\delta$ oscillates between $\delta_{min}$ and $\delta_{max}$. It can be shown that

$$W_2(\delta_{min}) + W_2(\delta_{max}) = 0$$

Note that when $\delta > \delta_{u1}$, the rotor accelerates and the KE continues to increase until $\delta = \delta_s + 2\pi$. However, the KE gained is of such magnitude that $\delta$ will continue to increase even after it crosses $\delta_{u1} + 2\pi$. Hence, the system cannot regain stability. We say that the system trajectory becomes unbounded.

Now, let us consider a situation where the fault or disturbance causes deceleration of the rotor (this is an unlikely event, but possible if we assume that the fault resistance is such that electrical power during the fault is greater than the initial value prior to the fault). If the fault is cleared when $(\delta_s - 2\pi) < \delta_{cl} < -(\pi + \delta_s)$, the rotor angle $\delta$ will continue to decrease even if $\dot\delta_{cl} = 0$ and it can be shown (from EAC) that the system will settle at the second SEP $= (\delta_s - 2\pi)$. This implies that the system will be stable after the pole slipping.

Now assume that $\delta_{cl} > \delta_{u2}$, where $\delta_{u2}$ is the angle corresponding to the second UEP $(\delta_{u2} = -(\pi + \delta_s))$. Since $P_m > P_e$, the rotor accelerates during the postfault period until $\delta$ reaches $\delta_s$. The rotor starts decelerating after $\delta_s$ is crossed and it is likely that the system becomes unstable in the backswing. Thus, in answer to the questions, we can state

1. The first UEP $(\pi - \delta_s)$ determines transient stability
2. Pole slipping is possible and the system can settle down to a stable equilibrium

## Remarks

1. Since the power angle curve is periodic, it is also possible to restrict the range of $\delta$ to $(-\pi, \pi)$. In this case, the state space is the surface of a cylinder instead of a plane (where the angle can take any value). With the restriction on the angle (within a period), it is obvious that there is only one SEP $(\delta_s)$ and one UEP $(\pi - \delta_s)$. The trajectories in the angle space lie on the circle shown in Figure 2.8 where the SEP is

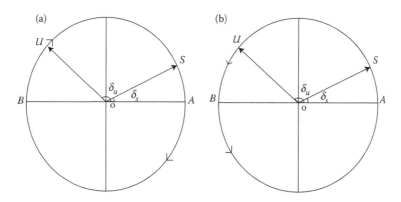

**FIGURE 2.8**
Trajectories in angle space for a SMIB system. (a) Stable case and (b) unstable case.

marked as point S and UEP is marked as point U. Note that point B corresponds to both $\pi$ and $-\pi$ values of the angle. If the faulted trajectory (moving in the clockwise direction in the angle space) crosses the UEP toward the SEP, then the system is stable; otherwise, it is unstable (see Figure 2.8a). Similarly, if the faulted trajectory (moving anticlockwise in the angle space) crosses UEP, then the system is unstable; otherwise, it is stable (see Figure 2.8b).

2. We can extend the analysis to determine the region of stability based on UEP to a multimachine system. However, there are a large number of UEPs in a multimachine power system. According to one estimate, these can be (up to) $(2^n - 2)$ UEPs. The question then is which UEP is relevant in computing the critical energy.

### 2.7.2.2 Multimachine System

With $n$ machines, we have to consider the $(n - 1)$-dimensional space of the relative angles $(\delta_i - \delta_n)$, $i = 1, 2, \ldots, (n - 1)$. The PE is minimum at SEP and starts increasing as we move away from SEP. The determination of stability in a multimachine system, based on transient energy, can be explained, using the mechanical analogy of a ball rolling on the PE surface in a multidimensional bowl. The ball is initially at rest at the bottom of the bowl (where the PE is minimum). A disturbance results in imparting KE to the ball that can move upward and toward the rim of the bowl over which it can escape. The rim of the bowl is called PEBS. The rim of the bowl is not even and has several saddle points (UEPs, typically of type 1). The PE at the UEPs varies. The UEP with the lowest energy (measured with respect to SEP) is called the closest UEP. In the past, the closest UEP was considered to define the region of stability. This gave very conservative results. In addition, it was computationally burdensome as all possible UEPs had to be found and their energies compared. The determination of UEP is numerically complex and approximations were used to estimate them.

### 2.7.3 Potential Energy Boundary Surface

The concept of PEBS was first mooted by Kakimoto et al. (1978) as a set of curves passing through UEPs and orthogonal to the equipotential curves. The significance of PEBS is in the context of suggestions made by Kakimoto et al. (1978) and Athay et al. (1979) that it is better to use the CUEP instead of the closest UEP to estimate the stability domain. The CUEP lies on the PEBS through which the unstable trajectory leaves the stability region. The point of intersection of the unstable trajectory with the PEBS is called the "exit point."

Mathematically, PEBS is defined by the points that satisfy the relation that the directional derivative of the PE function along the rays emanating from SEP is zero. Thus, the following equation characterizes PEBS (Athay et al. 1979):

$$\sum_{i=1}^{n-1} f_i(\theta)(\theta_i - \theta_i^s) = 0 \qquad (2.79)$$

The advantage of PEBS as originally defined by Kakimoto et al. (1978) is that a simple procedure for the computation of stability boundary can be given. The computation of the CCT is given below:

*Step 1*: Integrate the fault-on trajectory until $W_{PE}$ reaches a local maximum. This value is an estimate of the true $W_{cr}$.

*Step 2*: From the fault-on trajectory, determine the instant when $W = W_{cr}$. This is the CCT ($t_{cr}$). Figure 2.9 shows this graphically.

It is meaningful to view PEBS as the stability boundary of a reduced gradient system (in the angle space only) given by

$$\dot{\theta}_i = f_i(\theta) = -\frac{\partial W_{PE}}{\partial \theta_i} \qquad (2.80)$$

Chiang et al. (1988) show that PEBS defined as the stability boundary of the reduced dynamical system is consistent with the geometric construction procedure suggested in the paper by Kakimoto et al. (1978). This is in view

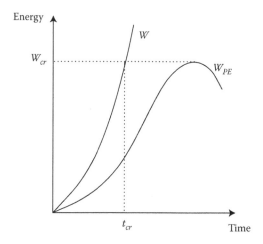

**FIGURE 2.9**
Determination of critical clearing time using PEBS.

of the fact that the stability boundary for the system described by Equation 2.80 is given by

$$\partial A(\theta^s) = \bigcup_i M^s(\theta^i) \tag{2.81}$$

where $\theta^i$ is the type 1 UEP lying on the stability boundary and $M^s$ indicates the stable manifold of $\theta^i$. It is easy to see that PEBS defined above intersects the level surface $W_{PE}(\theta) = C$ orthogonally as the vector field is orthogonal to the level surface.

However, PEBS is not identical to the projection of the stability boundary of the original system in the subspace characterized by $\omega = 0$. The PEBS intersects the projection of the stability boundary of the original system at the UEPs that lie on the stability boundary.

An important result given in Chiang et al. (1988) is stated below.

**Theorem**

Let $(\theta^u, 0)$ be a UEP on the stability boundary $\partial A(\theta^s, 0)$ of the original system. Then, the connected constant energy surface $\partial W(\theta^u, 0)$ intersects the stable manifold $M^s(\theta^u, 0)$ only at the point $(\theta^u, 0)$; moreover, the connected component of the set $W(\theta, \omega) < W(\theta^u, 0)$ does not contain any point that belongs to the stable manifold $M^s(\theta^u, 0)$.

Chiang et al. (1988) also show that the PEBS method can give either optimistic or slightly pessimistic results depending upon the point of intersection of the fault-on trajectory with the PEBS. They propose a modified PEBS method as follows:

*Step 1*: From the fault-on trajectory, detect the point $\theta^*$ at which the projected trajectory $\theta(t)$ crosses the PEBS.

*Step 2*: Find the EP of the system whose stable manifold contains the point $\theta^*$, say $\theta^u$. The value of $W_{PE}(\theta^u)$ is the critical energy $W_{cr}$.

This modification is the basis of the boundary of stability region-based controlling unstable equilibrium point (BCU) method described later.

### 2.7.4  Controlling UEP Method

In this method, the critical energy is determined as

$$W_{cr} = W(x_u^*) = W_2(\delta_u^*)$$

where $x_u^*$ is the type 1 UEP lying on the stability boundary and whose stable manifold is intersected by the fault-on trajectory. As the rotor velocity deviation, $\omega$, is zero at any EP, the critical energy is the same as the PE $W_2$ evaluated at the value $\delta_u^*$ corresponding to the CUEP.

The determination of CUEP can be complex. There are several approaches to the determination of CUEP. Two prominent approaches that have been applied for large systems are

1. Mode of disturbance (MOD) (Fouad and Vittal 1992)
2. BCU method (Chiang et al. 1994)

Here, we consider the second approach, which is related to the PEBS method as mentioned earlier.

### 2.7.5 BCU Method

The BCU method was proposed by Chiang et al. (1994). BCU stands for boundary of stability region-based controlling unstable equilibrium point. This method is based on the relationship between the stability boundary of the (postfault) classical power system model and the stability boundary of the following (postfault) reduced system model defined as

$$\dot{\theta} = f(\theta) \tag{2.82}$$

The state variables of this reduced system are rotor angles only with dimension of $n$ while the dimensions of the original system is $2n$. It is easy to see that if $\hat{\theta}$ is an EP of the system defined by Equation 2.82, then $(\hat{\theta},0)$ is an EP of the original system. Under the condition of small transfer conductances, it can be shown that

1. $(\hat{\theta}^s)$ is an SEP of the reduced system if and only if $(\hat{\theta}^s,0)$ is an SEP of the original system.
2. $(\hat{\theta}^u)$ is a type-k EP of the reduced system if and only if $(\hat{\theta}^u,0)$ is a type-k EP of the original system.
3. If the one-parameter transversality condition is satisfied, then $(\hat{\theta})$ is on the stability boundary $\partial A(\hat{\theta}^s)$ of the reduced system if and only if $(\hat{\theta},0)$ is on the stability boundary $\partial A(\hat{\theta}^s,0)$ of the original system.

The above results establish a relationship between the stability boundary $\partial A(\hat{\theta}^s)$ and the stability boundary $\partial A(\hat{\theta}^s,0)$ and suggest a method of finding the CUEP of the original system via the location of CUEP of the reduced system.

One version of the BCU method involves the following steps for the determination of the CUEP:

*Step 1*: From the fault-on trajectory, detect the exit point $\theta^*$, which is the point where the projected trajectory $\theta(t)$ exits the stability boundary of the reduced system. This exit point corresponds to the point at which the first local maximum of the PE $W_{PE}$ is reached.

*Step 2*: Use the point $\theta^*$ as the initial condition and integrate the postfault reduced system of Equation 2.82 to find the first local minimum of

$$\sum_{i=1}^{n} \|f_i(\theta)\|, \quad \text{say at } \theta_0^*.$$

*Step 3*: Use the point $\theta_0^*$ as the initial guess to solve

$$\sum_{i=1}^{n} \|f_i(\theta)\| = 0, \quad \text{say at } \theta_{c0}^*.$$

The CUEP of the original system is given by $x_u = (\theta_{c0}^*, 0)$.

The reduced system equations may be stiff and a suitable integration method is to be used in step 2.

Tests on realistic systems indicate that the MOD method is not always reliable. In comparison, the BCU method, when properly tuned to the system, provides a good compromise in reliability and computational speed.

## Remarks

1. Step 1 is the PEBS method and with the addition of two more steps, the method is refined to provide improved accuracy.

2. Step 2 of the algorithm is based on the fact that every trajectory on the stability boundary, say $\partial A(x_s)$ of the region of attraction of the SEP, $x_s$, converges to one of the EPs on $\partial A(x_s)$ as time increases. The basic result on the stability boundary is that it is contained in the set that is a union of the stable manifolds of the UEP on the $\partial A(x_s)$ (Chiang et al. 1987, Tsolas et al. 1985).

3. The BCU method is primarily applicable for classical generator models with reduced network representation. This has the drawback of having to consider path-dependent terms to account for transfer conductances.

Zou et al. (2003) have attempted to provide a theoretical foundation of the CUEP method for the analysis of network-preserving power system models. However, the extended CUEP method tends to be more conservative in stability evaluation.

## 2.8 Extended Equal Area Criterion

A simple yet reasonably reliable approach for direct stability evaluation is suggested in Xue et al. (1989, 1992). This is based on the conjecture that "the loss of synchronism of a multimachine system, whenever it occurs, is triggered off by the machine's irrevocable separation into two groups; hence, the idea of subdividing the system machines into the 'critical group,' generally comprising a few machines, and the remaining group, comprising the majority of machines." This conjecture is supported by theoretical analysis of Chiang et al. (1987) where it is shown that the CUEP lying on the stability boundary is usually of type 1.

On the basis of the conjecture, it is further assumed that the system stability can be assessed by replacing the machines of each group by an equivalent and finally the two equivalent machines replaced by an SMIB system. The EAC is then applied to this SMIB system for stability assessment.

The formulation of the EEAC method is given below. The generators are represented by classical models and loads by constant impedances. There is no need for the use of the COI reference frame.

### 2.8.1 Formulation

The machine equations are given by

$$M_i \frac{d^2\delta_i}{dt^2} = P_{mi} - P_{ei} \tag{2.83}$$

where

$$P_{ei} = E_i^2 Y_{ii} \cos\beta_{ii} + \sum_{j=1, j\neq i}^{m} E_i E_j Y_{ij} \cos(\delta_i - \delta_j - \beta_{ij}) \tag{2.84}$$

$Y_{ij}\angle\beta_{ij} = Y(i, j)$ where $Y$ is the reduced bus admittance matrix. The other symbols are as defined earlier.

Denoting $S$ to be the set of critical machines and $A$ the set of remaining machines, we assume that

$$\delta_k = \delta_s \,\forall\, k \in S, \quad \delta_l = \delta_a \,\forall\, l \in A \tag{2.85}$$

On the basis of Equation 2.85, we can derive two swing equations for the two equivalent machines representing the two groups $S$ and $A$. These are

$$M_s \ddot{\delta}_s = \sum_{k\in S}(P_{mk} - P_{ek}), \quad M_a \ddot{\delta}_a = \sum_{l\in A}(P_{ml} - P_{el}) \tag{2.86}$$

where

$$M_s = \sum_{k \in S} M_k, \quad M_a = \sum_{l \in A} M_l$$

Finally, setting

$$\delta = \delta_s - \delta_a \tag{2.87}$$

and using Equation 2.86, we can derive

$$M\ddot{\delta} = P_m - [P_c + P_{max}\sin(\delta - \alpha)] = P_m - P_e \tag{2.88}$$

If the rotor angles in each group are assumed to be identical to the respective COA (according to Equation 2.85), then the parameters $P_c$, $P_{max}$ and $\alpha$ assume constant values. This is termed as static EEAC. In contrast, dynamic extended equal area criterion (DEEAC) permits the parameters of the OMIB system described by Equation 2.88 to vary by considering variations in the rotor angles.

## 2.8.2 Approximation of Faulted Trajectory

The use of the Taylor series helps to compute CCT based on critical clearing angle without having to integrate faulted system equations.

The Taylor series contains only even derivatives of $\delta$. Truncating the series after $t^4$ term yields

$$\delta = \delta_0 + \frac{1}{2}\gamma t^2 + \frac{1}{24}\ddot{\gamma} t^4 \tag{2.89}$$

where $\gamma$ denotes the second-order derivates of $\delta$ at $t = 0^+$ (immediately after the fault) and $\ddot{\gamma}$ its fourth derivative also at $t = 0^+$.

Athay et al. (1979) have proposed an alternative approximation of faulted trajectory (although not in the context of EEAC). They assume that during the fault, the swing equation can be approximated as

$$M\ddot{\delta} = f = a + b\cos\eta t \tag{2.90}$$

where the constants $a$, $b$, and $\eta$ can be determined from initial conditions.

### 2.8.3 Identification of Critical Cluster

The procedure for the identification of the critical cluster is given below:

1. Draw up a list of candidate critical machines.

   This can be done using initial acceleration criterion that consists of (a) classification of machines in a decreasing order of their initial accelerations and (b) selection of machines that have accelerations close to that of the top machines.

2. Consider candidate critical clusters composed of one, two, or more machines and obtained by successively combining the candidate critical machines.

3. Compute the corresponding candidate CCT: The smallest one is the actual CCT; the actual critical cluster is precisely that which furnishes the CCT.

The reliable identification of critical clusters is not a simple task and improvements in the above procedure have been suggested (Xue and Pavella 1993).

### Remarks

1. The approach suggested by Xue et al. (1993) is similar to that described earlier in the MOD method. In both methods, the CUEP is related to the generators that have advanced rotor angles from the rest.

2. By using EEAC, the computation of the critical energy is simplified. However, the problem of accurately identifying the critical cluster results in introducing some empirical approaches that may not work always.

# 3

## Structure Preserving Energy Functions for Systems with Nonlinear Load Models and Generator Flux Decay

### 3.1 Introduction

Most of the literature on Lyapunov-like energy functions (reviewed in the previous chapter) assume linear voltage-dependent load models that enable the reduction of power system network, retaining only the generator internal nodes. The generators are also modeled by the simplest model, neglecting saliency ($x'_d \neq x_q$). Field flux decay and damper windings are neglected and only the swing equations are included for the analysis of synchronous transient stability. The transfer conductances (TCs) in the reduced admittance matrix of the network are neglected to construct the energy function. It is to be noted that although the transmission network can be reasonably modeled as lossless, the TCs are mainly introduced due to the load impedances with substantial resistive components.

A major disadvantage of using reduced bus admittance matrix is that the original network topology is lost. This has two adverse effects: (1) the network controllers such as HVDC converter and FACTS controllers cannot be satisfactorily modeled and (2) the application of electrical circuit or network theory concepts is not feasible. Tavora and Smith (1972) in their three-part paper have applied circuit theory in the analysis of equilibrium and stability in power systems. However, they have not retained load nodes. Although they have modeled the loads as constant torque or power type, they have clubbed it with the mechanical torque. This is possible only if the generator internal impedances are neglected.

### 3.2 Structure Preserving Model

Bergen and Hill (1981) proposed an SPM for stability analysis by retaining the identity of the load nodes, which also preserved the structure of the transmission network. The active loads were modeled as frequency dependent, but independent of the bus voltages. They assumed that the relationship with frequency is linear.

The formulation assumed that

1. The transmission network is lossless
2. The network has $n$ buses with constant voltage magnitudes, which are all assumed to be equal to 1
3. The load model is given by

$$P_{Di} = P_{Di}^0 + D_i \dot{\delta}_i, \quad i = m+1,\ldots,n \tag{3.1}$$

It is assumed that the first $m$ nodes represent the internal nodes of the $m$ generators.

The system equations are

$$M_i \ddot{\delta}_i + D_i \dot{\delta}_i + \sum_{\substack{j=1 \\ j \neq i}}^{n} b_{ij} \sin(\delta_i - \delta_j) = P_{mi} - P_{Di}^0 \triangleq P_i^0, \quad i = 1, 2, \ldots, n \tag{3.2}$$

where

$$M_i > 0, \quad i = 1, 2, \ldots, m$$

$$M_i = 0, \quad i = m+1, \ldots, n$$

$$D_i > 0, \quad i = 1, 2, \ldots, n$$

$$P_{Di}^0 = 0, \quad i = 1, 2, \ldots, m$$

$$P_{mi} = 0, \quad i = m+1, \ldots, n$$

Note that Equation 3.2 represents both the generator and loads with appropriate choice of the parameters. The inertia and the mechanical power ($P_m$) are zero at the load buses, whereas the load power ($P_D$) is zero at the generator internal buses. In steady state, $\sum_{i=1}^{n} P_i = 0$. However, following a fault or a disturbance, this equality may not apply.

By defining

$$\omega_0 = \frac{\sum_{i=1}^{n} P_i^0}{\sum_{i=1}^{n} D_i} \tag{3.3}$$

and

$$\bar{\omega}_i = \omega_i - \omega_0 \tag{3.4}$$

$$\bar{P}_i = P_i^0 - D_i \omega_0 \tag{3.5}$$

we can observe that $\sum_{i=1}^{n} \bar{P}_i = 0$ and the equilibrium value of angular velocity deviation, $\bar{\omega}_i = 0$.

Note that $\omega_i = d\delta_i/dt$.

The state space has a dimension of $n + m - 1$ as one of the angles (say $\delta_n$) can be chosen as reference. Thus, there are $(n - 1)$ internodal angles and $m$ rotor velocity deviations. The angles are defined by the vector $\underline{\alpha}$ where

$$\underline{\alpha} = [\alpha_1, \dots, \alpha_{n-1}]^T, \quad \alpha_i = \delta_i - \delta_n \tag{3.6}$$

We define the power flow in the branch $k$ by

$$p_k = g_k(\sigma_k) = b_k \sin \sigma_k \tag{3.7}$$

where $\sigma_k = \delta_i - \delta_j = \alpha_i - \alpha_j$ assuming that the branch $k$ is connected between nodes $i$ and $j$. We can obtain the power injections at $(n - 1)$ nodes as

$$\underline{P} = Ag(\sigma) = Ag(A^T\alpha) \triangleq \underline{f}(\alpha) \tag{3.8}$$

where $A$ is the reduced incidence matrix. The $i$th element of vector $\underline{f}(\alpha)$ is given by

$$f_i(\alpha) = \sum_{\substack{k=1 \\ k \neq i}}^{n-1} b_{ik} \sin(\alpha_i - \alpha_k) + b_{in} \sin \alpha_i, \quad i = 1, 2, \dots, (n-1) \tag{3.9}$$

The system Equation 3.2 can be rewritten as

$$M_g \dot{\omega}_g + D_g \omega_g + T_1^T \left[ \underline{f}(\alpha) - \underline{P}^0 \right] = 0 \tag{3.10a}$$

$$D_l \omega_l + T_2^T \left[ \underline{f}(\alpha) - P^0 \right] = 0 \tag{3.10b}$$

where subscripts $g$ and $l$ refer to the generators and loads, respectively. $M_g$, $D_g$, and $D_l$ are diagonal matrices of appropriate dimensions. The matrices $T_1$ and $T_2$ are defined by

$$T = [I_{n-1} \vdots -e] \triangleq [T_1 \vdots T_2]$$

where $I_{n-1}$ is the identity matrix of dimension $(n - 1)$ and $e$ is the $(n - 1)$-dimensional vector with unity entries. The matrix $T_1$ has $m$ columns and the matrix $T_2$ has $(n - m)$ columns.

Bergen and Hill (1981) term the energy function defined on the SPM as topological Lyapunov function given by

$$V(\alpha, \omega_g) = \frac{1}{2} \omega_g^T M_g \omega_g + W(\alpha, \alpha_0) \tag{3.11}$$

$$W(\alpha, \alpha_0) = \int_{\alpha_0}^{\alpha} \left[ \underline{f}(\underline{\xi}) - \underline{f}(\underline{\xi}_0) \right]^T d\underline{\xi} \tag{3.12}$$

It can be shown that the second term in the Lyapunov function defined in Equation 3.11 can be expressed as the sum of potential energies of individual branches of the network. Thus

$$W(\alpha, \alpha_0) = \sum_{k=1}^{nb} b_k \int_{\sigma_{k0}}^{\sigma_k} (\sin \beta - \sin \sigma_{k0}) d\beta \tag{3.13}$$

where $nb$ is the numbers of lines (branches of the network) and $\sigma_k$ is the angle across the $k$th branch. The first term in the Lyapunov function is the sum of kinetic energies in individual generator rotors given by

$$\frac{1}{2} \omega_g^T M_g \omega_g = \frac{1}{2} \sum_{i=1}^{m} M_i \bar{\omega}_i^2 \tag{3.14}$$

where $\bar{\omega}_i$ is the relative frequency defined in Equation 3.4.

**Remarks**

1. Note that $\omega_0$ is the equilibrium value of the center of inertia (COI) frequency as

$$M_T \dot{\omega}_0 = \sum_{i=1}^{m} M_i \dot{\omega}_i = \sum_{i=1}^{n} P_i^0 - \sum_{i=1}^{n} D_i \omega_i \tag{3.15}$$

   Since $\omega_0$ is defined from Equation 3.3, we have $\dot{\omega}_0 = 0$. Since $M_i > 0$, it is obvious that $\omega_0$ is also the equilibrium value of individual generator speeds (measured with respect to a synchronously rotating reference frame).

2. Equation 3.13 is an important result, based on network theory. This enables the definition of energy on the critical cutset, which is defined

as the cutset whose branches (transmission lines) are all zero valued (lines for which cos $\sigma_i = 0$ are defined are zero valued). The significance of the critical cutset (according to Bergen and Hill) is that the EP for the power system described by Equation 3.10 is asymptotically stable if the system network including the generator stator branches has no critical cutsets. Note that any generator branch of transient reactances can also be a critical cutset with a single branch. Thus, it is a necessary condition that all generator branches should not be zero valued to ensure that the critical cutset is made of transmission lines.

3. The system separation also occurs along a cutset and the energy in a vulnerable outset can be related to the critical transient energy. This concept will be discussed in detail in Chapter 6.

4. The power flow equations for a lossless network with PV nodes have been interpreted to be that of a nonlinear resistive network. However, this analogy fails to relate the energy functions to the energy stored in the network.

It will be shown later that the structure of the system model can be interpreted as that of $R-L-C$ network with nonlinear inductors.

## 3.3 Inclusion of Voltage-Dependent Power Loads

Modeling the load buses as $P-V$ type with constant voltage magnitudes is not realistic unless voltage-regulating devices are present. In general, the load buses are modeled as $P-Q$ type with voltage-dependent reactive powers.

Narasimhamurthi and Musavi (1984) proposed a load model with constant active power and voltage-dependent reactive power defined by

$$Q_i = f_{qi}(V_i), \quad i = m+1,\ldots,n \tag{3.16}$$

They showed that this introduces an additional term in the PE function given by

$$W(V) = \sum_{i=m+1}^{n} \int_{V_{io}}^{V_i} \frac{f_{qi}(x)}{x} dx \tag{3.17}$$

Padiyar and Sastry (1987) proposed a load model where both active and reactive powers of the loads are voltage dependent. This is more realistic and constant power loads can be viewed as a special case of the general voltage-dependent loads. In general, loads are modeled as a linear combination of constant-power-, constant-current-, and constant-impedance-type loads. Mathematically, this can be expressed as

$$\left.\begin{aligned}P_{li} &= a_0 + a_1 V + a_2 V^2 \\ Q_{li} &= b_0 + b_1 V + b_2 V^2\end{aligned}\right\} \tag{3.18}$$

## 3.4 SPEF with Voltage-Dependent Load Models

The application of the SPM proposed by Bergen and Hill is extended by relaxing the assumption of *P–V* nodes for all the buses in the system (other than the generator internal buses) and also allowing for nonconstant power loads. The network is assumed to be lossless as before to ensure that the system is conservative. However, the energy function has a path-dependent term unless the load active powers are assumed to be constants. The PE can be expressed as a sum of four or five components that can be easily computed (Padiyar and Sastry 1987).

### 3.4.1 Dynamic Equations of Generator

Consider an *n*-bus multimachine system having *m* machines supplying nonlinear voltage-dependent loads (see Figure 3.1). In the direct transient stability evaluation using energy-type Lyapunov functions, the following assumptions are usually made:

1. Each synchronous machine is represented by a classical model, namely, a constant voltage source behind transient reactance.

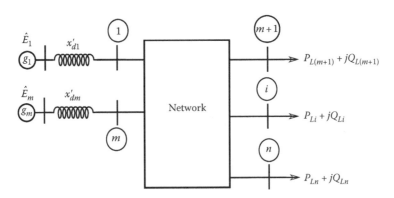

**FIGURE 3.1**
Multimachine system connected to nonlinear loads.

2. Governor action is not taken into account and thus the mechanical torque is assumed to be constant.

3. Machine damping is neglected.

4. The network is assumed to be lossless.

The last two assumptions can be relaxed without unduly complicating the analysis. However, these assumptions will lead to the system being conservative, as shown later.

Under the above assumptions, the motion of the $i$th machine is described by the following differential equations (in COI reference):

$$M_i \dot{\bar{\omega}}_i = P_{mi} - P_{ei} - \frac{M_i}{M_T} P_{COI} \tag{3.19}$$

$$\dot{\theta} = \bar{\omega}_i, \quad i = 1, 2, \ldots, m \tag{3.20}$$

where

$$P_{ei} = \frac{E_i V_i \sin(\theta_i - \phi_i)}{x'_{di}} \tag{3.21}$$

$$M_T = \sum_{i=1}^{m} M_i, \quad P_{COI} = \sum_{i=1}^{m} (P_{mi} - P_{ei}) \tag{3.22}$$

$\bar{\omega}_i = \omega_i - \omega_0 = \dot{\delta}_i - \dot{\delta}_0$. $\omega_0$ is the frequency deviation of COI defined by

$$M_T \dot{\omega}_0 = P_{COI} \tag{3.23}$$

In view of the definition of COI variables, we have

$$\sum_{i=1}^{m} M_i \theta_i = 0 \quad \text{and} \quad \sum_{i=1}^{m} M_i \bar{\omega}_i = 0 \tag{3.24}$$

### 3.4.2 Load Model

In the load model considered here, both active and reactive powers are assumed to be arbitrary functions of respective bus voltages. Thus, the equations for system loads can be written as

$$\left. \begin{array}{l} P_{li} = f_{pi}(V_i) \\ Q_{li} = f_{qi}(V_i) \end{array} \right\} \quad i = 1, 2, \ldots, n \tag{3.25}$$

The model considered here is more general than the polynomial load model of Equation 3.18.

### 3.4.3 Power Flow Equations

For a lossless transmission system, the following equations are applicable.
    Let

$$g_{1i} = \frac{E_i V_i \sin(\phi_i - \theta_i)}{x'_{di}}, \quad g_{2i} = \sum_{j=1}^{n} V_i V_j B_{ij} \sin \phi_{ij}$$

$$g_{3i} = \frac{V_i^2 - E_i V_i \cos(\theta_i - \phi_i)}{x'_{di}}, \quad g_{4i} = \sum_{j=1}^{n} V_i V_j B_{ij} \cos \phi_{ij}$$

The active power injected into the network from bus $i$ is

$$P_i = g_{1i} + g_{2i}, \quad \text{for } i = 1, 2, \ldots, m \tag{3.26}$$

$$= g_{2i}, \quad \text{for } i = m + 1, m + 2, \ldots, n \tag{3.27}$$

The reactive power injection at bus $i$ is

$$Q_i = g_{3i} - g_{4i}, \quad \text{for } i = 1, 2, \ldots, m \tag{3.28}$$

$$Q_i = -g_{4i}, \quad \text{for } i = m + 1, m + 2, \ldots, n \tag{3.29}$$

In the above expressions, $B_{ij} = \text{Im}[Y_{bus}(i,j)]$, where $Y_{bus}$ is the admittance matrix of the network (without including the machine reactances).
    The power flow equations at bus $i$ can be written as

$$P_i + P_{li} = P_i + f_{pi}(V_i) = 0 \tag{3.30}$$

and

$$Q_i + Q_{li} = Q_i + f_{qi}(V_i) = 0 \tag{3.31}$$

### 3.4.4 Structure Preserving Energy Functions

Athay et al. (1979) derived an energy function for the reduced system (after eliminating the load buses) using COI variables. The presence of TCs introduces a path-dependent term. This can be avoided by using the SPM. However, the energy function developed in Narasimhamurthy and Musavi (1984) and Tsolas et al. (1985) takes account only of constant active power loads and uses absolute rotor velocities in defining the KE. The latter would be accurate only if an infinite bus was considered in the system. The development of the energy function given here avoids both the drawbacks mentioned above. Any arbitrary voltage-dependent load characteristics are considered and the COI variables are used.

Consider the energy function defined for the postfault system.

$$W(\theta,\bar{\omega},V,\phi,t) = W_1(\bar{\omega}) + W_2(\theta,V,\phi,t) \tag{3.32}$$

where

$$W_1(\bar{\omega}) = \frac{1}{2}\sum_{i=1}^{m}M_i\bar{\omega}_i^2$$

$$W_2(\theta,V,\phi,t) = W_{21}(\theta) + W_{22}(t) + W_{23}(V) + W_{24}(V,\theta,\phi) + W_{25}(V,\theta)$$

$$W_{21}(\theta) = -\sum_{i=1}^{m}P_{mi}(\theta_i - \theta_{i0}), W_{22}(t) = \sum_{i=1}^{n}\int_{t_0}^{t}P_{li}\frac{d\phi_i}{dt}dt, W_{23}(V) = \sum_{i=1}^{n}\int_{V_{i0}}^{V_i}\frac{f_{qi}(x_i)}{x_i}dx_i$$

$$W_{24}(V,\theta,\phi) = \sum_{i=1}^{m}\left[(E_i^2 + V_i^2 - 2E_iV_i\cos(\theta_i - \phi_i))\right.$$

$$\left. - (E_{i0}^2 + V_{i0}^2 - 2E_{i0}V_{i0}\cos(\theta_{i0} - \phi_{i0}))\right]\frac{1}{2x_{di}'}$$

$$W_{25}(V,\phi) = -\frac{1}{2}\sum_{i=1}^{n}\sum_{j=1}^{n}B_{ij}(V_iV_j\cos\phi_{ij} - V_{i0}V_{j0}\cos\phi_{ij0})$$

$W_1$ is the KE and $W_2$ is the PE.

It can be proved, by direct verification, that the system defined by the Equations 3.19 through 3.31 is conservative.

**Proof**

Taking partial derivatives of $W$ with respect to $V_i$, $\theta_i$, $t$, $\phi_i$, and $\bar{\omega}_i$, respectively, one can easily derive (using system equations)

$$\frac{\partial W}{\partial V_i} = \frac{1}{V_i}(Q_{li} + Q_i) = 0 \quad \text{(from Equation 3.31)} \tag{3.33}$$

$$\frac{\partial W}{\partial \phi_i} = \frac{E_iV_i\sin(\theta_i - \phi_i)}{x_{di}'} + \sum_{j=1}^{n}B_{ij}V_iV_j\sin\phi_{ij}, \quad i = 1,2,...,m \tag{3.34}$$

$$= \sum_{j=1}^{n}B_{ij}V_iV_j\sin\theta_{ij}, \quad i = m+1,...,n$$

$$= P_i \text{ from Equations 3.26 and 3.27}$$

$$\frac{\partial W}{\partial t} = \sum_{i=1}^{n} P_{li} \frac{\partial \phi_i}{dt} \tag{3.35}$$

$$\frac{\partial W}{\partial \theta_i} = -P_{mi} + P_{ei} \tag{3.36}$$

and

$$\frac{\partial W}{\partial \bar{\omega}_i} = M_i \bar{\omega}_i \tag{3.37}$$

We have, from Equation 3.30

$$\sum_{i=1}^{n} \frac{\partial W}{\partial \phi_i} \frac{d\phi_i}{dt} + \frac{\partial W}{\partial t} = \sum_{i=1}^{n} (P_i + P_{li}) \frac{d\phi_i}{dt} = 0 \tag{3.38}$$

and

$$\sum_{i=1}^{m} \left[ \frac{\partial W}{\partial \bar{\omega}_i} \frac{d\bar{\omega}_i}{dt} + \frac{\partial W}{\partial \theta_i} \frac{d\theta_i}{dt} \right] = \sum_{i=1}^{m} (M_i \bar{\omega}_i - P_{mi} + P_{ei}) \bar{\omega}$$

$$= -\sum_{i=1}^{m} \frac{M_i}{M_T} P_{COI} \bar{\omega}_i = 0 \text{ (from Equations 3.19 and 3.24)} \tag{3.39}$$

Hence

$$\frac{dW}{dt} = \sum_{i=1}^{m} \left( \frac{\partial W}{\partial \omega_i} \frac{d\omega_i}{dt} + \frac{\partial W}{\partial \theta_i} \frac{d\theta_i}{dt} \right) + \sum_{i=1}^{m} \frac{\partial W}{\partial V_i} \frac{dV_i}{dt} + \sum_{i=1}^{n} \frac{\partial W}{\partial \phi_i} \frac{d\phi_i}{dt} + \frac{\partial W}{\partial t} = 0 \tag{3.40}$$

on substituting from Equations 3.33, 3.38, and 3.39.
    This shows that the total energy of the system is conserved.

**Comments**

1. It is assumed that the system models are well-defined in the sense
   that the voltages at the load buses can be solved in a continuous
   manner at any given time during the transient. This means that the
   system trajectories are smooth and that there are no jumps in the
   energy function.

2. Consider the term in the energy function that corresponds to the active
   power component of the load at bus $i$. Integrating by parts, we get

$$\int_{t_0}^{t} P_{li} \frac{d\phi_i}{dt} dt = P_{li}\phi_i \Big|_{t_0}^{t} - \int_{t_0}^{t} \frac{df_{pi}}{dV_i} \cdot \frac{dV_i}{dt} \phi_i \, dt$$

$$= (P_{li}\phi_i - P_{li0}\phi_{i0}) - \int_{t_0}^{t} \frac{df_{pi}}{dV_i} \frac{dV_i}{dt} \phi_i \, dt \tag{3.41}$$

The second term on the right-hand side of Equation 3.41 is path dependent. Approximation can be introduced by ignoring this term and evaluating the energy using only the first component, which is path independent. With this approximation, the energy function (3.32) can be modified by replacing

$$\int P_{li} \frac{d\phi}{dt} \, dt \text{ by } [P_{li}\phi_i - P_{li0}\phi_{i0}]$$

The time derivative of the modified energy function ($W_{mod}$) will not be zero and is given by

$$\frac{dW_{mod}}{dt} = \sum_{i=1}^{n} \frac{d}{dV_i} f_{pi}(V_i) \frac{dV_i}{dt} \phi_i \tag{3.42}$$

With constant voltage or constant active power loads, the right-hand side of Equation 3.42 is zero and the system is conservative. For other types of load, it is nonzero. However, if the derivative of the energy function is sufficiently small in magnitude, the use of modified energy function can give an accurate estimation of the stability region.

3. The terms of the energy function can be physically interpreted in the following way:

$W_1$: Total change in the rotor KE relative to COI

$W_{21}$: Change in the PE due to mechanical input relative to COI

$W_{22}$: Change in the PE due to voltage-dependent active power loads relative to COI

$W_{23}$: Change in the PE due to voltage-dependent reactive power loads

$W_{24}$: Change in the magnetic energy stored in the machine reactances

$W_{25}$: Change in the magnetic energy stored in the transmission lines

4. The last two terms in the energy function (3.32) represent the energy in the machine reactances and transmission line reactances. It is shown by Padiyar and Varaiya (1983) (see also Tsolas et al. 1985) that this energy can be expressed as half the sum of reactive power loss in each element of the network (including machine reactances). This energy is thus given by

$$\sum_{k=1}^{nb} \frac{1}{2} Q_k = \frac{1}{2} \left[ \sum_{i=1}^{m} Q_{gi} - \sum_{j=1}^{n} Q_{lj} \right] \tag{3.43}$$

where $Q_{gi}$ is the reactive power generation (at the internal bus) of generator $i = E_i^2 - E_iV_i\cos(\delta_i - \phi_i)/x'_{di}$; $Q_{lj}$ is the reactive power load at bus $i$; and $nb$ is the total number of elements in the network, including machine reactances.

The right-hand side of Equation 3.43 is easily calculated at the end of power flow solution at each step during the transient. Thus, the computation of the overall energy function is simplified.

5. If $P_{li}$ is not a constant, then $W_{22}(t)$ can also be obtained by numerical integration using the trapezoidal rule. Thus

$$W_{22}(t + h) = W_{22}(t) + \frac{1}{2}[P_{li}(t) + P_{li}(t + h)][\phi_i(t + h) - \phi_i(t)]$$

6. The PE component $W_{23}(V)$ is path independent as the integral can be obtained on the functions of bus voltage. For example, if

$$Q_l = b_0 + b_1V + b_2V^2$$

then

$$\int_{V_0}^{V} \frac{Q_l}{V} dV = b_0 \ln\frac{V}{V_0} + b_1(V - V_0) + \frac{b_2}{2}(V^2 - V_0^2)$$

where ln indicates a natural logarithm.

7. It is to be noted that the number (n) of buses here, does not include the generator buses (unlike in Section 3.2).

Note that if all the load buses have reactive power load characteristics with $b_0 = 0$, $b_1 = 0$, then the term $W_{23}$ can be merged with terms $W_{24}$ and $W_{25}$ and we can express

$$W_{23} + W_{24} + W_{25} = \frac{1}{2}\sum_{i=1}^{m} Q_g$$

### 3.4.5 Computation of Stability Region

Whenever a fault occurs in a power system, the total energy of the postfault system, whose stability is being examined, increases. After the fault is cleared, the total energy is nonincreasing. The key idea of the direct method is that the transient stability of the system, for a given contingency, can be determined directly by comparing the gain in the total system energy during the fault-on period with the "critical" energy. In the past, the critical energy was chosen to be the PE at the UEP closest (in terms of energy) to the postfault EP. This critical energy usually yields results that are very conservative. The concept of CUEP in determining the critical energy has removed much of the conservativeness in the results. Computational simplifications are possible with the use of the PEBS method.

The PEBS method is particularly attractive as it avoids the need for evaluating UEPs. In this method, it is assumed that the critically cleared trajectory goes near the UEP corresponding to the fault location. PEBS is defined as the surface formed in the angle space by the points corresponding to the first maxima of the PE (with respect to SEP). Thus, the instant at which the time derivative of the PE function changes its sign from positive to negative can be interpreted to be the instant when the trajectory crosses the PEBS. The value of the PE at this time is the critical energy.

### *Effect of Load Model*

Padiyar and Sastry show with examples of (i) four-machine system (El-Abiad and Nagappan 1966) and (ii) a seven-machine CIGRE system (Pai 1981) that the use of SPEF results in an accurate prediction of $T_{cr}$ in practically all the cases considered. It was also observed that the voltage dependence of active power characteristics has a significant effect on the CCT. It is maximum with constant impedance type of loads, minimum with the constant power loads, and intermediate with constant current loads. Thus, it may be optimistic to consider the loads to be of constant impedance type. The variation in the reactive power characteristics of loads has no significant effect on $T_{cr}$. These conclusions may not be valid in all cases but the results bring out the need for accurate load models.

Nonlinear characteristics of loads (particularly constant power type) can introduce convergence problems in the network solution during simulation. The nonconvergence of the solution implies either the nonexistence of the power flow solution or the failure of the solution algorithm.

### *Constant-Impedance-Type Loads*

It is possible to reduce the system network by eliminating load buses for the case of constant-impedance-type loads. The drawback in reducing the network is the creation of TCs (even with lossless lines). It was observed that ignoring TCs results in significant error. Also, the approximation of the path-dependent TC terms in the energy function introduces errors that can be eliminated using SPEF. Although the line resistances are neglected in the computation of SPEF, their effects are negligible.

## 3.5 Case Studies on IEEE Test Systems

Two case studies, one each on the two IEEE stability test systems (IEEE Committee Report 1992), are carried out and the results are presented in this section. SPEF is used and the critical energy is determined by using the PEBS (exit point) method.

There are two IEEE transient stability test systems:

a. 17-generator, 162-bus system
b. 50-generator, 145-bus system

The generators are represented by classical models and the loads are modeled as constant impedances.

In addition, the 10-generator, 39-bus New England test system is also investigated. The results for both classical and detailed models of the generators will be presented in the next chapter.

### 3.5.1 Seventeen-Generator System

All the machines are represented by classical models and loads are modeled as constant impedances. The disturbances considered are three-phase faults at different buses followed by line clearing. The critical energy is computed by the PEBS method, which is obtained as the peak of the PE during fault-on trajectory. (The energy is calculated for the postfault system.)

The comparison of CCT obtained by prediction and simulation is given in Table 3.1 for 15 different disturbances. For faults at bus #15 and bus #131, the postfault network is assumed to be the same as the prefault network. It is

**TABLE 3.1**

Seventeen-Generator IEEE Test System and Computation of $T_{cr}$ and $W_{cr}$

| Sl. No. | Fault Bus # | Line Cleared (from Bus # to Bus #) | Critical Clearing Time ($T_{cr}$) | | $W_{cr}$ |
| | | | Prediction | Simulation | |
|---|---|---|---|---|---|
| 1 | 75 | 75–9 | 0.34–0.35 | 0.35–0.36 | 25.43 |
| 2 | 95 | 95–97 | 0.30–0.31 | 0.30–0.31 | 7.11 |
| 3 | 52 | 52–79 | 0.34–0.35 | 0.35–0.36 | 8.46 |
| 4 | 93 | 93–91 | 0.24–0.25 | 0.24–0.25 | 9.66 |
| 5 | 110 | 110–141 | 0.27–0.28 | 0.26–0.27 | 14.02 |
| 6 | 15 | 15–11 | 0.25–0.26 | 0.25–0.26 | 17.72 |
| 7 | 1 | 1–2 | 0.21–0.22 | 0.21–0.22 | 18.39 |
| 8 | 124 | 124–109 | 0.34–0.35 | 0.36–0.37 | 40.98 |
| 9 | 112 | 112–120 | 0.19–0.20 | 0.20–0.21 | 14.52 |
| 10 | 70 | 70–149 | 0.23–0.24 | 0.23–0.24 | 13.61 |
| 11 | 126 | 126–127 | 0.25–0.26 | 0.26–0.27 | 68.72 |
| 12 | 130 | 75–130 | 0.30–0.31 | 0.30–0.31 | 11.29 |
| 13 | 27 | 27–25 | 0.35–0.36 | 0.35–0.36 | 72.58 |
| 14 | 15 | — | 0.21–0.22 | 0.22–0.23 | 12.69 |
| 15 | 131 | — | 0.25–0.26 | 0.25–0.26 | 20.80 |

interesting to observe that for all the disturbances considered, the predicted CCT are in very good agreement with those obtained from simulation. Thus, the use of the PEBS method with SPEF gives accurate results.

For the fault at bus #75, the swing curves for the critically stable and unstable cases are shown in Figure 3.2. The generator 16 separates from the rest. The variations of total ($W$), kinetic ($W_1$), and potential ($W_2$) energies with time are shown in Figure 3.3. Both stable and unstable cases are considered. It is interesting to observe that while $W$ is approximately constant (as expected) after the fault is cleared, the KE $W_1$ continues to increase when the system is unstable. The increase in $W_1$ is not monotonic but superimposed with an

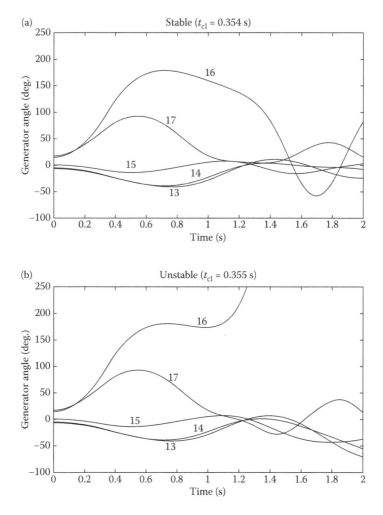

**FIGURE 3.2**
Swing curves (17-generator system). (a) Stable. (b) Unstable.

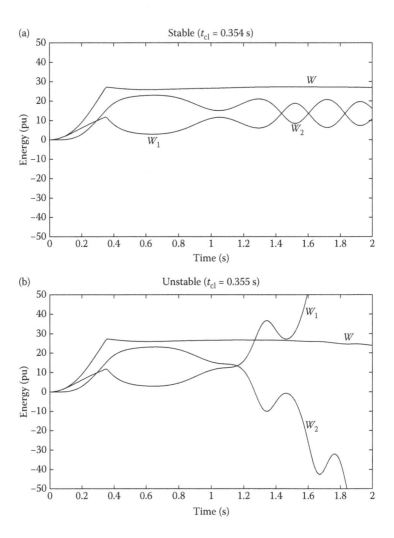

**FIGURE 3.3**
Variations of the potential, kinetic, and total energies. (a) Stable case. (b) Unstable case.

oscillatory component. The variation in PE, $W_2$, is of similar nature except that $W_2$ continues to decrease. If transmission line resistances (that are included in this example) are neglected, $W$ would be exactly constant after the fault is cleared.

The variations of the components of the PE are shown in Figure 3.4. These components are defined as

$$W_{21} = -\sum_{i=1}^{m} P_{mi}(\theta_i - \theta_{io})$$

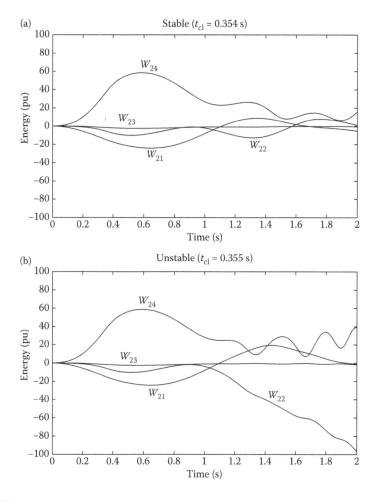

**FIGURE 3.4**
Variations of the potential energy components. (a) Stable case. (b) Unstable case.

$$W_{22} = -\sum_{i=1}^{n} \int_{to}^{t} P_{li}(V_i) \frac{d\phi_i}{dt} dt$$

$$W_{23} = \sum_{i=1}^{n} \int_{V_{io}}^{V_i} \frac{f_{qi}(x_i)}{x_i} dx_i$$

$$W_{24} = \frac{1}{2} \left[ \sum_{i=1}^{m} Q_{gi} - \sum_{j=1}^{n} Q_{lj} \right]$$

From Figure 3.4, it is observed that the components $W_{23}$ and $W_{24}$ are bounded. In this example, $W_{21}$ also appears to be bounded while $W_{22}$ is unbounded for the unstable case. The oscillations in $W_{24}$ are due to intermachine oscillations during the transient. The component $W_{23}$ is small compared to others.

### 3.5.2 Fifty-Generator System

Three-phase faults at various buses are considered. Both cases—with and without line resistances—are included. The line resistances are considered in the network solution; however, the SPEF does not account for them. The results are shown in Table 3.2.

The last entry in Table 3.2 is a benchmark test for system simulation. The swing curves for the critical generators (connected at buses 104 and 111) are shown in Figure 3.5. The disturbance considered is a three-phase fault at bus 7 cleared by opening the line between bus 7 and bus 6. The simulation shows that the CCT lies in the range 0.108–0.109 s. The results are in agreement with those given in the paper IEEE Committee Report (1992).

For the benchmark test case, there is a significant difference in the CCT obtained from prediction and simulation. This is mainly due to the fact that the mode of instability (MOI) changes when the resistances are neglected. While generators connected to buses 104 and 111 go out of step when the resistances are included, the generator connected to bus 137 goes out of step first when the resistances are neglected. However, it can be observed from Table 3.2 that the prediction of the CCT is not affected by the inclusion of resistances in most of the cases in spite of the presence of a few lines with high negative resistance (that occur due to the fact that the test system is derived from a larger system by network reduction). This observation in the case of the 50-generator system justifies the assumption of neglecting the small resistances of the EHV lines where $X/R$ ratio is around 10.

**TABLE 3.2**

Comparison of Prediction and Simulation for 50-Generator IEEE Test System

| Fault (Bus #) | Line Cleared (Bus # to Bus #) | Resistance Neglected | | Resistance Included | |
|---|---|---|---|---|---|
| | | $t_{cr}$ (pred) | $t_{cr}$ (sim) | $t_{cr}$ (pred) | $t_{cr}$ (sim) |
| 58 | 58–57 | 0.20–0.21 | 0.21–0.22 | 0.20–0.21 | 0.21–0.22 |
| 108 | 108–75 | 0.21–0.22 | 0.22–0.23 | 0.21–0.22 | 0.21–0.22 |
| 91 | 91–75 | 0.18–0.19 | 0.19–0.20 | 0.19–0.20 | 0.19–0.20 |
| 90 | 90–92 | 0.16–0.17 | 0.18–0.19 | 0.18–0.19 | 0.19–0.20 |
| 96 | 96–73 | 0.08–0.09 | 0.09–0.10 | 0.11–0.12 | 0.11–0.12 |
| 98 | 98–72 | 0.11–0.12 | 0.12–0.13 | 0.11–0.12 | 0.12–0.13 |
| 111 | 111–66 | 0.16–0.17 | 0.18–0.19 | 0.18–0.19 | 0.18–0.19 |
| 105 | 105–73 | 0.18–0.19 | 0.20–0.21 | 0.18–0.19 | 0.20–0.21 |
| 7 | 6–7 | 0.05–0.06 | 0.06–0.07 | 0.05–0.06 | 0.10–0.11 |

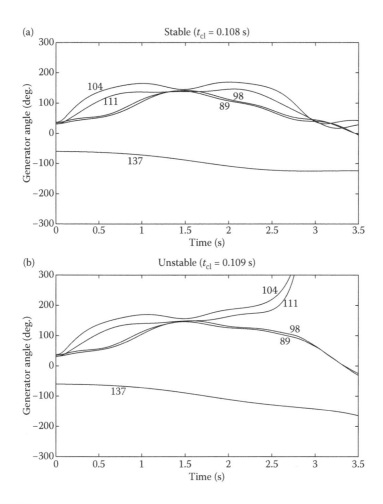

**FIGURE 3.5**
Swing curves (50-generator system). (a) Stable. (b) Unstable.

It is to be noted that although the SPEF is calculated for the postfault system, the datum (reference) for the energy function can be any convenient point (not necessarily the postfault SEP, $x_s$). It is convenient to take the initial (prefault) EP ($x_0$) as the datum. There is no loss of generality in doing this as the critical value of the energy will also change accordingly. To explain this, consider the stability criterion

$$W(t = t_{c1}) \le W_{cr} \qquad (3.44)$$

where $W$ is defined as

$$W(x) = W^*(x) - W^*(x_s)$$

such that $W(x_s) = 0$ and $W_{cr}$ is the value of $W$ computed at the CUEP.

Let

$$W'(x) = W^*(x) - W^*(x_0)$$

and

$$W'_{cr} = W_{cr} + W^*(x_s) - W^*(x_0)$$

The criterion (3.44) is transformed to

$$W'(t = t_{cl}) \le W'_{cr} \tag{3.45}$$

(Note that $W^*$ denotes the energy expression involving only the vector $x$. $W'_{cr}$ is the value of $W'$ computed at the CUEP.)

If there are no path-dependent terms in the energy function, there is no difference between the use of the stability criterion (3.44) or (3.45). However, when path-dependent terms are present, it is convenient (and accurate) to use (3.45) rather than (3.44) when $W_{cr}$ is determined from the PEBS method (Pai et al 1993).

## 3.6  Solution of System Equations during a Transient

The system equations during a transient can be expressed as a set of differential and algebraic equations given by

$$\dot{x} = f(x, y) \tag{3.46}$$

$$0 = g(x, y, p) \tag{3.47}$$

where $y$ consists of generator terminal voltages (expressed in either rectangular or polar coordinates) and $p$ is a set of parameters such as load powers. It is assumed that the generator voltages are sinusoidal and represented as slowly varying phasors, which are determined from network (including stator) algebraic Equation 3.47. Also, as the network changes due to switching or fault, Equation 3.47 changes in structure. It is to be noted that there are as many equations as the number of unknowns $y$. Thus, Equation 3.47 refers to power balance in the network and $y$ is the vector of bus voltages, including generator terminal voltages.

There are two major issues concerning the solution of DAE (3.46) and (3.47), namely

a. Existence of a solution of algebraic Equation 3.47
b. Methods of solution

It is usually assumed that the solution of differential Equation 3.46 exists for the initial conditions: $x = x_0$, $y = y_0$ at $t = t_0$. $y_0$ may be determined from power flow analysis (steady-state solution) and $x_0$ is computed from the knowledge of $y_0$ and power outputs of generators.

If it is assumed that the solution of the algebraic Equation 3.47 exists and $y$ can be solved in terms of $x$ as

$$y = h(x,p) \tag{3.48}$$

Substituting Equation 3.48 in Equation 3.46 results in a vector differential equation:

$$\dot{x} = f(x,h(x,p)) = F(x,p) \tag{3.49}$$

The solution of the above equation is relatively straightforward. Unfortunately, when $g$ is nonlinear, an exact solution of $y$ is not feasible. However, conceptually, $y$ can be expressed as in Equation 3.48, using implicit function theorem if the Jacobian matrix $(\partial g/\partial y)$ is nonsingular. In this case, the solution can be obtained by iterative methods. However, lack of existence of a solution creates problem as the simulation of the system comes to a halt. In such cases, it is fruitful to note that the algebraic equations for the network are actually obtained as approximations to the network models with fast (parasitic) dynamics given by

$$\epsilon\dot{y} = g(x,y,p) \tag{3.50}$$

where $\epsilon$ is small and approximated to zero.

It must be noted that multiple solutions of Equation 3.47 is not a major problem as the required solution is determined from considerations of continuity. For example, if $y_0$ is the solution at $t = t_0$, the solution at $t = t_0 + h$ must be close to $y_0$ as $h \to 0$.

The details of the network solution for transient stability analysis using simulation are given in Appendix C.

---

## 3.7 Noniterative Solution of Networks with Nonlinear Loads

For a class of nonlinear voltage-dependent load models, it is possible to get an exact (noniterative) solution (if it exists) when the load buses are not connected (Padiyar 1984). The following assumptions are made in the analysis:

1. The network is lossless.

2. The load buses are unconnected. Each load bus may be connected to one or more generators, but not to another load. An approximate

method of handling the power flow equations, using orthogonal transformation, when the load buses are connected is given in Padiyar and Sastry (1986a).

3. The loads are assumed to be voltage dependent with the following characteristics:

$$P_L = a_0 + a_1 V_L + a_2 V_L^2 = \left[ c_0 + c_1 \left( \frac{V_L}{V_{L0}} \right) + c_2 \left( \frac{V_L}{V_{L0}} \right)^2 \right] P_L^0$$

$$Q_L = b_0 + b_1 V_L + b_2 V_L^2 = \left[ d_0 + d_1 \left( \frac{V_L}{V_{L0}} \right) + d_2 \left( \frac{V_L}{V_{L0}} \right)^2 \right] Q_L^0$$

where $P_L^0$ and $Q_L^0$ are the initial steady-state values of load active and reactive power, respectively. It is obvious that

$$c_0 + c_1 + c_2 = 1$$
$$d_0 + d_1 + d_2 = 1$$

4. For any generator, the following assumptions apply:

    i.   The stator is represented by a constant voltage source $E$ behind the transient reactance $x_d'$.

    ii.  Mechanical torque is assumed to be constant, that is, the governor action is ignored.

    iii. Damping and transient saliency are neglected.

    (Note that the above assumptions have been used in the previous analysis.)

### 3.7.1 System Equations

Consider the system having $m$ generators and $n$ nonlinear voltage-dependent loads as shown in Figure 3.6. For convenience, only the generator internal buses and load buses are shown in Figure 3.6. The intermediate buses (where the load is not connected) are assumed to be eliminated by network reduction.

### 3.7.2 Dynamic Equations of Generators

The swing equation of the $i$th generator in COI variables can be expressed as

$$M_i \dot{\omega}_i = P_{mi} - P_{ei} - \frac{M_i}{M_T} P_{COI}$$

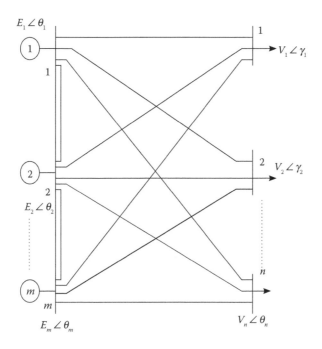

**FIGURE 3.6**
Multimachine system connected to uncoupled loads.

$$\dot{\theta}_i = \bar{\omega}_i, P_{COI} = \sum_{i=1}^{m}(P_{mi} - P_{ei})$$

where

$$P_{ei} = E_i\left[\sum_{j=1}^{m}E_jB_{ij}\sin\theta_{ij} + \sum_{k=1}^{n}V_kB_{il}\sin(\theta_i - \gamma_k)\right], l = k + m \qquad (3.51)$$

### 3.7.3 Power Flow Equations during a Transient

It will be shown that when the load buses are unconnected, it is possible to obtain load bus voltage magnitudes as solutions to a set of uncoupled quartic equations. The phase angles are determined uniquely once the voltage magnitudes are known.

For the system shown in Figure 3.6, the power flow equations at load bus $k$ can be expressed as

$$Q_{Lk} = V_k\left[\sum_{i=1}^{m}E_iB_{ik}\cos(\theta_i - \gamma_k) + V_kB_{kk}\right] \qquad (3.52)$$

$$P_{Lk} = V_k \left[ \sum_{i=1}^{m} E_i B_{ik} \sin(\theta_i - \gamma_k) \right] \tag{3.53}$$

By defining

$$S_{1k} = \frac{Q_{Lk}}{V_k} - B_{kk} V_k \tag{3.54}$$

$$S_{2k} = \frac{P_{Lk}}{V_k} \tag{3.55}$$

we can derive

$$\hat{S}_k = S_{1k} + jS_{2k} = \sum_{i=1}^{m} E_i B_{ik} e^{j(\theta_i - \gamma_k)} = e^{-j\gamma_k} \hat{R}_k \tag{3.56}$$

where

$$\hat{R}_k = R_{1k} + jR_{2k} = \sum_{i=1}^{m} E_i B_{ik} e^{j\theta_i} \tag{3.57}$$

The active and reactive power loads at bus $k$ are given by

$$P_{Lk} = a_{0k} + a_{1k} V_k + a_{2k} V_k^2 \tag{3.58}$$

$$Q_{Lk} = b_{0k} + b_{1k} V_k + b_{2k} V_k^2 \tag{3.59}$$

Substituting Equations 3.58 and 3.59 in Equations 3.54 and 3.55, we get

$$S_{1k} = (b_{2k} - B_{kk}) V_k + b_{1k} + \frac{b_{0k}}{V_k} \tag{3.60}$$

$$S_{2k} = a_{2k} V_k + a_{1k} + \frac{a_{0k}}{V_k} \tag{3.61}$$

From Equation 3.56, we get

$$|S_k| = |R_k| \tag{3.62}$$

$$\gamma_k = \tan^{-1} \frac{R_{2k}}{R_{1k}} - \tan^{-1} \frac{S_{2k}}{S_{1k}} \tag{3.63}$$

Expanding Equation 3.62, we get a quartic equation in $V_k$ given below

$$A_{1k}V_k^4 + A_{2k}V_k^3 + A_{3k}V_k^2 + A_{4k}V_k + A_{5k} = 0 \qquad (3.64)$$

where

$$A_{1k} = a_{2k}^2 + (b_{2k} - B_{kk})^2, \quad A_{2k} = 2[a_{2k}a_{1k} + (b_{2k} - B_{kk})b_{1k}]$$

$$A_{3k} = 2[a_{2k}a_{0k} + (b_{2k} - B_{kk})b_{0k}] + a_{1k}^2 + b_{1k}^2 - \alpha_k^2$$

$$A_{4k} = 2(a_{1k}a_{0k} + b_{1k}b_{0k}), \quad A_{5k} = a_{0k}^2 + b_{0k}^2, \quad \alpha_k = |R_k|$$

### 3.7.4 Special Cases

1. When the load is of constant power type. Here

$$a_{1k} = a_{2k} = 0, \quad b_{1k} = b_{2k} = 0$$

In this case, we get a quadratic equation in $V_k^2$

$$a_{0k}^2 + B_{kk}^2 V_k^4 + b_{0k}^2 + 2B_{kk}b_{0k}V_k^2 = \alpha^2 V_k^2 \qquad (3.65)$$

A real solution for $V_k^2$ can exist if and only if

i. $\alpha^2 > 2B_{kk}b_{0k}$

ii. $\alpha^2(\alpha^2 - 4B_{kk}b_{0k}) \geq 4B_{kk}^2 a_{0k}^2$

Actually, there are two positive solutions for $V_k$ if the above conditions are satisfied.

2. When the load is a combination of constant current and constant impedance type. Here

$$a_{0k} = b_{0k} = 0$$

We get a quadratic equation in $V_k$

$$\left[a_{2k}^2 + (b_{2k} - B_{kk})^2\right]V_k^2 + 2\left[a_{2k}a_{1k} + b_{2k}b_{1k} - B_{kk}b_{1k}\right]V_k + b_{1k}^2 + a_{1k}^2 = \alpha^2 \qquad (3.66)$$

A positive real solution for the above equation exists if

$$\alpha^2 \geq (a_{1k}^2 + b_{1k}^2)$$

under the assumption that all the coefficients on the left-hand side (LHS) of Equation 3.66 are positive. Moreover, if the solution exists, it is also unique. It is to be noted that generally, $B_{kk} < 0$.

### 3.7.5 Solutions of the Quartic Equation

The solutions of the quartic Equation 3.64 can be obtained from the coefficients of the fourth-order polynomial. As a matter of fact, if we express the nonlinear equation $f(x) = 0$, where $f(x)$ is a polynomial of order $n$, then exact solutions (as functions of the coefficients) can be computed only when $n \leq 4$. Although a quartic equation has four solutions, only positive real solutions are of interest as the load bus voltages are positive real numbers.

We can show that there can be two, one, or zero solutions of interest (positive real). We will demonstrate the existence of the positive real solutions of Equation 3.64 by a graphical analysis.

It is not difficult to verify that $S_{1k}$ and $S_{2k}$ (defined in Equations 3.60 and 3.61) reach minimum values at (i) $V_k = V_{k1}$ and (ii) $V_k = V_{k2}$, respectively, where $V_{k1} = [b_{0k}/(b_{2k} - B_{kk})]^{1/2}$ and $V_{k2} = [a_{0k}/a_{2k}]^{1/2}$. The locus of $(S_{1k} + jS_{2k})$ as a function of $V_k$ is shown in Figure 3.7 for the case when $V_{k2} < V_{k1}$. The direction of locus of $S_k$ shown is drawn as the voltage $V_k$ is increased. The direction of the locus will reverse when $V_{k2} > V_{k1}$.

Since $|R_k| = |S_k|$, we can observe that if $|R_k| < OC$, there is no solution, whereas if $|R_k| > OC$, there are two solutions $OA$ and $OB$ for the bus voltage. When $|R_k| = OC$, the two solutions coincide.

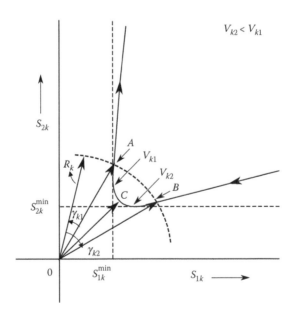

**FIGURE 3.7**
Locus of $S_k$ as $V_k$ is varied.

**Remarks**

1. There is only one feasible solution at point $C$ when $OC = |R_k|$. At this point, the current magnitude supplied to the load is minimum. Note that $S_{1k} = I_{rk}$ and $S_{2k} = I_{pk}$, where $I_{rk}$ and $I_{pk}$ are the reactive current and the active current, respectively, at bus $k$. Hence, $|S_k|$ is the magnitude of the current supplied to the load at bus $k$ (depending on the load characteristics).

2. Since the voltage magnitude at point $C$ lies between $V_{k1}$ and $V_{k2}$, a sufficient condition for the existence of a solution is that the voltage at bus $k$ lies outside the range $(V_{k1} - V_{k2})$.

### 3.7.6 Network Transformation for Decoupling of Load Buses

Since the assumption of unconnected buses is rather restrictive, the question can be posed whether it is possible to transform a given network to a network with unconnected load buses.

Consider a lossless system having $m$ generators supplying nonlinear voltage-dependent loads described earlier. The buses are numbered first with $m$ internal nodes of generators and then the $n$ load buses.

The network equations can be expressed as

$$jB_{GG}\underline{E} + jB'_{GL}\underline{V}' = \underline{I}_G \tag{3.67}$$

$$jB'_{LG}\underline{E} + jB'_{LL}\underline{V}' = \underline{I}'_L \tag{3.68}$$

Let $T$ be a transformation matrix such that

$$\underline{V}' = T\underline{V}, \quad \underline{I}'_L = T\underline{I}_L \tag{3.69}$$

where $\underline{V}$ and $\underline{I}_L$ are the load bus voltage and current vectors of the transformed network. The equations of the transformed network are given by

$$jB_{GG}\underline{E} + jB_{GL}\underline{V} = \underline{I}_G \tag{3.70}$$

$$jB_{LG}\underline{E} + jB_{LL}\underline{V} = \underline{I}_L \tag{3.71}$$

where

$$B_{GL} = B'_{GL}T, \quad B_{LG} = T^{-1}B'_{LG}, \quad \text{and} \quad B_{LL} = T^{-1}B'_{LL}T$$

The matrix $T$ is chosen such that the matrix $B_{LL}$ is diagonal. Since $B'_{LL}$ is a real, symmetric matrix, the matrix $T$ will be real and orthogonal. The columns of the matrix $T$ are the eigenvectors of $B'_{LL}$. By choosing normalized eigenvectors, we get a power invariant transformation, such that

$$(\underline{V}')^T(\underline{I}'_L)^* = (\underline{V})^T(\underline{I}_L)^* \tag{3.72}$$

The diagonal matrix $B_{LL}$ implies that the load buses in the transformed network are unconnected.

### 3.7.7 Transformation of the Load Characteristics

In the original network, we consider the load at a particular bus to be dependent only on the voltage at the bus. However, the load bus voltages in the transformed network are a linear combination of the voltages of the load buses in the original network. This complicates the characterization of the loads as dependent on the voltage magnitudes of their respective bus voltages. This problem can be resolved if we assume that the voltages of the connected load buses in the original system are coherent. Mathematically, this can be expressed as

$$V'_j = \beta_j V_{\text{ref}} \tag{3.73}$$

where $\beta_j$ is a complex constant and $V_{\text{ref}}$ is a reference (complex) voltage.

There is no loss of generality, if we assume $V_{\text{ref}} = 1.0 + j0$. The use of Equation 3.73 results in the initial load currents $I'_{Lj}$ ($j = 1, 2, \dots, n$) becoming independent of $V'_j$. For example, if the load is of constant power type, we get

$$I'_{Lj} = \frac{P'_{Lj} - jQ'_{Lj}}{(V'_j)^*} = \frac{P'_{Lj} - jQ'_{Lj}}{\beta_j^*} \tag{3.74}$$

Since $P'_{Lj}$, $Q'_{Lj}$, and $\beta_j$ are constants, $I'_{Lj}$ is independent of $V'_j$ and can be easily transformed to $I_{Lj}$ by the use of Equation 3.69.

### Remarks

1. It is necessary to consider each type of load separately. Thus, constant current type of load components can be transformed in a similar fashion.

2. For constant impedance type of loads, it can be shown that no assumptions are required. Since

$$\underline{I}'_L = [y'_L]V' \tag{3.75}$$

where $[y'_L]$ is a diagonal matrix of load admittances, we can show that (since $T$ is an orthogonal matrix)

$$\underline{I}_L = T^{-1}[y'_L]T\underline{V} = [y_L]\underline{V} \tag{3.76}$$

where $[y_L]$ is also a diagonal matrix.

3. When the generators connected to the loads remain coherent (maintain synchronous stability), the assumption about coherent load buses is valid. Synchronous stability implies that $\theta_i$ ($i = 1, 2,..., m$) reach a constant value after a disturbance and thus, $V'_j$ (which depends on $\theta_j$) tend to reach constant steady-state values following a disturbance.

The above-stated observations lead to the following proposition.

## Proposition

In a coherent area, it is feasible to transform the network using power invariant transformation such that the load buses in the transformed network are unconnected.

---

## 3.8 Inclusion of Transmission Losses in Energy Function

### 3.8.1 Transformation of a Lossy Network

It was mentioned earlier that Lyapunov-like energy function is accurate when the transmission losses are neglected. It is generally true that transmission losses in EHV lines can be neglected in the evaluation of transient stability. However, it is possible to define SPEF for a lossy network with a constant $G/B$ ratio for its elements. The following lemma (Padiyar 1984) can be used to formulate the SPEF:

## Lemma

A lossy network with constant $G/B = \alpha$ for all its elements is equivalent to a lossless network with a new set of power injections. The two sets of injections are related by

$$P_i = P'_i + \alpha Q'_i \tag{3.77a}$$

$$Q_i = -\alpha P'_i + Q'_i \tag{3.77b}$$

where $P_i$ and $Q_i$ refer to injections for the original network and $P'_i$ and $Q'_i$ refer to the injections for the transformed (lossless) network. The susceptance matrices for both networks are identical.

*Proof.*

For an element $k$ (connected across the buses $i$ and $j$) in the network, the power and reactive power at node $i$ are given by

$$P_{ij} = (V_i^2 - V_i V_j \cos \delta_{ij}) g_{ij} + b_{ij} V_i V_j \sin \delta_{ij} \tag{3.78}$$

$$Q_{ij} = (V_i^2 - V_i V_j \cos \delta_{ij}) b_{ij} - g_{ij} V_i V_j \sin \delta_{ij} \tag{3.79}$$

From the above, we can derive

$$P_{ij} - \alpha Q_{ij} = (1 + \alpha^2) b_{ij} V_i V_j \sin \delta_{ij}$$

$$\frac{\alpha}{1 + \alpha^2} P_{ij} + \frac{1 - \alpha^2}{1 + \alpha^2} Q_{ij} = b_{ij}(V_i^2 - V_i V_j \cos \delta_{ij})$$

where $\alpha = g_{ij}/b_{ij}$ for all elements in the network.
Defining

$$P'_{ij} = b_{ij} V_i V_j \sin \delta_{ij} \tag{3.80}$$

$$Q'_{ij} = b_{ij}(V_i^2 - V_i V_j \cos \delta_{ij}) \tag{3.81}$$

we have

$$P'_{ij} = \frac{1}{1 + \alpha^2} P_{ij} - \frac{\alpha}{1 + \alpha^2} Q_{ij} \tag{3.82}$$

$$Q'_{ij} = \frac{\alpha}{1 + \alpha^2} P_{ij} + \frac{1}{1 + \alpha^2} Q_{ij} \tag{3.83}$$

It is easy to see that the above equations are still applicable when $P_{ij}$ and $Q_{ij}$ are replaced by $P_i$ and $Q_i$ and similarly when $P'_{ij}$ and $Q'_{ij}$ are replaced by $P'_i$ and $Q'_i$. Thus, we have

$$\begin{bmatrix} P'_i \\ Q'_i \end{bmatrix} = \frac{1}{1 + \alpha^2} \begin{bmatrix} 1 & -\alpha \\ \alpha & 1 \end{bmatrix} \begin{bmatrix} P_i \\ Q_i \end{bmatrix} \tag{3.84}$$

From Equation 3.84, we can easily derive Equation 3.77.

### 3.8.2 Structure Preserving Energy Function Incorporating Transmission Line Resistances

Consider the system with $m$ generators and $(m + n)$ buses (including the internal buses of generators represented by classical models) as shown in Figure 3.8a. This shows that the first $m$ buses are the internal nodes of the generators where voltages $(E\angle\theta)$ are specified and the remaining buses where active and reactive powers are specified. (It is possible that at some nodes, e.g., the generator terminal nodes, there is no load. But this is equivalent to specifying $P = Q = 0$.) The network admittance matrix $Y$ is assumed to have the following structure:

$$Y = G + jB = \alpha B + jB$$

The transformed network that is lossless is shown in Figure 3.8b. Here, the loads are specified as $P' + jQ'$ (as defined in Equation 3.84). The power outputs of generator $i$ at the internal bus, in the transformed network, are obtained as

$$P'_{ei} = b_i E_i V_i \sin(\theta_i - \phi_i), \quad Q'_{ei} = b_i[E_i^2 - E_i V_i \cos(\theta_i - \phi_i)], \quad i = 1, 2, \ldots, m \quad (3.85)$$

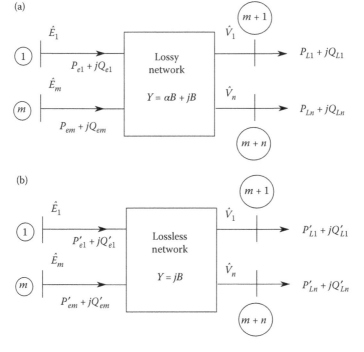

**FIGURE 3.8**
System diagram. (a) Original network. (b) Transformed (lossless) network.

where $b_i = 1/x'_{di}$ and $V_i \angle \phi_i$ are the terminal voltages of generator $i$ in the transformed network. Note that both $\theta_i$ and $\phi_i$ are referred to the COI. The swing equations for the original network are given by

$$\dot{\theta}_i = \bar{\omega}_i \qquad (3.86a)$$

$$M_i \dot{\bar{\omega}}_i = P_{mi} - P_{ei} - \frac{M_i}{M_T} P_{COI} \qquad (3.86b)$$

Since $P_{ei}$ is related to $P'_{ei}$ and $Q'_{ei}$ by

$$P_{ei} = P'_{ei} + \alpha Q'_{ei} \qquad (3.87)$$

we can express (3.86b) as

$$M_i \dot{\bar{\omega}}_i = P'_{mi} - P'_{ei} - \frac{M_i}{M_T} P'_{COI} \qquad (3.88)$$

where

$$P'_{COI} = \sum_{l=1}^{m} P'_{mi} - P'_{ei}, \quad P'_m = P_{mi} - \alpha Q'_{ei}$$

The SPEF for the transformed network is given by

$$W = W_1 + W_2 = W_1 + \sum_{i=1}^{5} W_{2i}$$

The KE component $W_1$ is given by

$$W_1 = \frac{1}{2} \sum_{i=1}^{m} M_i \bar{\omega}_i^2$$

The five components of the energy function are defined in a similar fashion as in Section 3.4 using the transformed network. Assuming constant active and reactive power loads, we obtain

$$W_{21} = -\sum_{i=1}^{m} \int_{t_0}^{t} P'_{mi} \frac{d\theta_i}{dt} \, dt$$

$$W_{22} = \sum_{i=m+1}^{n+m} P'_{Li}(\phi_i - \phi_{i0}), \quad W_{23} = \sum_{i=m+1}^{m+n} \int_{V_{i0}}^{V_i} \frac{Q'_{Li}}{V_i} \, dV_i$$

$$W_{24} = \sum_{i=1}^{m} b_i \left[ E_i^2 + V_i^2 - 2E_i V_i \cos(\theta_i - \phi_{i0}) - (E_i^2 + V_{i0}^2 - 2E_i V_{i0} \cos(\theta_i - \phi_i)) \right]$$

$$W_{25} = -\frac{1}{2} \sum_{i=m+1}^{m+n} \sum_{j=m+1}^{m+n} B_{ij}(V_i V_j \cos\phi_{ij} - V_{i0} V_{j0} \cos\phi_{ij0})$$

It can be shown that

$$W_{24} + W_{25} = \sum_{k=1}^{nb} \frac{1}{2} Q'_k = \frac{1}{2} \left[ \sum_{i=1}^{m} Q'_{ei} - \sum_{j=m+1}^{m+n} Q'_{Lj} \right]$$

Since the specified quantities at the load buses $(i = m + 1,\ldots, m + n)$ are $P_{Li}$ and $Q_{Li}$, we have to express $P'_{Li}$ and $Q'_{Li}$ in terms of $P_{Li}$ and $Q_{Li}$.
Thus, we can rewrite $W_{22}$ and $W_{23}$ as

$$W_{22} = \sum_{i=m+1}^{m+n} [(P_{Li} - \alpha Q_{Li})(\phi_i - \phi_{i0})]/(1 + \alpha^2)$$

$$W_{23} = \sum_{i=m+1}^{m+n} \int_{V_{i0}}^{V_i} \frac{(\alpha P_{Li} + Q_{Li})}{(1 + \alpha^2)V_i} dV_i$$

$W_{21}$ can be expressed as

$$W_{21} = -\sum_{i=1}^{m} P_{mi}(\phi_i - \phi_{i0}) + W_{26}$$

where

$$W_{26} = \sum_{i=1}^{m} \alpha \int_{t_0}^{t} Q'_{ei} \frac{d\theta_i}{dt} dt = \sum_{i=1}^{m} \alpha b_i E_i^2(\theta_i - \theta_{i0}) - \sum_{i=1}^{m} \left[ \alpha b_i E_i \int_{t_0}^{t} V_i \cos(\theta_i - \phi_i)\dot{\theta}_i \, dt \right]$$

The second term of $W_{26}$ is path dependent and has to be numerically integrated.

## Remarks

1. For constants $P_L$ and $Q_L$ and for the original lossy network with the special structure ($\alpha$ is a constant), the energy function has only one path-dependent term in $W_{26}$ as defined above.

2. For voltage-dependent active or reactive power, $W_{22}$ becomes path dependent and has to be computed numerically (using the trapezoidal rule).

3. $W_{23}$ is a path-independent term even with voltage-dependent active power loads.

On the basis of a case study of the 10-generator (New England test system), Padiyar and Ghosh (1988) claim that the path-dependent component of $W_{26}$ is negligible initially (following a disturbance) and does not affect the accuracy of the prediction of the CCT.

The results of the case studies presented in Padiyar and Ghosh (1988) indicate that the critical energy and CCT increase with increase in $\alpha$. Since the special structure of the network admittance matrix ($G = \alpha B$) is not realistic, they have also considered a case where the results of a network with actual line resistances which are compared to those with an average value of $\alpha$ ($G/B$). These results for a 10-generator system show that the transformation of the network (to a lossless network with altered power injections) using the average value of $\alpha$ can give accurate results for a lossy network (Padiyar and Ghosh 1988). The disturbances considered are three-phase faults at the terminals of various generators. Both constant power and constant impedance type of active power loads are considered whereas the reactive power loads are assumed to be of constant impedance type only.

## 3.9 SPEF for Systems with Generator Flux Decay

Bergen et al. (1986) developed a Lyapunov function for systems with generator flux decay and voltage-dependent reactive power loads (in addition to frequency-dependent active power loads) based on the SPM. Tsolas et al. (1985) proposed an SPEF taking into account the one-axis model of the generators (considering field flux delay) and transient saliency. The loads are represented by constant real ($P$) and reactive ($Q$) power demand. This paper also presents results for the region of attraction (for the asymptotic stability of a system with classical models of generators) based on the SPM.

In what follows, the system model and the formulation of the SPEF are taken from Sastry (1984) and Padiyar and Sastry (1986b).

### 3.9.1 System Model

Consider the system with $m$ generators supplying $n$ voltage-dependent loads. The generator model considered is a one-axis model with field flux delay and transient saliency. The damper windings are neglected along with the mechanical damping terms. The mechanical inputs to the generators are assumed to be constant.

### 3.9.1.1 Generator Model

The generators are described by differential and algebraic equations using angle variables defined with respect to the center of inertia reference frame.
The swing equations are (for $i = 1, 2,..., m$)

$$\dot{\theta}_i = \bar{\omega}_i \tag{3.89}$$

$$M_i \dot{\bar{\omega}}_i = P_{mi} - P_{ei} - \frac{M_i}{M_T} P_{COI} \tag{3.90}$$

where

$$P_{ei} = [E'_{qi} + (x'_{di} - x_{qi})i_{di}]i_{qi} \tag{3.91}$$

$$P_{COI} = \sum_{i=1}^{m} (P_{mi} - P_{ei}) \tag{}$$

The equation for the field winding is

$$T'_{doi} \frac{dE'_{qi}}{dt} = E_{fdi} - E'_{qi} + (x_{di} - x'_{di})i_{di} \tag{3.92}$$

The algebraic equations for the stator are

$$V_{qi} = E'_{qi} + x'_{di}i_{di} \tag{3.93}$$

$$V_{di} = -x_{qi}i_{qi} \tag{3.94}$$

where

$$(V_{qi} + jV_{di})e^{j\theta_i} = V_i e^{j\phi_i} \tag{3.95}$$

From Equation 3.94, we can derive

$$V_{qi} = V_i \cos(\theta_i - \phi_i) \tag{3.96}$$

$$V_{di} = -V_i \sin(\theta_i - \phi_i) \tag{3.97}$$

From Equations 3.93 through 3.97, we can derive expressions for $i_{di}$ and $i_{qi}$ as

$$i_{di} = \frac{V_i \cos(\theta_i - \phi_i) - E'_{qi}}{x'_{di}} \tag{3.98}$$

$$i_{qi} = \frac{V_i \sin(\theta_i - \phi_i)}{x_{qi}} \tag{3.99}$$

Substituting Equations 3.98 and 3.99 in Equation 3.91, we get

$$P_{ei} = \frac{E'_{qi} V_i \sin(\theta_i - \phi_i)}{x'_{di}} + \frac{V_i^2 (x'_{di} - x_{qi}) \sin 2(\theta_i - \phi_i)}{2 x'_{di} x_{qi}} \tag{3.100}$$

Substituting for $i_{di}$ in Equation 3.92, we obtain

$$T'_{doi} \dot{E}'_{qi} = E_{fdi} - \frac{x_{di}}{x'_{di}} E'_{qi} + \frac{V_i \cos(\theta_i - \phi_i)}{x'_{di}} (x_{di} - x'_{di}) \tag{3.101}$$

### 3.9.1.2 Load Model

Each load at bus $i$ is represented as an arbitrary function of the voltage at bus $i$. Thus

$$P_{li} = f_{pi}(V_i) \tag{3.102}$$

$$Q_{li} = f_{qi}(V_i) \tag{3.103}$$

### 3.9.1.3 Power Flow Equations

For the lossless system, the power injected at any bus $i$ can be expressed as

$$P_i = -P_{ei} + \sum_{j=1}^{n} B_{ij} V_i V_j \sin \phi_{ij}, \quad i = 1, 2, \ldots, m \tag{3.104}$$

$$= \sum_{j=1}^{n} B_{ij} V_i V_j \sin \phi_{ij}, \quad i = (m+1), (m+2), \ldots, n \tag{3.105}$$

$$Q_i = \frac{V_i^2 - E'_{qi} V_i \cos(\theta_i - \phi_i)}{x'_{di}} - \frac{V_i^2 (x'_{di} - x_{qi})}{2 x'_{di} x_{qi}} [\cos(2\theta_i - 2\phi_i) - 1]$$

$$- \sum_{j=1}^{n} B_{ij} V_i V_j \cos \phi_{ij}, \quad i = 1, 2, \ldots, m \tag{3.106}$$

$$= -\sum_{j=1}^{n} B_{ij} V_i V_j \cos\phi_{ij}, \quad i = (m+1), (m+2), \ldots, n \tag{3.107}$$

In the above expressions, $B_{ij} = \text{Im} [Y(i, j)]$, where $Y$ is the $n \times n$ bus admittance matrix. (Note that the internal nodes of generators are not included. The generator terminal buses are labeled first.)

The power flow equations at any bus $i$ can be expressed as

$$P_i + P_{li} = P_i + f_{pi}(V_i) = 0 \tag{3.108}$$

$$Q_i + Q_{li} = Q_i + f_{qi}(V_i) = 0 \tag{3.109}$$

### 3.9.2 Structure Preserving Energy Function

For the system considered, we can define an SPEF as

$$W(\bar{\omega}, E'_q, \theta, V, \phi, E_{fd}, t) = W_1 + W_2 = W_1 + \sum_{k=1}^{8} W_{2k} \tag{3.110}$$

where $W_1$ is the KE and $W_2$ is the PE defined with respect to an EP. The PE has eight components defined below:

$$W_{21}(\theta) = -\sum_{i=1}^{m} P_{mi}(\theta_i - \theta_{i0}) \tag{3.111}$$

$$W_{22}(t) = \sum_{i=1}^{n} \int_{t_0}^{t} P_{li} \frac{d\phi_i}{dt} dt \tag{3.112}$$

$$W_{23}(V) = \sum_{i=1}^{n} \int_{V_{i0}}^{V_i} \frac{f_{qi}(V_i)}{V_i} dV_i \tag{3.113}$$

$$W_{24}(E'_q, V, \theta, \phi) = \sum_{i=1}^{m} \left[ \begin{array}{l} E'^2_{qi} + V_i^2 - 2E'_{qi} V_i \cos(\theta_i - \phi_i) - \\ (E'^2_{qi0} + V_{i0}^2 - 2E'_{qi0} V_{i0} \cos(\theta_{i0} - \phi_{i0})) \end{array} \right] \frac{1}{2x'_{di}} \tag{3.114}$$

$$W_{25}(V, \phi) = -\frac{1}{2} \sum_{i=1}^{n} \sum_{j=1}^{n} B_{ij} (V_i V_j \cos\phi_{ij} - V_{i0} V_{j0} \cos\phi_{ij0}) \tag{3.115}$$

$$W_{26}(t) = -\sum_{i=1}^{m}\left[\int_{t_0}^{t}\frac{E_{fdi}(t)}{(x_{di} - x'_{di})}\frac{dE'_{qi}}{dt}dt\right] \tag{3.116}$$

$$W_{27}(E'_q) = \sum_{i=1}^{m}\frac{(E'^2_{qi} - E'^2_{qi0})}{2(x_{di} - x'_{di})} \tag{3.117}$$

$$W_{28}(V,\theta,\phi) = -\sum_{i=1}^{m}\left[V_i^2\{\cos(2\theta_i - 2\phi_i) - 1\} - V_{i0}^2\{\cos(2\theta_{i0} - 2\phi_{i0}) - 1\}\right]\frac{(x'_{di} - x_{qi})}{4x'_{di}x_{qi}} \tag{3.118}$$

The KE is given by

$$W_1(\bar{\omega}) = \frac{1}{2}\sum_{i=1}^{m}M_i\bar{\omega}_i^2 \tag{3.119}$$

**Theorem**

The derivative of the energy function is given by

$$\frac{dW}{dt} = -\sum_{i=1}^{m}\frac{T'_{doi}(\dot{E}_{qi})^2}{(x_{di} - x'_{di})} < 0 \tag{3.120}$$

*Proof.*
Partially differentiating $W_2$ with respect to $E'_{qi}$, $V_i$, $\phi_i$, $\theta_i$, and $t$, we get

$$\frac{\partial W_2}{\partial E'_{qi}} = \frac{-T'_{doi}\dot{E}'_{qi} + E_{fdi}}{(x_{di} - x'_{di})} \text{ from Equation 3.101} \tag{3.121}$$

$$\frac{\partial W_2}{\partial V_i} = [Q_i + f_{qi}(V_i)]\frac{1}{V_i} = 0 \text{ from Equation 3.109} \tag{3.122}$$

$$\frac{\partial W_2}{\partial \phi_i} = P_i \text{ from Equations 3.104 and 3.105} \tag{3.123}$$

$$\frac{\partial W_2}{\partial \theta_i} = -P_{mi} + P_{ei} \text{ from Equation 3.91} \tag{3.124}$$

$$\frac{\partial W_2}{\partial t} = \sum_{i=1}^{n} P_{li} \frac{d\phi_i}{dt} - \sum_{i=1}^{m} \frac{E_{fdi}}{(x_{di} - x'_{di})} \frac{dE'_{qi}}{dt} \tag{3.125}$$

From Equation 3.119, we get

$$\frac{\partial W_i}{\partial \bar{\omega}_i} = M_i \bar{\omega}_i \tag{3.126}$$

From Equations 3.124 and 3.126, we can derive

$$\frac{\partial W_1}{\partial \bar{\omega}_i} \frac{d\bar{\omega}_i}{dt} + \frac{\partial W_2}{\partial \theta_i} \frac{d\theta_i}{dt} = -\frac{M_i \bar{\omega}_i}{M_T} P_{\text{COI}} \text{ from Equation 3.90}$$

Thus

$$\sum_{i=1}^{m} \left( \frac{\partial W}{\partial \bar{\omega}_i} \dot{\bar{\omega}}_i + \frac{\partial W_2}{\partial \theta_i} \bar{\omega}_i \right) = \frac{-P_{\text{COI}}}{M_T} \sum_{i=1}^{m} M_i \bar{\omega}_i = 0$$

$$\sum_{i=1}^{n} \frac{\partial W_2}{\partial V_i} \frac{dV_i}{dt} = 0 \text{ from Equation 3.122}$$

From Equations 3.108, 3.123, and 3.125, we can obtain

$$\sum_{i=1}^{n} \left( \frac{\partial W_2}{\partial \phi_i} \frac{d\phi_i}{dt} \right) + \frac{\partial W_2}{\partial t} = -\sum_{i=1}^{m} \frac{E_{fdi}}{(x_{di} - x'_{di})} \frac{dE'_{qi}}{dt} \tag{3.127}$$

Finally, using Equations 3.121 and 3.127, we can derive

$$\frac{dW}{dt} = \sum_{i=1}^{n} \left( \frac{\partial W_2}{\partial \phi_i} \frac{d\phi_i}{dt} + \frac{\partial W_2}{\partial t} \right) + \sum_{i=1}^{m} \frac{\partial W_2}{\partial E'_{qi}} \frac{dE'_{qi}}{dt} = -\sum_{i=1}^{m} \frac{T'_{doi} (\dot{E}'_{qi})^2}{(x_{di} - x'_{di})}$$

**Remarks**

1. From Equation 3.120, it is evident that $dW/dt$ is negative along a system trajectory as $dE'_{qi}/dt$ is not identically zero. Hence, the energy function satisfies the conditions for being a Lyapunov function and the system is asymptotically stable in the neighborhood of the EP.
2. For the classical model of the generator neglecting saliency, $E'_{qi} = E_i$ = constant. In this case, the terms $W_{26}$, $W_{27}$, and $W_{28}$ are

identically zero and the PE $W_2$ (with five components) is identical to that defined in Section 3.4.

3. The sum of the terms $W_{26}$ and $W_{27}$ corresponds to the change in the energy stored due to flux decay. In general, $E_{fdi}$ is a function of time in the presence of AVR and the term $W_{26}$ is path dependent. However, when $E_{fdi}$ is a constant, the integral term $W_{26}$ can be expressed as

$$W_{26} = \sum_{i=1}^{m} E_{fdi} \frac{(E'_{qi} - E'_{qi0})}{(x_{di} - x'_{di})}$$

4. The term $W_{22}$ is path independent only if $P_{li}$, $i = 1, 2, \ldots, m$ is constant. However, the integral can be numerically evaluated using the trapezoidal rule.

5. It can be shown that the sum of terms $W_{24}$ and $W_{28}$ is the energy stored in the machine reactances $x'_d$ and $x_q$. That is

$$W_{24} + W_{28} = \frac{1}{2} \sum_{i=1}^{m} (i^2_{di} x'_{di} + i^2_{qi} x_{qi})$$

$$= \frac{1}{2} \sum_{i=1}^{m} [i_{qi}(E'_{di} - V_{di}) - i_{di}(E'_{qi} - V_{qi})]$$

Thus, $W_{24} + W_{28}$ is equal to half the reactive power losses in the machine reactances. Note that $E'_{di} = 0$ for the one-axis model of the generators.

6. $W_{28}$ accounts for transient saliency ($x'_d \neq x_q$). In the absence of transient saliency, this term is zero.

7. For the classical models, $dW/dt = 0$, whereas $dW/dt < 0$ with one-axis model.

### 3.9.3 Example

The four-machine power system (El-Abiad and Nagappan 1966) is taken up for the stability analysis. It is assumed that a three-phase fault occurs at bus 3 and the postfault system is same as the prefault system. The generators 2 and 3 that are close to the fault location are represented by one-axis models whereas generators 1 and 4 are represented by classical models. The load characteristics are assumed to be of constant power type for active power and constant impedance type for reactive power. The CCT obtained by prediction (using the PEBS method) and digital simulation for different cases of the generator models are shown in Table 3.3.

**TABLE 3.3**

CCT for Different Generator Models

| Generator Model | Prediction | Simulation |
|---|---|---|
| 1. Classical model | 0.455–0.460 | 0.455–0.460 |
| 2. With flux decay | | |
|     a. With saliency | 0.470–0.475 | 0.460–0.465 |
|     b. Without saliency | 0.445–0.450 | 0.455–0.460 |
| 3. With AVR | | |
|     a. With saliency | 0.455–0.460 | 0.470–0.475 |
|     b. Without saliency | 0.435–0.440 | 0.465–0.470 |

**TABLE 3.4**

Comparison of Predicted CCT

| Generator Model | Predicted from Sasaki's Method | CCT Predicted from SPEF | Digital Simulation |
|---|---|---|---|
| Classical model | 0.40 | 0.485–0.490 | 0.485–0.490 |
| With flux decay | 0.42 | 0.490–0.495 | 0.485–0.490 |

Table 3.3 also shows the effect of modeling an excitation system with continuously acting AVR (IEEE type 1). The inclusion of AVR results in a path-dependent term for $W_{26}$ that is evaluated by numerical integration. The results indicate that the application of SPEF enables a fairly accurate prediction of CCT in general, although the results in the case with AVR are slightly in the pessimistic side. This could be due to the fact that the energy function decays with time $((dW/dt) < 0)$. For the classical model, the system is conservative (with constant energy).

To compare the results obtained from SPEF with earlier methods, CCT obtained by Sasaki (1979) (who considered flux decay in an approximate manner using an RNM with constant impedance type load models) are presented in Table 3.4 along with SPEF results (Sastry 1984).

It is to be noted that AVR has not been considered in the previous methods using reduced system models.

## 3.10 Network Analogy for System Stability Analysis

It was pointed out in Chapter 2 that the electrical power $(P_{ei})$ output of a generator (Equation 3.91) is strictly an expression for the electrical torque. In steady state, when the system is operating at constant (rated) frequency, the per-unit torque is numerically equal to the per-unit power. However, during a transient, the power and torque have different values.

It is meaningful to assume that a lossless transmission line permits the transfer of a constant torque from the sending end to the receiving end. We can propose the following network analogy to interpret and derive SPEF:

Torque ($T$) ~ current ($I$)
Velocity ($\omega$) ~ voltage ($V$)
Inertia ($M$) ~ capacitance ($C$)
Angle ($\delta$) ~ flux linkage ($\lambda$)
Damping ($D$) ~ conductance ($G$)

The swing equation for generator $i$ (represented by the classical model) is given by

$$M_i \frac{d\omega_i}{dt} + D_i\omega_i = T_{mi} - T_{ei} \tag{3.128}$$

where

$$T_{ei} = \frac{E_i V_i \sin(\delta_i - \delta_i^t)}{x_{di}'} = \frac{E_i V_i \sin \sigma_i}{x_{di}'}$$

where $\delta_i^t$ is the angle of the generator terminal bus. Note that $\omega_i = \dot{\delta}_i$ is the generator rotor velocity measured with respect to a synchronously rotating reference. Also

$$\delta_i - \delta_i^t = \theta_i - \phi_i$$

where $\theta_i$ and $\phi_i$ are the angles with respect to COI.

Equation 3.128 represents an equivalent circuit shown in Figure 3.9. Here, $T_{mi}$ is a constant current source at the internal bus ($g_i$) of generator $i$. $T_{ei}$ represents the electrical torque in $i$th generator and is similar to the current flowing in the nonlinear inductor that represents the stator. $\omega_i$ is similar to the voltage at the internal bus ($g_i$). From the equivalent circuit, we can derive the following equation:

$$\int_{t_0}^{t} (T_{mi} - T_{ei})\omega_i \, dt = \int_{\delta_{i0}}^{\delta_i} (T_{mi} - T_{ei}) \, d\delta_i = \frac{1}{2} M_i(\omega_i^2 - \omega_{i0}^2) + \int_{t_0}^{t} D_i\omega_i^2 dt$$

By defining an energy function, $W_i$ ($\delta_i$, $\omega_i$), given by

$$W_i(\delta_i, \omega_i) = W_{1i}(\omega_i) + W_{2i}(\delta_i, \delta_i^t, V_i) = \frac{1}{2} M_i(\omega_i^2 - \omega_{i0}^2) + \int_{\delta_{i0}}^{\delta_i} (T_{ei} - T_{mi}) d\delta_i$$

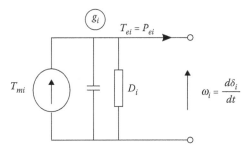

**FIGURE 3.9**
Equivalent circuit for a generator.

$$\frac{dW_i}{dt} = -D_i\omega_i^2$$

Note that $W_i$ is the transient energy associated with the $i$th generator and it has two components: $W_{1i}$ (KE associated with the $i$th generator) and $W_{2i}$ (PE associated with the $i$th generator).

It is possible to express the total energy as the sum of the energies for $m$ generators in the system. If one of the buses in the network is an infinite bus, then there is no need to define COI and rotor frequencies (in radians per second) can be referred (measured) with respect to the synchronously rotating reference frame. The total energy is given by

$$W(\delta, \omega) = W_1(\omega) + W_2(\delta)$$

$$= \sum_{i=1}^{m} \frac{1}{2} M_i(\omega_i - \omega_{i0})^2 + \sum_{i=1}^{m} \int_{\delta_{i0}}^{\delta_i} (T_{ei} - T_{mi})d\delta_i \qquad (3.129)$$

It is to be noted that we can use the symbols $P_{mi}$ and $P_{ei}$ in place of $T_{mi}$ and $T_{ei}$ as is normally done. However, conceptually, they are torques that are analogous to currents.

Equation 3.129 represents an alternative formulation of the energy function, particularly, the PE. It would appear that the PE is path dependent. However, for classical models of generators, it will be shown that the PE can also be expressed as path independent if load active powers are constants.

According to the Bergen and Hill model (1981), the load model at bus $j$ is assumed to be of type

$$P_{Lj} = P_{Lj0} + D_j\omega_j \qquad .$$

(Note that the use of symbol $P$ is for convenience, which actually represents load torque.) The load model can be represented by the analogous equivalent circuit shown in Figure 3.10.

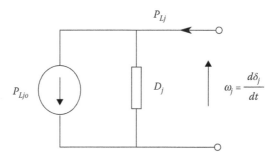

**FIGURE 3.10**
Equivalent circuit for a load.

The transmission line $k$, connecting buses $i$ and $j$, can be viewed as a non-linear inductor defined by the equation

$$T_k = b_k V_i V_j \sin \lambda_k = T_{km} \sin \lambda_k \tag{3.130}$$

where $T_k$ is the torque (or power) in line $k$.

The equivalent circuits of the $m$ generators and $n$ load buses can be clubbed together with the equivalent circuits of the transmission network (consisting of $nl$ series branches). The analogous network consists of linear capacitors, resistors, nonlinear inductors, and current sources. The combined network has $(n + m)$ nodes and $nb = (nl + m)$ series-connected branches. Note that in a lossless network, the shunt branches of the electrical network need not be included in the analogous network (see Figure 3.11) where the currents represent torques (or powers). (The resistors are not shown.)

The analogous network in general, consists of linear capacitors, resistors, and nonlinear inductors. We can choose a tree consisting of capacitors (inertias) at $m$ generator internal buses (numbered $g_1$ through $g_m$) and resistors

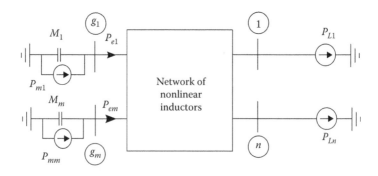

**FIGURE 3.11**
Analogous network for the system.

(damping) at load buses (numbered 1 through $n$). The reference bus for the network is the infinite bus (numbered $(n + 1)$).

Using the analogous network shown in Figure 3.11, we can prove the following theorem.

**Theorem**

The PE $W_2(\delta)$ defined in Equation 3.129 can be expressed as

$$W_2(\delta) = \sum_{i=1}^{m} \int_{\delta_{i0}}^{\delta_i} (P_{ei} - P_{mi}) d\delta_i = W_{21} + W_{22} + \sum_{i=1}^{m} \int_{\sigma_{i0}}^{\sigma_i} P_{ei} d\sigma_i + \sum_{k=1}^{nl} \int_{\phi_{k0}}^{\phi_k} P_k d\phi_k \quad (3.131)$$

where

$$W_{21} = -\sum_{i=1}^{m} P_{mi}(\delta_i - \delta_{i0}), \quad W_{22} = \sum_{j=1}^{n} P_{Lj}(\delta_j - \delta_{j0}), \quad \sigma_i = \delta_i - \delta_i^t$$

and

$$\frac{dW}{dt} = -\sum_{i=1}^{m+n} D_i \omega_i^2 \quad (3.132)$$

*Proof.*
Applying Tellegen's theorem (Tellegen 1952, Penfield et al. 1970) for the circuit shown in Figure 3.11, we get

$$\sum_{i=1}^{m} M_i \omega_i \dot{\omega}_i - \sum_{i=1}^{m} P_{mi} \frac{d\delta_i}{dt} + \sum_{j=1}^{n} P_{Lj} \frac{d\delta_j}{dt} + \sum_{i=1}^{m} P_{ei} \frac{d\sigma_i}{dt} + \sum_{k=1}^{nl} P_k \frac{d\phi_k}{dt} = -\sum_{i=1}^{n} D_i \omega_i^2$$

$$(3.133)$$

Integrating the LHS, with respect to time $t$, we get

$$W = W_1 + W_2 = -\sum_{i=1}^{m+n} \int D_i \omega_i^2 dt$$

where $W_1 = (1/2)\sum_{i=1}^{m} M_i \omega_i^2$ is the KE and $W_2$ is the PE defined in Equation 3.131. Note that $D_i$ includes damping terms at both generator (internal) and load buses.

The LHS of Equation 3.133 is $dW/dt$. This proves Equation 3.132.

Bergen and Hill (1981) assume that the bus voltages are maintained at constant values. With this assumption, the following theorem can be stated.

**Theorem**

The total transient energy stored in the network elements (including the generator transient reactances) is equal to half the sum of the reactive power loss in individual elements. Mathematically, this can be expressed as

$$\sum_{i=1}^{m} P_{ei} d\sigma_i + \sum_{j=1}^{nl} \int_{\phi_{k0}}^{\phi_k} P_k d\phi_k = W_{24} + W_{25} = \frac{1}{2}\sum_{k=1}^{nb}(Q_k - Q_{k0}) \qquad (3.134)$$

where $Q_k$ is the reactive power loss in the branch $k$.

*Proof.*

$$W_{24} = \sum_{i=1}^{m} \int_{\sigma_{i0}}^{\sigma_i} P_{ei} d\sigma_i$$

Substituting $P_{ei} = E_i V_i \sin\sigma_i / x'_{di}$, we can obtain

$$W_{24} = \sum_{i=1}^{m} -\frac{E_i V_i}{x'_{di}}(\cos\sigma_i - \cos\sigma_{i0})$$

Adding and subtracting $(E_i^2 + V_i^2)/2$ to each term in the sum, we get

$$W_{24} = \sum_{i=1}^{m} \frac{1}{2}(Q_i - Q_{i0}) \qquad (3.135)$$

Note that the reactive power loss in the $i$th generator transient reactance is given by

$$Q_i = \frac{E_i^2 + V_i^2 - 2E_i V_i \cos\sigma_i}{x'_{di}}$$

Similar arguments can be used to show that

$$\sum_{k=1}^{nl} \int_{\phi_{k0}}^{\phi_k} P_k d\phi_k = W_{25} = \frac{1}{2}\sum_{k=1}^{nl}(Q_k - Q_{k0}) \qquad (3.136)$$

where $Q_k = (V_i^2 + V_j^2 - 2V_iV_j \cos \phi_k)/x_k$, $\phi_k = (\delta_i - \delta_j)$ is the reactive power loss in the branch $k$ with reactance $x_k$. It is assumed that the branch is connected across nodes $i$ and $j$.

From Equations 3.135 and 3.136, we can derive Equation 3.134.

Note that the sum of reactive power losses in the network branches is equal to the algebraic sum of reactive power injections in the network. Thus, we get

$$W_{24} + W_{25} = \frac{1}{2}\left[\sum_{i=1}^{m} Q_{gi} - \sum_{j=1}^{n} Q_{Lj}\right] \qquad (3.137)$$

## Remarks

1. Tavora and Smith (1972a) interpret the swing equations of the generators as a mass-spring system using mechanical (system) analogy. However, it is well known that a mechanical system can be modeled as an electrical network with capacitors analogous to inertias, springs represented as inductors, and damping as conductances. There is a large body of network theory that can be applied to derive important results.

2. Bergen and Hill (1981) use the analogy of power and bus angle as analogous to current and voltage, respectively. This results in the nonlinear resistive network. However, this analogy fails to relate the energy function to the energy stored in the network.

3. The analogous network of linear capacitors and nonlinear inductors helps to define energy for individual elements of the network. This fact can be used to define critical cutsets that separate the network into two parts, each of which maintains synchronous stability. There is also a one-to-one correspondence between the MOI and the critical cutset (defined as one whose elements experience unbounded increase in angles across them as they connect two subsystems that separate from each other). These issues are discussed in Chapter 6. Quick identification of the critical cutset and using an adaptive protection scheme can prevent widespread blackouts.

4. The application of energy functions for control of system stability has not been discussed in the previous literature except in a cursory manner. Controlling the power flow in the lines belonging to the critical cutset can assist in maintaining stability in the postfault system. Thus, even if the system energy is greater than the critical energy at the time of clearing the fault, the control action in the postfault state can help maintain system stability. The location of the network controllers (FACTS) and the control laws are the major issues

in the design of self-healing grids (which do not require active intervention by the system operator).

5. The energy stored in a three-phase transmission line with balanced
   (positive sequence) currents flowing through the conductors is
   given by

$$W_L = \frac{1}{2}L_1(i_a^2 + i_b^2 + i_c^2) = \frac{1}{2}L_1(I_d^2 + I_q^2) = \frac{1}{2}L_1I^2$$

where $L_1$ is the positive sequence series inductance of the line. The
current $i_d$ and $i_q$ are the $d$–$q$ components (Park or Kron) of the currents flowing through the line conductors. In per-unit quantities,
$L_1 = X_1$, where $X_1$ is the series reactance of the line. Thus, the magnetic energy stored in the line $W_L = Q_L/2$, where $Q_L$ is the reactive
power loss in the line. Note that in the electrical circuit of a linear
inductor $W_L = (1/2)LI^2 = Q_L/2$. Of course, it is not surprising that
we get identical answers whether we consider the electrical network with linear inductors or analogous network with a nonlinear
inductor.

6. Since we have assumed that the electrical network has only $P$–$V$
   nodes (with constant bus voltages), the shunt susceptances or load
   reactive power do not contribute to the energy function and thus

$$W_{23} = \sum_{j=1}^{n} \int_{V_{j0}}^{V_j} \frac{Q_j}{V_j} dV_j = 0$$

In the next chapter, we will consider realistic models for both generators and loads. The energy functions will have additional components, which can be easily derived by relaxing the assumptions
about $E'$ and $V$. However, Equation 3.129 still applies for systems
with detailed models.

7. Example 2 in Chapter 1 is the application of the network analogy for
   direct transient stability evaluation of a lossless SMIB.

# 4

# Structure Preserving Energy Functions for Systems with Detailed Generator and Load Models

## 4.1 Introduction

In the previous chapter, we presented SPEF for a power system with classical generator models and nonlinear voltage-dependent load models with examples and case studies on the two IEEE test systems—17-generator and 50-generator systems. We also considered the extension of SPEF for one-axis generator models with flux decay. In this chapter, we will consider detailed generator models with two-axis machine models and excitation control systems with AVR. The formulations of SPEF will be presented in different ways and aimed at simplifying the computations in practical systems. The role of dynamic loads such as induction motors on voltage stability is emphasized. We will also outline a composite framework for synchronous and voltage stability using energy functions. The applications of SPEF will be illustrated using several examples and a case study of the New England 10-generator system.

Tsolas et al. (1985) and Bergen et al. (1986) applied SPM to develop Lyapunov-like energy functions for the stability analysis of multimachine systems with modeling of field flux decay and voltage-dependent reactive loads. Padiyar and Sastry (1986b) proposed a "topological" energy function to account for one-axis model with transient saliency and excitation controllers with AVR. However, in all these papers using SPM, the SPEF has a time derivative that is negative semidefinite and the results tend to be conservative. In this chapter, a new SPEF is proposed using COI variables, which remains constant along the postfault trajectory. This may introduce some terms that are path dependent and have to be numerically evaluated but leads to better accuracy in the prediction of transient stability using the PEBS method. Simpler and more general expressions for the energy functions are derived and it is shown that the use of classical models is a special case.

This chapter also presents the reformulation of the SPEF (based on the network analogy introduced in Chapter 3) that is also applicable for detailed models. The reformulation based on SPM enables defining energy function as the sum of the energies defined for each component of the network. This helps in developing network controllers to improve system stability.

## 4.2 System Model

### 4.2.1 Generator Model

We consider a system with $m$ generators supplying up to $n$ nonlinear voltage-dependent loads (see Figure 4.1). The network has a total of $n + m$ buses with $m$ internal buses of generators. The internal buses are labeled as $g_1, g_2, \ldots, g_m$. The terminal buses are labeled as $1, 2, \ldots, m$ and there could be loads connected at these buses. The generator model considers two rotor circuits: (i) field winding on the direct axis and (ii) a damper winding on the quadrature axis. This is a standard two-axis model (Anderson and Fouad 1977). The excitation control system is represented by standard IEEE models (IEEE Committee Report 1981b).

The generator equations are given below:

For $i = 1, 2, \ldots, m$

$$\dot{\theta}_i = \bar{\omega}_i \tag{4.1}$$

$$M_i \dot{\bar{\omega}}_i = P_{mi} - P_{ei} - (M_i / M_T) P_{\text{COI}} \tag{4.2}$$

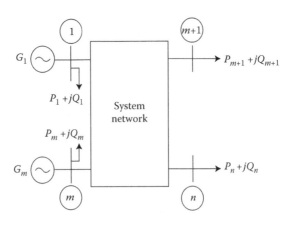

**FIGURE 4.1**
A multimachine system with detailed generator models.

$$T'_{doi}\dot{E}'_{qi} = E_{fdi} - E'_{qi} + (x_{di} - x'_{di})i_{di} \tag{4.3}$$

$$T'_{qoi}\dot{E}'_{di} = -E'_{di} - (x_{qi} - x'_{qi})i_{qi} \tag{4.4}$$

$$V_{qi} = E'_{qi} + x'_{di}i_{di} \tag{4.5}$$

$$V_{di} = E'_{di} - x'_{qi}i_{qi} \tag{4.6}$$

$$V_i e^{j\phi_i} = (V_{qi} + jV_{di})e^{j\theta_i} \tag{4.7}$$

From Equations 4.5 through 4.7, we obtain

$$i_{di} = \frac{V_i \cos(\theta_i - \phi_i) - E'_{qi}}{x'_{di}} \tag{4.8}$$

$$i_{qi} = \frac{E'_{di} + V_i \sin(\theta_i - \phi_i)}{x'_{qi}} \tag{4.9}$$

The expression for the electrical power ($P_{ei}$) is

$$P_{ei} = [E'_{qi} + (x'_{di} - x'_{qi})i_{di}]i_{qi} + E'_{di}i_{di} \tag{4.10}$$

Substituting for $i_{di}$ and $i_{qi}$ from Equations 4.8 and 4.9, we get

$$P_{ei} = \frac{E'_{qi}V_i \sin(\theta_i - \phi_i)}{x'_{di}} + \frac{E'_{di}V_i \cos(\theta_i - \phi_i)}{x'_{qi}} + \frac{V_i^2(x'_{di} - x'_{qi})\sin(2\theta_i - 2\phi_i)}{2x'_{di}x'_{qi}} \tag{4.11}$$

## 4.2.2 Excitation System Model

From the block diagram representation of various types of excitation control systems, we can derive the equations in the standard state space form. Considering a simplified model of a static type of excitation system shown in Figure 4.2, we get the following equations:

$$\dot{V}_{Ri} = \frac{K_A(V_{refi} - V_i + V_{si}) - V_{Ri}}{T_{Ai}} \tag{4.12}$$

$$\left.\begin{aligned}
E_{fdi} &= V_{Ri} \quad \text{if } E_{fdi\,min} \le V_{Ri} \le E_{fdi\,max} \\
&= E_{fdi\,min} \quad \text{if } E_{fdi\,min} > V_{Ri} \\
&= E_{fdi\,max} \quad \text{if } E_{fdi\,max} < V_{Ri}
\end{aligned}\right\} \tag{4.13}$$

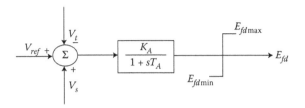

**FIGURE 4.2**
A static excitation system.

Note that $V_s$ is the output of an auxiliary controller (PSS). The subscript $i$ in the equations refers to the generator $i$.

### 4.2.3 Load Model

The load at any bus (except the internal bus of a generator) is represented by an arbitrary function of the voltage at that bus. Thus

$$P_{Li} = f_{pi}(V_i), \quad i = 1, \ldots, n \tag{4.14}$$

$$Q_{Li} = f_{qi}(V_i), \quad i = 1, \ldots, n \tag{4.15}$$

### 4.2.4 Power Flow Equations

For the lossless network, we can write the following equations for the injected power and reactive power:

$$P_i = -P_{ei} + \sum_{j=1}^{n} B_{ij} V_i V_j \sin\phi_{ij}, \quad i = 1, 2, \ldots, m$$

$$= \sum_{j=1}^{n} B_{ij} V_i V_j \sin\phi_{ij}, \quad i = m+1, \ldots, n \tag{4.16}$$

$$Q_i = \frac{V_i^2}{x'_{di}} - \frac{E'_{qi} V_i \cos(\theta_i - \phi_i)}{x'_{di}} + \frac{E'_{di} V_i \sin(\theta_i - \phi_i)}{x'_{qi}} - \frac{V_i^2(x'_{di} - x'_{qi})}{2x'_{di}x'_{qi}}\left[\cos(2\theta_i - 2\phi_i) - 1\right]$$

$$- \sum_{j=1}^{n} B_{ij} V_i V_j \cos\phi_{ij}, \quad i = 1, 2, \ldots, m$$

$$Q_i = -\sum_{j=1}^{n} B_{ij} V_i V_j \cos\phi_{ij}, \quad i = m+1, \ldots, n \tag{4.17}$$

In the above expressions, $B_{ij} = \text{Im } [Y_{ij}]$, where $Y$ is the admittance matrix of the network (excluding the generator stator reactances). The power flow equations are

$$P_i + f_{pi}(V_i) = 0, \quad i = 1, 2, \ldots, n \tag{4.18}$$

$$Q_i + f_{qi}(V_i) = 0, \quad i = 1, 2, \ldots, n \tag{4.19}$$

## 4.3 Structure Preserving Energy Function with Detailed Generator Models

In the previous chapter, an SPEF was formulated for systems with classical models for generators. Simplified (classical) generator models are acceptable only if the transient stability could be determined within the time frame of less than a second (before AVR can respond). However, for large systems with low frequency, interarea modes of oscillations, this premise is not correct. Hence, it becomes necessary to have the formulation of energy function with more detailed generator models. In what follows, an energy function is presented and derived for the two-axis machine model (see Appendix A) including AVR. The presence of AVR contributes a component of energy that is path dependent.

### 4.3.1 Structure Preserving Energy Function

Consider the following function defined for the postfault system:

$$W(\theta, \bar{\omega}, E_q', E_d', V, \phi, t) = W_1(\bar{\omega}) + \sum_{i=1}^{11} W_{2i} \tag{4.20}$$

where

$$W_1(\bar{\omega}) = \frac{1}{2} \sum_{i=1}^{m} M_i \bar{\omega}_i^2, \quad W_{21}(\theta) = -\sum_{i=1}^{m} P_{mi}(\theta_i - \theta_{i0})$$

$$W_{22}(t) = \sum_{i=1}^{n} \int_{t_0}^{t} f_{pi}(V_i) \frac{d\phi_i}{dt} dt, \quad W_{23}(V) = \sum_{i=1}^{n} \int_{V_{i0}}^{V_i} \frac{f_{qi}(x_i)}{x_i} dx_i$$

$$W_{24}(\theta, E_q', V, \phi) = \sum_{i=1}^{m} \left[ E_{qi}'^2 + V_i^2 - 2E_{qi}' V_i \cos(\theta_i - \phi_i) \right.$$

$$\left. - (E_{qi0}'^2 + V_{i0}^2 - 2E_{qi0}' V_{i0} \cos(\theta_{i0} - \phi_{i0})) \right] \left( \frac{1}{2x_{di}'} \right)$$

$$W_{25}(V,\phi) = -\frac{1}{2}\sum_{i=1}^{n}\sum_{j=1}^{n} B_{ij}(V_i V_j \cos\phi_{ij} - V_{i0}V_{j0}\cos\phi_{ij0})$$

$$W_{26}(\theta,V,\phi) = -\sum_{i=1}^{m}\left[V_i^2(\cos 2(\theta_i - \phi_i) - 1) - V_{i0}^2(\cos 2(\theta_{i0} - \phi_0) - 1)\right] \times \frac{(x'_{di} - x'_{qi})}{4x'_{di}x'_{qi}}$$

$$W_{27}(\theta,E'_d,V,\phi) = \sum_{i=1}^{m}\left[E'^2_{di} + V_i^2 + 2E'_{di}V_i\sin(\theta_i - \phi_i)\right.$$

$$\left. - (E'^2_{dio} + V_{io}^2 + 2E'_{dio}V_{i0}\sin(\theta_{i0} - \phi_{i0}))\right]\left(\frac{1}{2x'_{qi}}\right)$$

$$W_{28}(V) = -\sum_{i=1}^{m}\frac{(V_i^2 - V_{i0}^2)}{2x'_{qi}}, \quad W_{29}(t) = -\sum_{i=1}^{m}\int_{t0}^{t}\left[\frac{E_{fdi}}{(x_{di} - x'_{di})}\right]\frac{dE'_{qi}}{dt}dt$$

$$W_{210}(E'_q) = \sum_{i=1}^{m}\left[\frac{(E'^2_{qi} - E'^2_{qi0})}{2(x_{di} - x'_{di})}\right], \quad W_{211}(E'_d) = \sum_{i=1}^{m}\left[\frac{(E'^2_{di} - E'^2_{di0})}{2(x_{qi} - x'_{qi})}\right]$$

The subscript 0 in the above expressions indicates quantities at initial equilibrium (operating point).

It can be proved that the time derivative of $W$ is nonpositive along the post-fault trajectory. That is

$$\frac{dW}{dt} = -\sum_{i=1}^{m}\left[\frac{T'_{doi}}{(x_{di} - x'_{di})}\left(\frac{dE'_{qi}}{dt}\right)^2 + \frac{T'_{qoi}}{(x_{qi} - x'_{qi})}\left(\frac{dE'_{di}}{dt}\right)^2\right] \tag{4.21}$$

*Proof*

Partial differentiation of $W_2$ with respect to $E'_{qi}$, $E'_{di}$, $V_i$, $\phi_i$, $\theta_i$, and $t$ yields the following expressions after some algebraic manipulations:

$$\frac{\partial W_2}{\partial E'_{qi}} = -\frac{T'_{doi}\cdot(dE'_{qi}/dt)}{(x_{di} - x'_{di})} + \frac{E_{fdi}}{(x_{di} - x'_{di})} \tag{4.22}$$

$$\frac{\partial W_2}{\partial E'_{di}} = -\frac{T'_{qoi}\cdot(dE'_{di}/dt)}{(x_{qi} - x'_{qi})} \tag{4.23}$$

$$\frac{\partial W_2}{\partial V_i} = 0 \tag{4.24}$$

$$\frac{\partial W_2}{\partial \phi_i} = P_i \tag{4.25}$$

$$\frac{\partial W_2}{\partial \theta_i} = -P_{mi} + P_{ei} \tag{4.26}$$

$$\frac{\partial W_2}{\partial t} = \sum_{i=1}^{n} f_{pi}(V) \frac{d\phi_i}{dt} - \sum_{i=1}^{m} \frac{E_{fdi}}{(x_{di} - x'_{di})} \frac{dE'_{qi}}{dt} \tag{4.27}$$

Since

$$\frac{\partial W_1}{\partial \bar{\omega}_i} = M_i \bar{\omega}_i,$$

$$\frac{\partial W_1}{\partial \bar{\omega}_i} \frac{d\bar{\omega}_i}{dt} + \frac{\partial W_2}{\partial \theta_i} \frac{d\theta_i}{dt} = (M_i \dot{\bar{\omega}}_i - P_{mi} + P_{ei})\bar{\omega}_i = -\frac{M_i \bar{\omega}_i}{M_T} P_{COI} \tag{4.28}$$

Thus

$$\sum_{i=1}^{m} \left( \frac{\partial W_1}{\partial \bar{\omega}_i} \frac{d\bar{\omega}_i}{dt} + \frac{\partial W_2}{\partial \theta_i} \frac{d\theta_i}{dt} \right) = -\sum_{i=1}^{m} \frac{M_i \bar{\omega}_i}{M_T} P_{COI} = 0 \tag{4.29}$$

It can be shown using Equations 4.22 through 4.27 that

$$\frac{dW}{dt} = \sum_{i=1}^{m} \left( \frac{\partial W_1}{\partial \bar{\omega}_i} \frac{d\bar{\omega}_i}{dt} + \frac{\partial W_2}{\partial \theta_i} \frac{d\theta_i}{dt} \right) + \sum_{i=1}^{n} \left[ \left( \frac{\partial W_2}{\partial V_i} \frac{dV_i}{dt} + \frac{\partial W_2}{\partial \phi_i} \frac{d\phi_i}{dt} \right) \right] + \frac{\partial W}{\partial t}$$

$$+ \sum_{i=1}^{m} \left[ \frac{\partial W_2}{\partial E'_{qi}} \frac{dE'_q}{dt} + \frac{\partial W_2}{\partial E'_{di}} \frac{dE'_{di}}{dt} \right] = -\sum_{i=1}^{m} \left[ \frac{T'_{doi}(dE'_{qi}/dt)^2}{(x_{di} - x'_{di})} + \frac{T'_{qoi}(dE'_{di}/dt)^2}{(x_{qi} - x'_{qi})} \right]$$

**Remarks**

1. The first five terms in the PE ($W_{21}$ through $W_{25}$) are the same as those defined in Padiyar and Sastry (1987), except that in term $W_{24}$, $E'_{qi}$ is replaced by $E_i$ (for the classical model).

2. The term $W_{26}$ accounts for transient saliency and is identically zero if $x'_d = x'_q$.

3. The terms $W_{27}$, $W_{28}$, and $W_{211}$ arise due to the presence of the damper windings in the quadrature axis. If this winding is neglected, then $E'_{di} = 0$ and

$$W_{27} + W_{28} = 0 = W_{211}$$

4. If $E_{fd}$ = constant (AVR is neglected), then

$$W_{29} = -\sum_{i=1}^{m} E_{fdi} \frac{(E'_{qi} - E'_{qio})}{(x_{di} - x'_{di})} \tag{4.30}$$

becomes a path-independent function of $E'_q$. If AVR is considered, then $E_{fd}$ is variable and the term $W_{29}$ has to be computed by numerical integration. Using trapezoidal rule, we have

$$W_{29}(t + h) = W_{29}(t) - \frac{1}{2} \sum_{i=1}^{m} [E_{fdi}(t + h) + E_{fdi}(t)] \cdot \frac{[E'_{qi}(t + h) - E'_{qi}(t)]}{(x_{di} - x'_{di})}$$

Note that any type of excitation system can be considered as $E_{fdi}$ is treated as a variable parameter in the energy function.

5. If $E_{fd}$ and active load $P_l$ are constants, then the energy function defined is a general (global) Lyapunov function.

6. As mentioned in the previous chapter, we neglect the effects of variations in the rotor frequencies during a transient, in the computation of $P_m$ and $P_e$ and assume $T_m \simeq P_m$ and $T_e \simeq P_e$.

7. The governor and prime mover dynamics have generally minor effects on the first swing stability. However, if these effects are included, the mechanical power (torque) will be time varying and the term $W_{21}$ is modified to

$$W_{21} = -\sum_{i=1}^{m} \int_{t_0}^{t} P_{mi}(t) \left( \frac{d\theta_i}{dt} \right) dt \tag{4.31}$$

### 4.3.2 Simpler Expression for SPEF

For the analysis of first swing stability, it is convenient to introduce two extra (path-independent) terms in the PE such that the total energy $W$ remains constant along the postfault trajectory. Thus, a modified energy function, defined below, is introduced:

$$W' = W_1 + W'_2 = W_1(\omega) + \sum_{i=1}^{13} W_{2i} \tag{4.32}$$

where

$$W_{212} = \sum_{i=1}^{m} \int_{t_0}^{t} [T'_{doi}/(x_{di} - x'_{di})] \left(\frac{dE'_{qi}}{dt}\right)^2 dt, \quad W_{213} = \sum_{i=1}^{m} \int_{t_0}^{t} [T'_{qoi}/(x_{qi} - x'_{qi})] \left(\frac{dE'_{di}}{dt}\right)^2 dt$$

Although the number of terms is increased by two, some simplifications are possible by grouping the terms. It can be shown that

$$W_{24} + W_{25} + W_{26} + W_{27} + W_{28} = \frac{1}{2} \sum_{k=1}^{nb} (Q_k - Q_{k0}) = W'_{24} \tag{4.33}$$

$$W_{29} + W_{210} + W_{212} = \sum_{i=1}^{m} \int_{t_0}^{t} i_{di} \left(\frac{dE'_{qi}}{dt}\right) dt = W'_{25} \tag{4.34}$$

$$W_{211} + W_{213} = -\sum_{i=1}^{m} \int_{t_0}^{t} i_{qi} \left(\frac{dE'_{di}}{dt}\right) dt = W'_{26} \tag{4.35}$$

where $nb = nl + m$ is the total number of series-connected branches in the network, including machine reactances. Note that $nl$ is the number of lines (including the transformers) in the network.

Thus, a simpler expression for the energy function is obtained as

$$W = W_1(\overline{\omega}) + W_{21} + W_{22} + W_{23} + W'_{24} + W'_{25} + W'_{26} \tag{4.36}$$

where $W'_{24}$, $W'_{25}$, and $W'_{26}$ are defined above. Note that for the classical model, both $W'_{25}$ and $W'_{26}$ are zero. For one-axis models, $W'_{26} = 0$.

It is to be noted that $W'_{25}$ accounts for both the field coil and effects of AVR, while $W'_{26}$ accounts for the damper winding on the $q$-axis. The term $W'_{24}$ accounts for the energy stored in all series-connected reactances, including the transient reactances of the machine. With transient saliency considered, the energy stored in the machine reactances is given by

$$\frac{1}{2}(x'_{di}i_{di}^2 + x'_{qi}i_{qi}^2)$$

If $x'_{di} = x'_{qi} = x'_i$, then this component reduces to $(1/2)x'_i I_{ai}^2$, where $I_{ai}^2 = i_{di}^2 + i_{qi}^2$.

The expressions for SPEF given in this and the previous chapters have considered the datum (reference) at the value corresponding to the initial (prefault) operating point. There is no loss of generality in doing so as the critical value of $W$ will also change appropriately.

## 4.4 Numerical Examples

### 4.4.1 SMIB System

The single-line diagram of the system is shown in Figure 4.3. The system data in per unit (except for time constants) on a 1000 MVA base are given below:

*Generator:* $x_d = 1.7572$, $x_q = 1.5845$, $x'_d = 0.4245$; $x'_q = 1.04$, $T'_{do} = 6.66s$,

$$T'_{qo} = 0.44s; H = 3.542, f_B = 50 \text{ Hz}, D = 0.0$$

*Transformer:* $x_t = 0.1364$

*Transmission line (per circuit):* $R_l = 0.0$, $x_l = 0.8125$ (impedance between nodes 1 and 2)

*Excitation system:* Static excites with single time constant AVR is used $K_A = 400$, $T_A = 0.025$ s

$$E_{fd\,max} = 6.0, \quad E_{fd\,min} = -6.0$$

*Operating data:* $E_b = 1.0$, $V_t = 1.05$, $P_t = 0.6$, $X_{Th} = 0.13636$

A three-phase fault occurs at the high-voltage bus at the sending end. Find the critical clearing time by simulation and prediction. Assume the postfault system identical to the prefault system.

**Solution**

The value of the critical energy is found to be 0.633 (by the PEBS method) and the critical clearing time is 0.231 s.

By simulation, the system is stable when the clearing time is 0.235 s and is unstable when the clearing time is 0.236 s. The swing curves for both cases are shown in Figure 4.4.

The energy function for the system is given by

$$W = W_1 + W_2 = W_1 + W_{21} + W_{24} + W_{25} + W_{26}$$

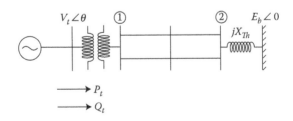

**FIGURE 4.3**
An SMIB system.

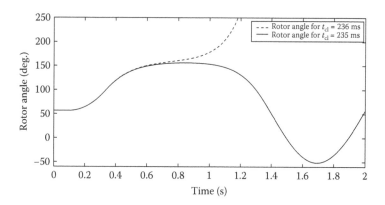

**FIGURE 4.4**
Swing curves for stable and unstable cases.

where

$$W_1 = \frac{1}{2} M \left( \frac{d\delta}{dt} \right)^2$$

$$W_{21} = -P_m (\delta - \delta_0)$$

$$W_{24} = \frac{1}{2} \left[ (i_d^2 - i_{d0}^2)x_d' + (i_q^2 - i_{q0}^2)x_q' + (I_a^2 - I_{ao}^2)^2 x_e \right] = \frac{1}{2} [Q_g + Q_b - (Q_{g0} + Q_{b0})]$$

$$W_{25} = \int_{E_{q0}'}^{E_q'} i_d dE_q'$$

$$W_{26} = -\int_{E_{d0}'}^{E_d'} i_q dE_d'$$

$I_a^2 = i_d^2 + i_q^2$, $x_e$ is the effective external reactance between the generator and the infinite bus. $Q_g = E_d' i_q - E_q' i_d$ and $Q_b$ is the reactive power injected by the infinite bus.

The total energy ($W$), KE ($W_1$), and the PE ($W_2$) for the stable and unstable cases are shown in Figure 4.5. The components of the PE for the stable and unstable cases are shown in Figure 4.6.

Note that the PE components $W_{22}$ and $W_{23}$ are zero in this example as there is no load bus.

## 4.4.2 Ten-Generator, 39-Bus New England Test System

The data for this system are given in Appendix D.

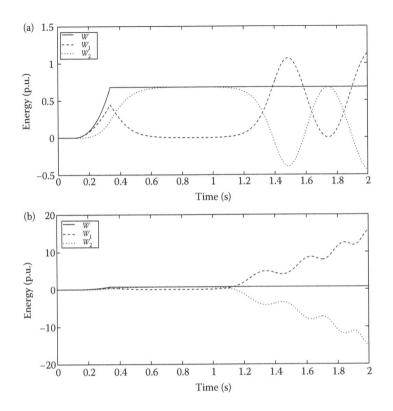

**FIGURE 4.5**
Potential, kinetic, and total energies: (a) stable case and (b) unstable case.

There are six cases of faults considered at various buses followed by tripping of faulted lines in four cases (see Table 4.1). All the cases are investigated
with the following generator model:

1. Classical
2. Detailed
   a. With damper winding (two-axis model) and transient saliency
      neglected
   b. Without damper winding (one-axis model) and transient saliency
      included

For both cases (a) and (b), the effects of modeling AVR were investigated.
Line resistances were neglected and loads are modeled as constant impedances. Tables 4.2 and 4.3 show the critical energy, the critical clearing time by
prediction, and the simulation for various cases.

The prediction was based on the PEBS method where sustained fault
trajectory is used to compute critical energy. To improve the accuracy in

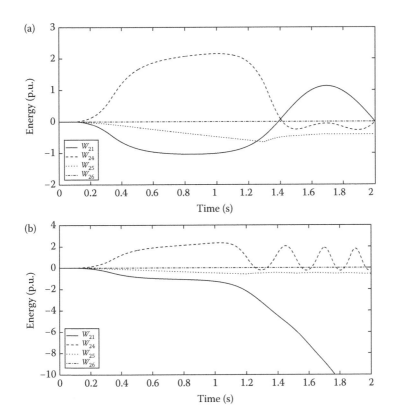

**FIGURE 4.6**
Components of potential energy: (a) stable case and (b) unstable case.

**TABLE 4.1**

Cases Considered for 10-Generator System

| Case | Fault at Bus # | Line Tripped (from Bus # to Bus #) |
|------|----------------|-------------------------------------|
| 1 | 37 | No line tripped |
| 2 | 8 | No line tripped |
| 3 | 26 | 26–28 |
| 4 | 24 | 24–23 |
| 5 | 14 | 14–34 |
| 6 | 10 | 12–25 |

**TABLE 4.2**

Results for Cases 1, 2, and 3 (10-Generator System)

| | Critical Energy and Critical | Classical | One-Axis Model | | Two-Axis Model | |
|---|---|---|---|---|---|---|
| | | | Without | | Without | |
| Case | Clearing Time | Model | AVR | With AVR | AVR | With AVR |
| 1 | Critical energy (predicted) | 10.84 | 3.83 | 38.06 | 4.53 | 22.39 |
| | Critical energy (simulated) | 11.40 | 4.03 | 30.17 | 4.17 | 21.81 |
| | CCT (predicted) | 0.21–0.22 | 0.12–0.13 | 0.32–0.33 | 0.13–0.14 | 0.29–0.30 |
| | CCT (simulated) | 0.21–0.22 | 0.12–0.13 | 0.31–0.32 | 0.13–0.14 | 0.29–0.30 |
| 2 | Critical energy (predicted) | 9.35 | 5.12 | 9.81 | 3.07 | 6.45 |
| | Critical energy (simulated) | 8.75 | 4.63 | 9.64 | 3.46 | 7.57 |
| | CCT (predicted) | 0.23–0.24 | 0.18–0.19 | 0.22–0.23 | 0.17–0.18 | 0.21–0.22 |
| | CCT (simulated) | 0.23–0.24 | 0.18–0.19 | 0.23–0.24 | 0.17–0.18 | 0.21–0.22 |
| 3 | Critical energy (predicted) | 1.43 | 0.15 | 4.78 | 0.35 | 4.23 |
| | Critical energy (simulated) | 1.16 | 0.17 | 4.80 | 0.29 | 4.33 |
| | CCT (predicted) | 0.08–0.09 | 0.03–0.04 | 0.15–0.16 | 0.04–0.05 | 0.14–0.15 |
| | CCT (simulated) | 0.08–0.09 | 0.03–0.04 | 0.15–0.16 | 0.04–0.05 | 0.15–0.16 |

the presence of damper winding (which introduces fast dynamics), $W_{26}$ is approximated as

$$W_{26} \simeq \frac{1}{2} \sum_{i=1}^{m} \left[ (x_{qi} - x'_{qi})(i_{qi}^2 - i_{qi0}^2) \right]$$

This modification is based on the paper by Sauer et al. (1989).

## Discussion

1. The prediction of critical clearing times for different generator models is quite accurate except for case 6, which will be discussed separately. All the 10 generators are modeled similarly when different models are considered.

2. In considering various models, it is observed that the provision of static excitation systems improves the transient stability in practically all the cases (except case 2). The classical model with generator reactance assumed to be $x'_d$ gives results on CCT that are better than the cases when flux decay is considered.

**TABLE 4.3**

Results for Cases 4, 5, and 6 (10-Generator System)

| Case | Critical Energy and Critical Clearing Time | Classical Model | One-Axis Model Without AVR | One-Axis Model With AVR | Two-Axis Model Without AVR | Two-Axis Model With AVR |
|---|---|---|---|---|---|---|
| 4 | Critical energy (predicted) | 7.78 | 2.82 | 28.44 | 2.08 | 20.63 |
| | Critical energy (simulated) | 8.87 | 2.76 | 24.61 | 2.38 | 19.12 |
| | CCT (predicted) | 0.20–0.21 | 0.11–0.12 | 0.31–0.32 | 0.11–0.12 | 0.29–0.30 |
| | CCT (simulated) | 0.20–0.21 | 0.11–0.12 | 0.31–0.32 | 0.11–0.12 | 0.30–0.31 |
| 5 | Critical energy (predicted) | 11.27 | 5.10 | 27.59 | 4.06 | 18.75 |
| | Critical energy (simulated) | 11.97 | 4.64 | 27.45 | 4.78 | 18.66 |
| | CCT (predicted) | 0.25–0.26 | 0.16–0.17 | 0.35–0.36 | 0.16–0.17 | 0.34–0.35 |
| | CCT (simulated) | 0.26–0.27 | 0.16–0.17 | 0.36–0.37 | 0.17–0.18 | 0.34–0.35 |
| 6 | Critical energy (predicted) | 16.56 | — | 4.40 | — | 4.28 |
| | Critical energy (simulated) | 2.81 | — | 6.24 | — | 5.24 |
| | CCT (predicted) | 0.49–0.50 | <0.01 | 0.46–0.47 | <0.01 | 0.44–0.45 |
| | CCT (simulated) | 0.24–0.25 | <0.01 | 0.57–0.58 | <0.01 | 0.57–0.58 |
| | CCT (modified PEBS method) | 0.34–0.35 | | | | |

The presence of damper winding in the quadrature axis has little effect on the transient stability.

3. The effect of transient saliency ($x_d' \neq x_q$) in one-axis models is to marginally reduce the critical clearing time. However, the provision of AVR has beneficial effect in the presence of transient saliency.

4. Table 4.4 compares the critical clearing times obtained from prediction and simulation for cases 1 through 5 with different numbers of generators represented in detail. Generally, generators close to the fault location are modeled in detail, while other generators are represented by classical models. The results shown in Table 4.4 indicate that, except for cases 2 and 3 in all other cases, there is an increase in CCT as more generators are represented in detail. This is due to the beneficial effects of providing AVRs in more generators (in improving transient stability).

5. Case 6 is the solitary case where the predicted CCT are very different from the results obtained by simulation. The AVR has a pronounced effect in this case as the inclusion of AVR gives CCT of 0.57–0.60 s depending on the model used. Although the fault occurs

**TABLE 4.4**

CCT and $W_{cr}$ with Different Number of Generators Modeled in Detail
(10-Generator System)

| Case | Number of Generators Modeled in Detail | Prediction | | Simulation | |
|------|----------------------------------------|------------|--------|------------|--------|
|      |                                        | CCT        | $W_{cr}$ | CCT      | $W_{cr}$ |
| 1 | 8, 9, and 10 | 0.22–0.23 | 13.24 | 0.22–0.23 | 12.00 |
| 1 | All | 0.29–0.30 | 22.39 | 0.29–0.30 | 21.81 |
| 2 | 8 and 10 | 0.21–0.22 | 8.34 | 0.21–0.22 | 7.07 |
| 2 | All | 0.21–0.22 | 6.45 | 0.21–0.22 | 7.57 |
| 3 | 8, 9, and 10 | 0.14–0.15 | 3.60 | 0.15–0.16 | 4.30 |
| 3 | All | 0.14–0.15 | 4.23 | 0.15–0.16 | 4.33 |
| 4 | 6 and 7 | 0.23–0.24 | 12.98 | 0.23–0.24 | 11.42 |
| 4 | 4, 5, 6, and 7 | 0.25–0.26 | 16.88 | 0.25–0.26 | 13.44 |
| 4 | All | 0.29–0.30 | 20.63 | 0.30–0.31 | 19.12 |
| 5 | 1 and 3 | 0.27–0.28 | 14.27 | 0.27–0.28 | 14.23 |
| 5 | 1, 3, 8, and 10 | 0.30–0.31 | 15.32 | 0.30–0.31 | 16.90 |
| 5 | All | 0.34–0.35 | 18.75 | 0.34–0.35 | 18.66 |

at the terminals of generator 10, the generator 10 is stable and the generators 8 and 9 accelerate and separate from the rest of the generators. The PEBS method fails to predict CCT as the sustained fault trajectories indicate that all nine generators accelerate and separate from generator 2, which has a very high inertia.

6. The failure of prediction of the critical energy based on sustained fault trajectory can be explained from the fact that in the postfault system, the generators 8 and 9 are connected to the rest of the system only through the line 26–27, which happens to be the critical cutset (the set of lines (elements) that separate the system into two parts; the angle across each element of the set becomes unbounded). Thus, generators 8 and 9 accelerate and separate from the rest of the generators.

It was observed that the fault at bus 12 followed by tripping of the line 12–25 also leads to a similar result—generators 8 and 9 separate from the rest. The importance of the critical cutset in determining the MOI is discussed in Chapter 6.

The modified PEBS method involves the first two steps of the BCU method. The first step of the BCU method is the same as the PEBS method. The second step involves using the exit point (obtained in the first step) as the initial condition and integration of the postfault, reduced gradient system to find the first local minimum of $\sum_{i=1}^{m} \|f_i(\delta)\|$. The application of the modified PEBS method did not give accurate results for the classical models in case 6.

7. It was observed that the mode of instability is unaffected by the generator model.

### 4.4.3 Variation of Total Energy and Its Components

The swing curves for case 5 (three-phase fault at bus #14 followed by tripping the line connecting bus 14 and bus 34) are shown in Figure 4.7 for the stable and unstable cases, considering the detailed models of all 10 generators. The critical clearing time lies between 0.34 and 0.35 s. The instability is due to the separation of generator 2, which decelerates while other generators accelerate. The variations of the total, kinetic, and potential energies with time for stable and unstable cases are shown in Figure 4.8. The six components of the potential energies for the stable and unstable cases are shown in Figures 4.9 and 4.10. It is observed that $W_{23}$ has negligible contribution while $W_{24}$ has a major contribution. $W_{21}$ and

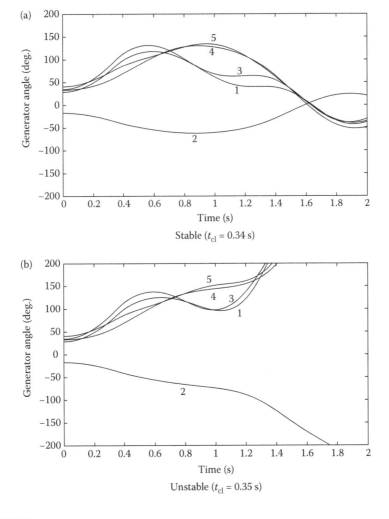

**FIGURE 4.7**
Swing curves for case 5 (10-generator system): (a) critically stable and (b) critically unstable.

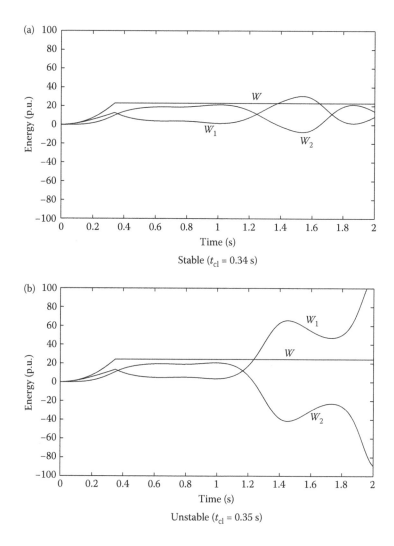

**FIGURE 4.8**
Potential, kinetic, and total energies: (a) stable case and (b) unstable case.

$W_{22}$ tend to cancel each other. $W_{26}$ has a small positive component while $W_{25}$ has a negative component with slightly higher magnitude.

## 4.5 Modeling of Dynamic Loads

As discussed in Chapter 2, the aggregate load at any bus can be modeled as functions of the bus voltage and frequency. However, the dynamics of the induction motor can be explicitly modeled if required. The action of the

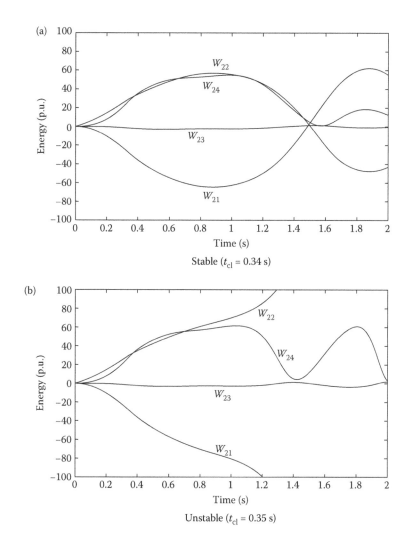

**FIGURE 4.9**
First four components of potential energy: (a) stable case and (b) unstable case.

OLTC or any other type of voltage regulator can also be modeled as part of the dynamic load. The load modeling is significant whenever voltage stability is to be investigated. Although voltage collapse (instability) is usually analyzed by ignoring the synchronous generator dynamics, the phenomenon of transient voltage stability involves dynamics in a time frame of a few seconds or less. In such cases, dynamics of load needs to be considered. In addition to the induction motor loads, HVDC converter control dynamics also affects transient voltage stability. The shunt FACTS devices such as SVC and STATCOM (Hingorani and Gyugyi 2000, Padiyar 2007) can be applied to

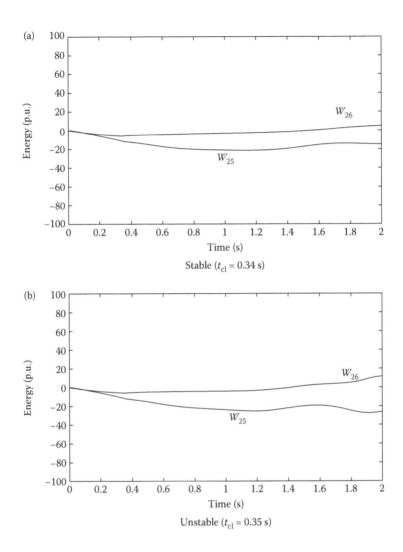

**FIGURE 4.10**
Last two components of potential energy: (a) stable case and (b) unstable case.

improve transient voltage stability in addition to enhancing synchronous or angle stability. The modeling of HVDC converter and FACTS controllers will be described in Chapter 5. Here, we will consider the induction motor load.

### 4.5.1 Induction Motor Model

The detailed model of an induction motor is a fifth-order model considering stator transients. However, since network transients (including synchronous generator stator transients) are generally neglected in stability analysis, a third-order model is adequate and is described by the following equations (Brereton et al. 1957, Padiyar 2002):

$$2H_m \frac{dS}{dt} = T_m - T_e \tag{4.37}$$

$$\frac{dE'}{dt} = -j2\pi fSE' - \frac{1}{T_0}[E' - j(X - X')I_t] \tag{4.38}$$

where

$$E' = E'_Q + jE'_D, \quad I_t = \frac{V_t - E'}{R_s + jX'} = I_Q + jI_D \tag{4.39}$$

$$T_0 = \frac{X_r + X_m}{2\pi fR_r} \tag{4.40}$$

$$X = X_s + X_m, \quad X' = X_s + \frac{X_m X_r}{X_m + X_r} \tag{4.41}$$

Note that $E'_Q$ and $E'_D$ are components of the complex voltage source $E'$ in the stator equivalent circuit shown in Figure 4.11. They are related to the flux linkages in the rotor coils. $f$ is the operating frequency in hertz and $S$ is the motor slip defined as

$$S = \frac{\omega_0 - \omega_m}{\omega_0} = \frac{f - f_m}{f} \tag{4.42}$$

where $\omega_m = 2\pi f_m$ is the rotational frequency (radians per second) of the rotor. The reactances $X_s$, $X_m$, and $X_r$ and the resistances $R_s$ and $R_r$ are defined in Figure 4.12, which shows the steady-state equivalent circuit

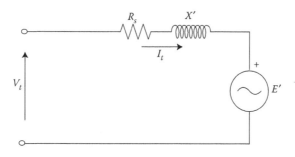

**FIGURE 4.11**
Stator equivalent circuit of an induction motor.

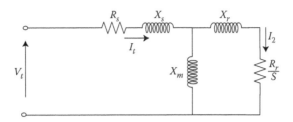

**FIGURE 4.12**
Steady-state equivalent circuit of an induction motor.

of the induction motor (neglecting the transients in the rotor coils). The electrical torque developed in the motor is given by the expression

$$T_e = E'_D I_D + E'_Q I_Q \tag{4.43}$$

It can be shown that Figure 4.11 is obtained from Figure 4.12 by replacing $R_r/S$ by the variable voltage source $E'((X_m + X_r)/X_m)$ and simplifying the circuit by applying the Thevenin theorem.
If $dE'/dt$ is assumed to be zero, we obtain

$$E' = I_2 \frac{R_r}{S} \frac{X_m}{(X_m + X_r)} \tag{4.44}$$

and the electrical torque $T_e$ is given by

$$T_e = \frac{I_2^2 R_r}{S} \tag{4.45}$$

### 4.5.2 Voltage Instability in Induction Motors

Consider an induction motor connected to a power system that can be represented by a Thevenin equivalent of a voltage source $E \angle 0$ in series with an impedance $Z$. It is assumed that the induction motor supplies a constant mechanical load torque. In steady state, the power drawn by the motor will be constant (independent of bus voltage) assuming that the losses in the stator coils and the magnetic circuits are negligible. However, during a transient caused by a disturbance in the voltage $E$, the motor speed can fluctuate before reaching a steady state (assuming the peak electrical torque in the postdisturbance state is greater than the mechanical torque). The final speed or slip can be different from the initial speed (or slip). Typically, during faults, there will be a dip in the voltage $E$, which reduces the motor terminal voltage. The motor decelerates as the electrical torque is not adequate to meet the required mechanical torque. Depending on the duration of the fault, it is possible that the motor may not reach a steady state after the fault is cleared and will continue to decelerate, leading to stalling of the motor. The stalled motor draws a large current and results in a further dip in the voltage. The undervoltage protection is usually provided to trip an induction motor. However, tripping of the motor can be avoided if the fault is cleared in a time less than the critical clearing time.

This phenomenon of instability of induction motors can be explained with reference to the torque slip characteristics of an induction motor shown in Figure 4.13. Under normal operating conditions, the peak torque is $T_{p1}$ and the motor operates stably at slip $S_1$ for a specified mechanical torque $T_m$. Note that there is another equilibrium value of slip ($S_2$). However, this is an UEP as $(dT_e/dS) < 0$.

During a fault, the peak torque $T_{p2}$ is less than the load torque $T_m$ and the motor decelerates. Assuming the postfault system identical to the prefault system, the motor will regain the original steady-state slip $S_1$ if the slip at the time of clearing the fault does not exceed $S_2$. If the motor slip goes beyond $S_2$, the motor continues to decelerate even after the fault is cleared and will stall eventually.

If the postfault system is different from the prefault system, we have to consider the torque slip characteristics of the postfault system to determine the stability. Assuming that the unstable slip corresponding to the postfault system is $S_2'$, the critical clearing time is obtained when the slip at the clearing of the fault is $S_2'$. The motor will settle at a steady-state slip of $S_1'$ if it is stable. In general, $S_2' < S_2$ and $S_1' > S_1$.

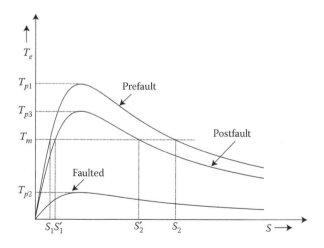

**FIGURE 4.13**
Torque slip characteristics of a grid connected induction motor.

### 4.5.3 Simpler Models of Induction Motors

A simplified model due to Walve (1986) specifies the real and reactive power demands $P_L$ and $Q_L$ of the motor in terms of load bus voltage $V_L$ and the frequency deviation $\dot{\delta}_L$. The expressions specifying $P_L$ and $Q_L$ are

$$P_L = P(\delta, V) = P_0 + K_{p\omega}\dot{\delta}_L + K_{pv}(V_L + T\dot{V}_L) \tag{4.46}$$

$$Q_L = Q(\delta, V) = Q_0 + K_{q\omega}\dot{\delta}_L + K_{qv}V_L + K_{qv2}V_L^2 \tag{4.47}$$

From the above, we can express the derivatives of $\delta_L$ and $V_L$ as follows (Dobson and Chiang 1989):

$$K_{q\omega}\dot{\delta}_L = -K_{qv2}V_L^2 - K_{qv}V_L + Q(\delta, V) - Q_0 \tag{4.48}$$

$$TK_{q\omega}K_{pv}\dot{V}_L = K_{p\omega}K_{qv2}V_L^2 + (K_{p\omega}K_{qv} - K_{q\omega}K_{pv})V_L$$

$$+ K_{q\omega}[P(\delta, V) - P_0] - K_{p\omega}[Q(\delta, V) - Q_0] \tag{4.49}$$

where $P(\delta, V)$ and $Q(\delta, V)$ are the active and reactive power supplied by the network. $P_0$ and $Q_0$ are the constant real and reactive power components of the motor.

An alternative and simpler model is utilized by Overbye et al. (1992) to discuss some aspects of the energy function approach to analyze angle and voltage stability. In this model, only reactive power ($Q_L$) is assumed to be dependent on the derivative of the load bus voltage specified by

$$Q_L + K_L\dot{V}_L = Q(\delta, V) \tag{4.50}$$

The frequency dependence of the real power and reactive power is neglected. However, $P_L$ and $Q_L$ may be modeled as voltage-dependent functions.

### 4.5.4 Energy Function Analysis of Synchronous and Voltage Stability

The simplest system that can exhibit both synchronous (angle) and voltage stability is a three-bus system with one generator, an infinite bus, and a load (shown in Figure 4.14). Bus 1 is the internal bus of the generator. If the aggregate load ($P_L + jQ_L$) is a second-order polynomial function, it is possible to express the load bus voltage magnitude ($V_L$) as real solutions of a quartic equation (as shown in Chapter 3). In general, there are two real solutions for the equation. If the generator is modeled by the classical model (using a swing equation) and the load is modeled by Equation 4.50, there are three

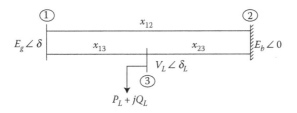

**FIGURE 4.14**
A three-bus system.

state variables $\delta, \dot{\delta}$, and $V_L$. For a specified generator power output and specified $P_L$ and $Q_L$, there can be one or two equilibrium values of $\delta$ for each solution of $V_L$. Since $\dot{\delta} = 0$ and $\dot{V}_L = 0$ is always satisfied at an equilibrium point, there can be two or four equilibrium points. We can pose the following questions concerning angle and voltage stability:

1. Assuming that there are more than one UEP, which UEP determines transient synchronous stability?
2. How do the loading conditions affect the nature of instability?
3. Does the nature of the models used affect the stability?

The answers to these questions will be answered by taking up a case study based on the example presented in Overbye et al. (1992).

## Case Study

Data: $H = 12\,\text{s}$, $X_d = 0.5$, $X_d' = 0.05$, $E_b = 1.0$, $E_g = 1.034$, $b_{12} = \dfrac{1}{x_{12}} = 3.1579$,

$b_{13} = \dfrac{1}{x_{13}} = 4.2105$, $b_{23} = \dfrac{1}{x_{23}} = 21.0526$, $K_L = 1$, $P_m = 3.0$

The dynamic equations of the system are

$$
\left.
\begin{aligned}
\frac{d\delta}{dt} &= \dot{\delta} = \omega \\[2mm]
M\dot{\omega} &= P_m - P_e \\[2mm]
K_L\dot{V}_L &= Q(\delta, V) - Q_L
\end{aligned}
\right\}
\tag{4.51}
$$

where

$$
Q(\delta, V) = V_L\left[ b_{13}E_g \cos(\delta - \delta_L) + b_{23}E_b \cos\delta_L + b_{33}V_L^2 \right]
$$

$$b_{33} = -(b_{13} + b_{23})$$

$$P_e = b_{12}E_gE_b \sin\delta + b_{13}E_gV_L \sin(\delta - \delta_L)$$

$$P_L = P(\delta, V) = V_L\left[b_{13}E_g \sin(\delta - \delta_L) - b_{23}E_b \sin\delta_L\right]$$

Since $P_L$ and $Q_L$ are considered as constants (independent of the load bus voltage), we can solve $V_L^2$ from the quadratic equation given below:

$$A_4V_L^4 + A_2V_L^2 + A_0 = 0$$

where

$$A_4 = b_{33}^2, \ A_2 = -b_{33}Q_L - \alpha^2, \ A_0 = P_L^2 + Q_L^2, \ \alpha = \left|b_{13}E_ge^{j\delta} + b_{23}E_b\right|$$

### 4.5.4.1 Computation of Equilibrium Points

For specified values of $P_m = P_e$ (generator power), $P_L$, and $Q_L$, we can find the equilibrium points. We assume $Q_L = 0.5P_L$ and $P_m = 3.0$, and vary $P_L$ from zero to the maximum feasible value. For (i) $P_L = 2.0$ and (ii) $P_L = 6.0$, the EP are given in Table 4.5 for $K_L = 1$. It is observed that for $P_L = 2.0$ (and $Q_L = 1.0$), there are four EPs. One of them is SEP, two are type 1 UEPs, and one is type 2 UEP (with two eigenvalues in the RHP). Note that the nature of the EP is found by computing the eigenvalues of the Jacobian (obtained by linearizing Equation 4.51) defined by

$$\begin{bmatrix} \Delta\dot{\delta} \\ \Delta\dot{\omega} \\ \Delta\dot{V}_L \end{bmatrix} = [J] \begin{bmatrix} \Delta\delta \\ \Delta\omega \\ \Delta V_L \end{bmatrix} \tag{4.52}$$

**TABLE 4.5**

Characterization of EPs (Three-Bus System)

| $P_L$ | EPs | Classical Model $(K_L = 0.0)$ | With Load Dynamics $(K_L = 1.0)$ |
|-------|-----|-------------------------------|----------------------------------|
|       |     | K | Eigenvalues |
| 2.0 | 1 | 6.17 (SEP) | $-0.04 \pm j10.31$, $-22.902$ (SEP) |
|     | 2 | $-6.12$ (UEP) | $9.15$, $-9.57$, $-13.45$ (type-1 UEP) |
|     | 3 | 2.06 (SEP) | $-0.15 \pm j5.80$, $8.05$ (type-1 UEP) |
|     | 4 | $-2.186$ (UEP) | $-5.89$, $6.08$, $4.42$ (type-2 UEP) |
| 6.0 | 1 | 5.94 (SEP) | $-0.09 \pm j9.98$, $-16.94$ (SEP) |
|     | 2 | 3.68 (SEP) | $-0.76 \pm j7.78$, $2.35$ (type-1 UEP) |

where

$$[J] = \begin{bmatrix} 0 & 1 & 0 \\ -\dfrac{C_1}{M} & 0 & -\dfrac{C_2}{M} \\ \dfrac{C_3}{K_L} & 0 & \dfrac{KC_4}{K_L} \end{bmatrix}$$

where

$$C_1 = b_{12}E_gE_b \cos\delta + b_{13}E_gV_L \cos(\delta - \delta_L), \quad C_2 = b_{13}E_g \sin(\delta - \delta_L)$$

$$C_3 = -b_{13}E_gV_L \sin(\delta - \delta_L), \quad C_4 = b_{13}E_g \cos(\delta - \delta_L) + b_{23}E_b \cos\delta + 2b_{33}V_L$$

For $P_L = 6.0$, there are only two EPs: one is an SEP and the other is a type 1 UEP. It was observed that the value of $K_L$ does not affect the nature of EP (whether it is stable or unstable and the type of UEP) unless $K_L = 0$. Note that, in this case, the dimension of the state space is reduced from 3 to 2 as the generator is modeled by the swing equation and the load dynamics is neglected. In this case, the stability is determined from the sign of $dP_e/d\delta = K$. If $K > 0$, the EP is stable, otherwise it is unstable. Also, for a classical model, the UEP is of type 1 as only one eigenvalue moves into RHP as $K$ changes its sign from positive to negative (Padiyar 2002). It is interesting to observe that with $K_L = 0$, the four EPs with $P_L = 2.0$ have different characteristics than the case when $K_L > 0$, two EPs in this case are stable, and the remaining two are UEPs. When $P_L = 6.0$, the UEPs disappear and both are SEPs.

To investigate the nature of EPs, the load bus voltage ($V_L$) is plotted against the generator angle ($\delta$) for $K_L = 1.0$ and different values of $P_L$ (see Figure 4.15). It is to be noted that when $P_L = Q_L = 0.0$, the load bus voltage is zero. One of

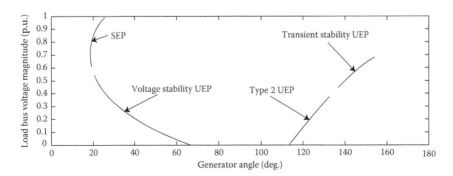

**FIGURE 4.15**
Variations of load bus voltage and rotor angle as load is varied.

the type 1 UEP tends to merge with SEP as $P_L$ is increased. Similarly, the remaining type 1 UEP merges with the type 2 UEP as the load is increased. The type 1 UEP with low voltages and normal values of angles can be labeled as the UEP associated with the voltage stability. The type 1 UEP associated with higher generator rotor angles is labeled as the UEP associated with loss of synchronous stability.

### 4.5.4.2 Computation of Energy at UEP

The critical energy that leads to the loss of synchronous stability is usually obtained from the PE evaluated at the CUEP. We define the SPEF for the three-bus system as follows:

$$W = W_1 + W_2 = W_1 + W_{21} + W_{22} + W_{23} + W_{24}$$

where

$$W_1 = \frac{1}{2}M\left(\frac{d\delta}{dt}\right)^2, \quad W_{21} = -P_m(\delta - \delta_0), \quad W_{22} = P_L(\delta_L - \delta_{L0})$$

$$W_{23} = Q_L \ln\left(\frac{V_L}{V_{L0}}\right), \quad W_{24} = \frac{1}{2}\sum_{k=1}^{3}(Q_k - Q_{k0})$$

$Q_k$ is the reactive power loss in line $k$. From Figure 4.14, we note there are three lines connecting the three buses.

The energies at the three UEPs are plotted as a function of $P_L$ as the load is increased (see Figure 4.16). The energy associated with transient stability UEP is much less than the energy associated with other two UEPs at reduced loads. As the load is increased, there is only one UEP (beyond $P_L = 4.0$) and the energy at that UEP (associated with voltage stability) keeps reducing until

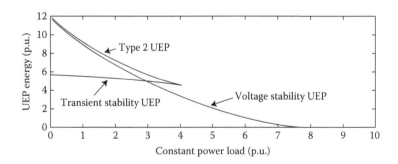

**FIGURE 4.16**
Potential energies at UEPs.

it becomes zero at maximum possible load. Note that the energy is computed relative to the SEP.

**Remarks**

1. Although a three-bus system with one generator is a very simple system, the results show correlation with what has been observed with large practical systems. Typically, loss of synchronous stability is a major issue in systems with moderate power transfers. On the other hand, voltage instability or collapse is observed in stressed power systems with high loadings.

2. It would appear that the CUEP for a particular disturbance can also be one of the UEPs associated with low-voltage solutions. Incidentally, energy methods have been applied previously for voltage stability assessment (Overbye and DeMarco 1991).

3. If $K_L = 0$, the application of the classical model of the generator does not predict loss of voltage stability. The load dynamics needs to be modeled to predict voltage collapse. It was observed that the inclusion of flux decay in the generator model also did not lead to the prediction of voltage instability. However, the modeling of the excitation system with AVR does help in identifying voltage instability. The analysis of a three-bus system shown in Figure 4.14 (with different data) is performed using a single time constant excitation system (Padiyar and Bhaskar 2002). With a fourth-order system model (for one-axis generator model and a static exciter with single time constant AVR), there are two complex pairs of eigenvalues, one corresponding to the swing mode (SM) and the other corresponding to the exciter mode. When the swing mode is stable, it was observed from simulation that while the power oscillations in the line connecting the generator to an infinite bus stabilize, the oscillations in the load bus voltage increase in magnitude if the exciter mode is unstable. The nature of load bus voltage variation changes when the voltage instability is caused by saddle-node bifurcation (SNB). There is a monotonic fall in the load bus voltage whereas the power oscillations in the line decay (see Figure 4.17). The disturbance considered is a pulse disturbance of 0.01 in $P_L$ for a duration of 0.5 s.

Based on the small signal stability analysis and simulation results, the following conclusions emerge:

1. The instability of power swings (due to angle instability) is caused by the instability of the swing mode that is affected by the fast excitation system. As the system loading is increased, with the normal

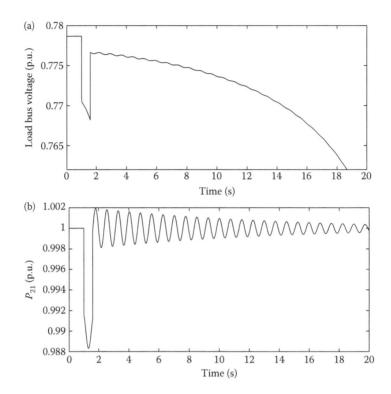

**FIGURE 4.17**
Simulation results for a three-bus system. (a) Load bus voltage. (b) Power flow in the line.

operation of AVRs, the initial system instability is due to the swing
mode that can be stabilized by damping controllers.

2. The voltage instability appears to result from the unstable exciter
   mode at higher loadings and is affected by the regulator gains and
   time constant. Low gains and high time constants can result in voltage instability. The nature of instability (monotonic or oscillatory)
   depends on the AVR gain. Below a critical gain, the instability is due
   to the SNB, while at higher gains, the instability is the oscillatory
   type (undamped oscillations) caused by Hopf bifurcation (HB).

3. If it is assumed that an equivalent generator can represent all generators (that remain in synchronism) in a multimachine system, the
   equivalent values of $K_A$ and $T_A$ are affected by the overexcitation limiters. Under abnormal conditions, the slowing down of the exciter
   response is equivalent to increase in $T_A$ or reduction in the effective
   value of $K_A$ or both.

4. While the angle instability affects both power swings and load voltage, the voltage instability has no significant effect on power swings.

### 4.5.5 Dynamic Load Models in Multimachine Power Systems

Davy and Hiskens (1997) propose a dynamic load model as follows:

$$P_{di} = P_{di}^0 + \omega_i D_i \tag{4.53}$$

$$Q_{di}(x_{qi}, V_i) = x_{qi} + Q_{ti}(V_i) \tag{4.54a}$$

$$T_{qi} \dot{x}_{qi} = -x_{qi} + Q_{si}^0 - Q_{ti}(V_i) \tag{4.54b}$$

where $P_{di}$ and $Q_{di}$ are the active and reactive load demands at bus $i$. Equation 4.53 is not new and has been proposed by Bergen and Hill (1981). Equations 4.53 and 4.54 describe the dynamics in the reactive power demand. In steady state, $Q_{di} = Q_{si}^0$. It is assumed that $Q_{si}^0$ is a constant although it can be a relaxed subject to some constraints. The load model is based on the paper by Hill (1993). $Q_{ti}$ is the transient change in the reactive power demand that depends on $V_i$. To establish a Lyapunov function, $Q_{ti}(V_i)$ must be of the form

$$Q_{ti}(V_i) = Q_{ti}^0 \ln\left(\frac{V_i}{\mu_i}\right) \tag{4.55}$$

where $Q_{ti}^0$ and $\mu_i$ are positive constants.

The contribution of the dynamic reactive power demands to the energy function is given by

$$\sum_{i=1}^{n}\left[ x_{qi} \ln\left(\frac{V_i}{V_i^s}\right) + \frac{Q_{ti}^0}{2}\left(\left(\ln\left(\frac{V_i}{\mu_i}\right)\right)^2 - \left(\ln\left(\frac{V_i^s}{\mu_i}\right)\right)^2\right) + \frac{x_{qi} - x_{qi}^s}{2Q_{ti}^0} \right] \tag{4.56}$$

where $n$ is the number of load buses. The superscript $s$ indicates equilibrium values.

### Remarks

1. The derivative of the total energy function (with the classical models of the generators) is given by

$$\frac{dW}{dt} = -\sum_{i=1}^{n} \frac{T_{qi}}{Q_{ti}^0} (\dot{x}_{qi})^2$$

2. If $Q_{si}^0 < Q_{ti}^0$, then the assumption about $Q_{si}$ being a constant can be relaxed.

3. The Lyapunov function is generated using a "first integral" analysis and compared with the Popov criterion.

4. Although load dynamics has been considered in voltage stability investigations, it is not in vogue in energy function analysis. Since the aggregate load models tend to be empirical, it would be adequate to view load as time varying with limits on the rate of change. However, step changes in the load can be considered as disturbances in the stability analysis.

## 4.6 New Results on SPEF Based on Network Analogy

In Chapter 3, we introduced a network analogy to derive energy functions. We will expand on the network analogy to derive new results that can be directly applied not only for simplification of the formulation of the energy functions but also for extending their application for the detection of loss of synchronism and improvement of stability. The applications of SPEF can be easily extended for the study of HVDC and FACTS controllers, which will be taken up in the next chapter.

In Chapter 3, we showed that the PE in a lossless system network without load damping can be expressed as

$$W_{PE} = W_2 = \sum_{i=1}^{m} \int_{\delta_{i0}}^{\delta_i} (P_{ei} - P_{mi})d\delta_i \tag{4.57}$$

This result is valid even when detailed generator and load models are considered (as shown later). We represent the lossless system by an analogous network of linear capacitors and nonlinear inductors. The loads are represented as current sources, which are part of the electrical network. $W_1(t_c)$, where $t_c$ is the clearing time of the fault, represents the KE that needs to be absorbed by the electrical network. However, since transmission lines and other series elements are modeled as nonlinear inductors, there is a limit on the energy that can be absorbed by the network.

Using Tellegen's theorem, we can also express the PE in terms of the energy stored in individual elements of the network. The following theorem gives the alternate representation of $W_2$ or $W_{PE}$:

## Theorem

For lossless systems with constant power loads, we can prove that

$$W_{PE} = \sum_{k=1}^{nb} \int_{t_0}^{t} (P_k - P_{k0})\frac{d\phi_k}{dt} dt = \sum_{k=1}^{nb} \int_{\phi_{k0}}^{\phi_k} (P_k - P_{k0})d\phi_k \quad (4.58)$$

where $nb$ is the total number of series branches in the network (consisting of transmission lines, transformers, and generator reactances).

*Proof.*

In a lossless network, the active power injection at a bus and the bus frequency $(d\phi/dt)$ satisfy Kirchhoff's current and voltage laws, respectively. Hence, the active powers (or torques) can be considered analogous to currents, bus frequencies to the voltages, and bus angles to the flux linkages in an equivalent electrical network. The power $(P_k)$ in the network branch $k$ (connected across buses $i$ and $j$) is given by

$$P_k = \frac{V_i V_j}{x_k} \sin \phi_k, \quad \phi_k = \phi_{ij}$$

This equation is similar to the equation of a nonlinear inductor given by $i = f(\lambda)$. Thus, the power system is analogous to the electrical network with nonlinear inductors (representing series reactances of the transmission lines and generators) (see Figure 4.18).

For any electrical network, Tellegen's theorem can be applied, which states that at any time, the sum of powers delivered to each branch of the network is zero (Tellegen 1952). Separating the (current) sources from other (passive) branches, we can apply Tellegen's theorem for the network shown in Figure 4.18 and obtain the following result:

$$\sum_{i=1}^{m} P_{ei}\frac{d\delta_i}{dt} - \sum_{j=1}^{n} P_{Lj}\frac{d\phi_j}{dt} = \sum_{k=1}^{nb} P_k\frac{d\phi_k}{dt} \quad (4.59)$$

where $P_{Lj}$ is the load (active) power at bus $j$. There are $n$ buses in the network (excluding the generator internal buses).

Tellegen's theorem is also valid when the branch voltage of one network and branch currents of another network are considered, provided the two networks have the same graph. If we consider the network shown in

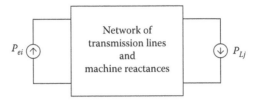

**FIGURE 4.18**
Analogous network for a lossless power system.

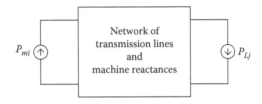

**FIGURE 4.19**
Analogous network in steady state.

Figure 4.18 in steady state, we get the source currents as $P_{mi}$ as $P_{ei} = P_{mi}$ (see Figure 4.19). The currents corresponding to loads remain unchanged in steady state as $P_{Lj}$ is assumed to be constant for $j = 1,2,\ldots,n$.

Considering branch currents in the (steady-state) network of Figure 4.19 and branch voltages of the network in Figure 4.18, we get the following result:

$$\sum_{i=1}^{m} P_{mi}\frac{d\delta_i}{dt} - \sum_{j=1}^{n} P_{Lj}\frac{d\phi_j}{dt} = \sum_{k=1}^{nb} P_{k0}\frac{d\phi_k}{dt} \tag{4.60}$$

where $P_{k0}$ is the steady-state power flow in branch $k$. Subtracting Equation 4.60 from Equation 4.59, we get

$$\sum_{i=1}^{m} (P_{ei} - P_{mi})\frac{d\delta_i}{dt} = \sum_{k=1}^{nb} (P_k - P_{k0})\frac{d\phi_k}{dt} \tag{4.61}$$

Integrating the above equation, we get the expression for $W_{PE}$ given by Equation 4.58.

## Remarks

1. It would appear from Equation 4.58 that the PE is path dependent. However, we can show that $W_{PE}$ is also equal to

$$W_{PE} = W_{PE1} + W_{PE2} + W_{PE3} \tag{4.62}$$

where

$$W_{PE1} = -\sum_{i=1}^{m} P_{mi}(\delta_i - \delta_{i0}) + \sum_{j=1}^{n} P_{Lj}(\phi_j - \phi_{j0}) \tag{4.63}$$

$$W_{PE2} = \frac{1}{2}\sum_{k=1}^{nb}(Q_k - Q_{k0}) \tag{4.64}$$

$$W_{PE3} = \sum_{j=1}^{n}\int_{V_{j0}}^{V_j}\frac{Q_{Lj}}{V_j}dV_j + \sum_{i=1}^{m}\int_{t_0}^{t}\left[-i_{qi}\frac{dE'_{di}}{dt} + i_{di}\frac{dE'_{qi}}{dt}\right]dt \tag{4.65}$$

where $Q_k$ is the reactive power absorbed by the branch $k$. $Q_{Lj}$ is the reactive power load at bus $j$. $i_{di}$ and $i_{qi}$ are the $d$- and $q$-axis components of the armature currents in the generator $i$. $E'_{di}$ and $E'_{qi}$ are the $d$- and $q$-components of the induced voltages in the stator windings of the generator $i$. Note that if we consider the simplified (classical) model of the generator, then the second term in the right-hand side (RHS) of Equation 4.65 is zero.

Since $\sum Q_k$ must be equal to the net injection of reactive power in the network, we get an alternative expression for $W_{PE2}$ as follows:

$$W_{PE2} = \frac{1}{2}\left[\sum_{i=1}^{m}(Q_{Gi} - Q_{Gi0}) - \sum_{j=1}^{n}(Q_{Lj} - Q_{Lj0})\right] \tag{4.66}$$

Note that the subscript "0" indicates the operating (steady state) value of the variable. $Q_{Gi}$ is the reactive power injected at the internal bus of generator $i$ and is calculated as

$$Q_{Gi} = E'_{di}i_{qi} - E'_{qi}i_{di}$$

**Proof of Equation 4.62**

The first term in the PE results from applying Equation 4.60 in the integral—$\int_{\phi_{k0}}^{\phi_k} P_{k0}d\phi_k$. The assumption here is that $P_{mi}$ is also a constant in addition to $P_{Lj}$. Nonconstancy of $P_{mi}$ and $P_{Lj}$ would result in path-dependent terms.

The second and third terms in the PE can be obtained from noting that

$$\int P_k d\phi_k = \int \frac{V_i V_j}{x_k}\sin\phi_k d\phi_k = -\frac{V_i V_j\cos\phi_k}{x_k} + \int \frac{\cos\phi_k}{x_k}(V_i dV_j + V_j dV_i) \tag{4.67}$$

Adding $((V_i^2 + V_j^2)/2x_k)$ and subtracting the same quantity, we can show that

$$\int P_k d\phi_k = \frac{V_i^2 + V_j^2 - 2V_iV_j \cos\phi_k}{2x_k} - \int (I_{ri}^k dV_i + I_{rj}^k dV_j) \tag{4.68}$$

where $I_{ri}^k$ and $I_{rj}^k$ are the reactive currents flowing into branch $k$ at the nodes $i$ and $j$, respectively. These are given by

$$I_{ri}^k = \frac{V_i - V_j \cos\phi_k}{x_k}, \quad I_{rj}^k = \frac{V_j - V_i \cos\phi_k}{x_k} \tag{4.69}$$

The first term in the RHS of Equation 4.68 is $Q_k/2$. After the summation over all branches and noting that the definite integral gives the increment over the steady-state values, we get the expression for $W_{PE2}$ as given in Equation 4.64.

Summing the second term in the RHS of Equation 4.68 gives

$$-\sum_{k=1}^{nb} \int (I_{ri}^k dV_i + I_{rj}^k dV_j) = \sum_{j=1}^{n} \int_{V_{j0}}^{V_j} \frac{Q_{Lj}}{V_j} dV_j \tag{4.70}$$

if we assume classical models for generators. In deriving Equation 4.70, we note that

$$\frac{Q_{Lj}}{V_j} = -\sum_{k \in j} I_{rj}^k \tag{4.71}$$

where the summation is over the branches that are incident on the bus $j$.

Note that Equation 4.70 includes the components corresponding to the generator terminal buses. The reactive current injected to the stator (reactance) branch at the generator internal bus also contributes to the RHS of Equation 4.70. However, the reactive current injected at the generator internal bus does not contribute to the term $W_{PE3}$ as the generator voltage is assumed to be a constant for the classical model.

### 4.6.1 Potential Energy Contributed by Considering the Two-Axis Model of the Synchronous Generator

If we consider the detailed two-axis model for the synchronous generators, $W_{PE3}$ is no longer given by the RHS of Equation 4.70.

The stator equations for a generator $i$ are given by

$$E_{qi}' + x_{di}'I_{di} = v_{qi} = V_i \cos(\delta_i - \phi_i) \tag{4.72}$$

$$E'_{di} - x'_{qi}i_{qi} = v_{di} = -V_i \sin(\delta_i - \phi_i) \tag{4.73}$$

The electrical power output ($P_{ei}$) of the generator $i$ is given by

$$P_{ei} = v_{di}i_{di} + v_{qi}i_{qi} \tag{4.74}$$

Substituting Equations 4.72 and 4.73 in 4.74, we get

$$P_{ei} = \frac{E'_{di}V_i \cos(\delta_i - \phi_i)}{x'_{qi}} + \frac{E'_{qi}V_i \sin(\delta_i - \phi_i)}{x'_{di}} + \frac{V_i^2 \sin\{2(\delta_i - \phi_i)\}}{2}\left(\frac{1}{x'_{qi}} - \frac{1}{x'_{di}}\right) \tag{4.75}$$

We can show that (Krishna 2003)

$$\int_{t_0}^{t} P_{ei} \frac{d(\delta_i - \phi_i)}{dt} dt = \int_{t_0}^{t} \left(i_{di}\frac{dE'_{qi}}{dt} - i_{qi}\frac{dE'_{di}}{dt}\right) dt + \int_{t_0}^{t} \frac{Q_{gi}}{V_i} dV_i + \frac{1}{2}(Q_{mi} - Q_{mi0}) \tag{4.76}$$

where $Q_{gi}$ is the reactive output of the generator (at the terminals) given by

$$Q_{gi} = v_{di}i_{qi} - v_{qi}i_{di} \tag{4.77}$$

$Q_{mi}$ is the reactive power loss in the stator given by

$$Q_{mi} = x'_{di}i_{di}^2 + x'_{qi}i_{qi}^2 \tag{4.78}$$

When we sum the PE in machine stator reactances, using the expression (4.76), the second term in the RHS of Equation 4.76 contributes to the first term in the RHS of Equation 4.65. The third term in the RHS of Equation 4.76 contributes to $W_{PE2}$. The only additional term contributing to the PE (by the two-axis models of the generators) is the second component of $W_{PE3}$ given by

$$W_{PE3}^m = \sum_{i=1}^{m} \int_{t_0}^{t} \left(i_{di}\frac{dE'_{qi}}{dt} - i_{qi}\frac{dE'_{di}}{dt}\right) dt \tag{4.79}$$

The above expression is path dependent. However, this includes the effect of the AVR and the excitation system control.

**Remarks**

1. The expression for the PE given in Section 4.3 has six components. From the network analogy, these can be reduced to three components as $W_{PE1} = W_{21} + W_{22}$ to the contribution of power drawn ($-P_m$ and $P_L$). $W_{PE2} = W'_{24}$ (defined in Equation 4.33) is the energy stored in the magnetic field of the series elements. $W_{PE3} = W_{23} + W'_{25} + W'_{26}$ (the last two components defined in Equations 4.34 and 4.35) is the magnetic energy in the shunt elements corresponding to the generators and loads.

2. The shunt susceptances in the lines do not carry power but contribute to $W_{PE3}$. Since the reactive current in the linear shunt susceptances is proportional to the voltage (across them), we obtain for the node $i$

$$-\int_{V_{i0}}^{V_i} I_{ri} dV_i = -\int_{V_{i0}}^{V_i} b_{si} V_i dV_i = -\frac{b_{si}}{2}(V_i^2 - V_{i0}^2) = -\frac{(Q_{si} - Q_{si0})}{2} \qquad (4.80)$$

where $Q_{si}$ is the reactive power loss in the shunt susceptance connected at bus $i$. Note that positive susceptance implies a capacitor that absorbs negative reactive power. The expression (4.80) can be clubbed with $W_{PE2}$.

3. We can express the energy stored in a lossless series element $k$ as $\int I_k dV_k$ as the current in the element, $I_k$, is lagging the voltage drop across the line, $V_k$. Since $V_k = x_k I_k$, we obtain the energy associated with line $k$ as $(1/2)x_k I_k^2 = (1/2)(V_k^2/x_k)$. Note that this expression is different from what is given in (4.68). However, if $V_i$ and $V_j$ are constants, the sum of the second term (in the RHS) of Equation 4.68 is zero. Note that the first term of $\int P_k d\phi_k$ is the same as $(1/2)(V_k^2/x_k)$.

4. If the voltages at nodes $i$ and $j$ (the terminal nodes of the series branch $k$) are varying, we can still express

$$\int I_k dV_k = \int \frac{V_k}{x_k} dV_k \qquad (4.81)$$

Since

$$V_k = V_i^2 + V_j^2 - 2V_i V_j \cos\phi_k$$

We can derive

$$dV_k = 2x_k[P_k d\phi_k + I^k_{ri}dV_i + I^k_{rj}dV_j] \tag{4.82}$$

Substituting Equation 4.82 in Equation 4.81, we obtain

$$\frac{(V_k^2 - V_{k0}^2)}{2x_k} = \int_{\phi_{k0}}^{\phi_k} P_k d\phi_k + \int (I^k_{ri}dV_i + IjdV_j) \tag{4.83}$$

Note that the above equation is equivalent to that of Equation 4.68.

5. The two components of the PE $W_{PE2}$ and $W_{PE3}$ (defined in Equations 4.64 and 4.65) are related to the reactive power/currents in the electrical network. For a lossless network, the concept of reactive current flow network (RCFN) is useful (Padiyar 1984, Padiyar and Suresh Rao 1996). This is a DC resistive network in which the currents are analogous to the reactive currents in the AC network and the voltages are analogous to the voltage magnitudes in the AC network. The analogous RCFN for a lossless transmission line represented by a $\pi$ equivalent network is shown in Figure 4.20. Here, $b_k = (1/x_k)$ and $b_i^s$ and $b_j^s$ are the equivalent shunt susceptances associated with the positive sequence capacitances of the line. It is seen that the resistances of the analogous network are a function of $\delta_{ij}$, which varies with the voltage magnitudes $V_i$ and $V_j$ to ensure the power flows in the line given by

$$P_k = b_k V_i V_j \sin\delta_{ij}, \quad Q_{ki} = b_k(V_i^2 - V_i V_j \cos\delta_{ij})$$

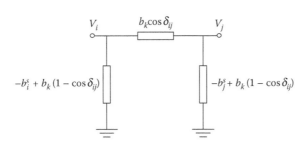

**FIGURE 4.20**
Reactive current flow network (RCFN) element for a lossless line.

where $Q_{ki}$ is the reactive power injection at bus $i$. Note that the resistances in the analogous network, representing line capacitances, are negative.

The power losses in the RCFN represent the reactive power losses in the power system (AC) network.

6. From integration by parts in the expression for $W_{PE3}$ given in Equation 4.65, we can derive

$$W_{PE3} = \sum_{i=1}^{m} \int_{t_0}^{t} \left[ E'_{di} \frac{di_{qi}}{dt} - E'_{qi} \frac{di_{di}}{dt} \right] dt - \sum_{j=1}^{n} \int_{t_0}^{t} V_j \frac{dI^q_{Lj}}{dt} dt$$

$$+ \sum_{j=1}^{n} (Q_{Lj} - Q_{Lj0}) - \sum_{i=1}^{m} (Q_{Gi} - Q_{Gi0}) \tag{4.84}$$

where $I^q_{Lj} = \dfrac{Q_{Lj}}{V_j}$

Combining the above expression with the expression $W_{PE2}$ given in Equation 4.66, we can derive,

$$W_{PE2} + W_{PE3} = \sum_{i=1}^{m} \int_{t_0}^{t} \left[ E'_{di} \frac{di_{qi}}{dt} - E'_{qi} \frac{di_{di}}{dt} \right] dt - \sum_{j=1}^{n} \int_{t_0}^{t} V_j \frac{dI^q_{Lj}}{dt} dt$$

$$+ \frac{1}{2} \left[ \sum_{j=1}^{n} (Q_{Lj} - Q_{Lj0}) - \sum_{i=1}^{m} (Q_{Gi} - Q_{Gi0}) \right] \tag{4.85}$$

Equation 4.85 is useful whenever the shunt reactive currents at the buses tend to be constants. This situation occurs due to limitations on the reactive current supply that can lead to voltage collapse following a major disturbance (rather than loss of synchronism).

## 4.7 Unstable Modes and Parametric Resonance

Chiang et al. (1993) predict the unstable modes in power systems, based on the determination of the unstable manifold of the corresponding CUEP (which may be found from the BCU method). They also show that the unstable modes due to different fault-on trajectories, with identical type-one CUEP, are identical. The unstable modes studied in Chiang et al. (1993) are related to the loss of transient stability.

However, there could be large disturbances that do not result in loss of transient stability. For example, generators are always subject to disturbances caused by rotor swings of other generators in the system and periodic variations in the system loads. Owing to system nonlinearities, there could be parametric resonances (Tamura and Yorino 1987, Yorino et al. 1989). Tamura and Yorino refer to two types of parametric resonances: (i) heteroparametric and (ii) autoparametric. The former is illustrated by the Mathieu equation, which has a periodically varying coefficient in the second-order differential equation given by (McLachlan 1964):

$$\frac{d^2y}{dx^2} + (a - 2q\cos 2x)y = 0$$

It is possible to express the swing equation of a generator connected to a quasi-infinite bus (whose voltage magnitude and phase angle are not constants, but have periodic components) in the form of Mathieu equation with a forcing term (Tamura and Yorino 1987). The autoparametric resonance in contrast refers to resonance among multiple modes in a multimachine power system. Vittal et al. (1991) and Thapar et al. (1997) apply the normal form of vector fields to predict the resonance.

### 4.7.1 Normal Forms

Consider a nonlinear system described by

$$\dot{X} = F(X), \quad X \in R^n$$

We can express the vector field $F$ (assumed to be analytic) in Taylor's series given by

$$\dot{X}_i = A_iX + (1/2)X^T H^i X + \text{high-order terms}$$

where $A_i$ is the *i*th row of the Jacobian $\partial F/\partial X$ (evaluated at the SEP) and $H^i = [\partial^2 F_i/\partial x_j\partial x_k]$ is the Hessian matrix. The linear part of the Taylor's series expansion is $\dot{X} = AX$, where $A$ is the Jacobian matrix that has $n$ eigenvalues given by $\lambda_i$ ($i = 1,2,\ldots,n$).

Defining the similarity transformation $X = VY$, where $V$ is the matrix of right eigenvectors, we can express the Taylor's series using the transformed variables $y_j$ ($j = 1,2,\ldots,n$) as

$$\dot{y}_j = \lambda_j y_j + \sum_{k=1}^{n} \sum_{l=1}^{n} C_{kl}^j y_k y_l$$

where $C^j$ is a matrix obtained by transforming the Hessian matrix.
  By transforming from $Y$ to $Z$ where

$$y_j = z_j + \sum_{k=1}^{n} \sum_{l=1}^{n} h_{kl}^j z_k z_l, \quad h_{kl}^j = \frac{C_{kl}^j}{\lambda_k + \lambda_l - \lambda_j}$$

it is possible to eliminate the nonlinear term in the series and derive

$$\dot{Z} = \Omega Z$$

where $\Omega$ is the diagonal matrix of eigenvalues.
  Note that if $\lambda_j = \lambda_k + \lambda_l$, we say that there is resonance and it is not possible to eliminate the nonlinear (quadratic) term. The resonance results in interaction among the modes of the linear system.
  In the conclusions of the paper by Thapar et al. (1997), it is suggested that spectral analysis could be carried out to verify the modal content in the time domain simulation results.

### 4.7.2 Fast Fourier Transform of Potential Energy

The total PE in a system subjected to a large disturbance encompasses the response of all the generators as $W_{PE}$ is defined as the negative of the sum of the integrals of the accelerating powers of all the generators in the system. We have observed that it can also be expressed as the sum of energies in each element of the analogous network. The three-phase faults at different locations that are critically cleared (resulting in critically stable trajectories) represent the severest disturbances in the system without affecting the system integrity. A case study on the IEEE 17-generator system is taken up to illustrate the analysis (Immanuel 1993).

#### 4.7.2.1 Results of the Case Study

The fast Fourier transform (FFT) of the PE ($W_{PE}$) was taken over more than 10 s, starting at $t = t_{cl}$ for the six cases shown in Table 4.6.
  The FFT and $W_{PE}$ for cases 5 and 6 are shown in Figures 4.21 and 4.22, respectively. The dominant modes of oscillation range from 0.98 to 3.42 Hz.

**TABLE 4.6**

Results of FFT Analysis of Potential Energy (17-Generator System)

| Case Number | Fault at Bus Number | Line Cleared | $t_{cr}$ (s) | Dominant Frequency (Hz) |
|---|---|---|---|---|
| 1 | 75 | 75–9 | 0.35 | 1.27 |
| 2 | 95 | 95–97 | 0.30 | 2.93 |
| 3 | 3 | 3–118 | 0.25 | 1.27 |
| 4 | 15 | — | 0.26 | 3.13 |
| 5 | 13 | 13–109 | 0.35 | 0.98 |
| 6 | 12 | — | 0.20 | 3.42 |

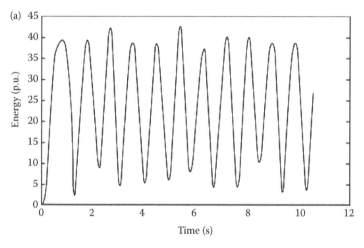

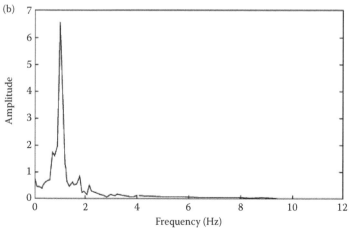

**FIGURE 4.21**

Potential energy (a) and FFT (b) for case 5 (10-generator system).

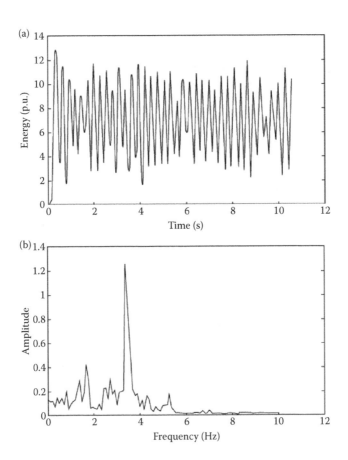

**FIGURE 4.22**
Potential energy (a) and FFT (b) for case 6 (10-generator system).

## Remarks

1. The spectral analysis is based on the PE function that shows sustained oscillations with more than one modal component. However, one mode is predominant that depends on the location and severity of the disturbance. The disturbance is quite severe as the system is critically stable.

2. The instability results from the SNB of the dominant mode as the system trajectory leaves the boundary of region of stability (or attraction). This behavior can be predicted from the observation that the initial loss of stability can be predicted from representing the system by a two-machine equivalent (Pavella et al. 2000). This has a single dominant oscillatory mode that undergoes an SNB as the fault clearing time is marginally increased.

# 5

# Structure Preserving Energy Functions for Systems with HVDC and FACTS Controllers

## 5.1 Introduction

In this chapter, we will consider the network controllers in the formulation of SPEF. It is not feasible (without making approximations) to model HVDC lines (with converter controls) and FACTS controllers (static var compensator (SVC), thyristor-controlled series capacitor (TCSC), static synchronous compensator (STATCOM), static synchronous series compensator (SSSC), and unified power flow controller (UPFC)) in stability studies. These controllers are increasingly being applied in power systems not only for flexible system operation in steady state but also to improve the dynamic performance under small or large disturbances. The introduction of HVDC links in the system can assist the system operator during a contingency in rescheduling power flows to avoid overloading of AC lines. FACTS controllers also serve a similar purpose. In the absence of the network controllers, generation rescheduling is required to prevent overloading of lines and this would increase the operating costs as economic dispatch is normally used to minimize operating costs.

In this chapter, we will concentrate on the dynamic performance of HVDC and FACTS controllers for improvement of stability. Since energy functions are used for direct stability evaluation, energy functions for these controllers will be presented with examples. SPMs of systems are utilized in developing the energy functions.

## 5.2 HVDC Power Transmission Links

### 5.2.1 HVDC Systems and Energy Functions

HVDC power transmission was introduced in 1954 with mercury arc valves for AC/DC conversion using Graetz bridge converters. There has been continuous updating of technology since then to improve system performance and reliability. Thyristor valves were introduced in the early 1970s, and in 1997, insulated gate bipolar transistor (IGBT) valves were introduced with voltage source converters (VSC) for improved performance. However, VSC-based HVDC links are currently rated at relatively low power levels (below 400 MW) and short distances using underground or submarine cables. In contrast, line commutated (current source) converters (LCC) using thyristor valves are now being applied for long distance, bulk power (up to 6000 MW) transmission at ±800 kV (Arrillaga 1998, Padiyar 2010, Sood 2004). HVDC transmission is also used for asynchronous interconnection when AC interconnection is not feasible. In such cases, typically back-to-back (BTB) HVDC links are used where there are no DC lines and conversion from AC to DC and back is performed at the same station.

The transient stability analysis of multimachine AC/DC power systems based on energy functions was first attempted by Pai et al. (1981) by using a simplified DC link model. The dynamic loads due to the HVDC link (along with other loads) were represented by current injections at the generator internal buses of the reduced system using distribution factors. Ni and Fouad (1987) have applied energy functions for AC/DC systems, neglecting DC link dynamics and assuming strong voltage support at the converter buses. DeMarco and Canizares (1992) have proposed a vector energy function for AC/DC systems with two components—one for AC and one for DC. Jing et al. (1995) presented the incorporation of HVDC and SVC models for implementation of direct transient stability assessment of a practical system. A study of DAE in AC–DC systems, focusing on the discontinuous solutions in the computation of the transient dynamics is reported in Susuki et al. (2008). The dynamic performance characteristics of North American HVDC systems were first reported by an IEEE Committee (1981a).

The application of SPEF was first proposed by Padiyar and Sastry (1984) and extended for multiterminal DC (MTDC) systems (1993). The SPEF incorporates the DC system as dynamic loads at converter buses. Detailed or simplified converter control models can be used. Voltage-dependent nonlinear loads in the AC network are modeled as described in Chapter 3.

### 5.2.2 HVDC System Model

The HVDC system consists of two or more converters (if multiterminal operation is to be considered). A converter terminal is shown in Figure 5.1. In general, there is more than one bridge connected in series. All the bridges at a terminal are identical and operate at the same value of the control angle $\theta$ (delay or extinction angle).

#### 5.2.2.1 Converter Model

In transient stability studies, it is adequate to represent the converter by a simplified model in which the valve switchings are ignored. This is equivalent to ignoring the AC and DC harmonics.

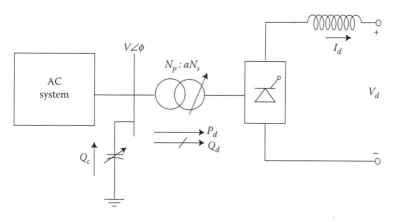

**FIGURE 5.1**
Single-line diagram of a converter station.

From converter theory (Padiyar 2010), the average DC voltage (pole to ground) in per unit at a converter terminal is given by

$$V_d = kaV\cos\theta - R_c I_d \tag{5.1}$$

where

$$k = \frac{3\sqrt{2}N_s n_b V_{acb}}{\left(\pi N_p V_{db}\right)}, \quad R_c = \pm\left(\frac{3n_b X_c}{\pi Z_{db}}\right), \quad Z_{db} = \left(\frac{V_{db}}{I_{db}}\right)$$

$V_{acb}$, $V_{db}$, $I_{db}$, and $Z_{db}$ are base AC voltage, DC voltage, DC current, and DC impedance, respectively. $n_b$ is the number of bridges per pole and $X_c$ is the converter transformer leakage reactance (on the valve side).

For a rectifier, $\theta = \alpha$ (delay angle), while for an inverter, $\theta = \gamma$ (extinction angle). $I_d$ and $R_c$ are assumed to be positive for the rectifier and negative for the inverter. The per-unit system used here is more general than that used earlier (e.g., in Fudeh and Ong 1981). This is because $V_{acb}$ can be chosen independently of $V_{db}$ and can assume different values for different terminals. Similarly, $n_b$, $N_s$, and $N_p$ can vary depending on the terminal. The effect of these parameters is included in a single parameter $k$, which can vary. From Equation 5.1, the converter terminal can be represented by Norton's equivalent of a current source $I$ in parallel with $R_c$ (see Figure 5.2). The power ($P_d$) and reactive power ($Q_d$) are given by

$$P_d = pV_d I_d \tag{5.2}$$

$$Q_d = |P_d|\tan\zeta \tag{5.3}$$

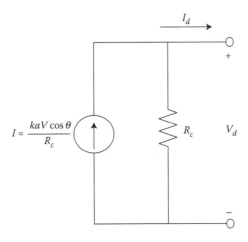

**FIGURE 5.2**
Norton's equivalent of a converter.

$$\cos\zeta = \frac{V_d}{kaV} \tag{5.4}$$

where $p$ is the number of poles (one or two).

### 5.2.2.2 DC Network Equations

Unlike in AC networks, transients in a DC network are sometimes taken into account. However, it is simpler to ignore fast transients and model only the resistive DC network. The use of Norton's equivalent circuits for converters enables the DC network equations to be expressed with the ground as a reference node. The resistive DC network is modeled as

$$[G]V_d = I \tag{5.5}$$

where $[G]$ is a conductance matrix of size $n_d$, where $n_d$ is the number of converter terminals. It is assumed that the nonconverter buses are absent or are eliminated.

### 5.2.2.3 Converter Control Model

The converter control is usually represented by block diagrams where the transfer function of each block is specified. A typical controller block diagram is shown in Figure 5.3a and b. In Figure 5.3b, the switch $S_1$ is closed in the case of equidistant pulse control (EPC) and the switch $S_2$ is closed in case the synchronizing circuit is provided. $\Delta\delta_V$ is the change in the phase of the converter bus voltage, and the transfer function of the synchronizing circuit is typically

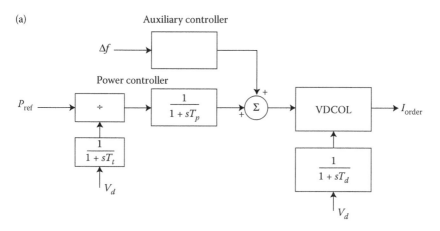

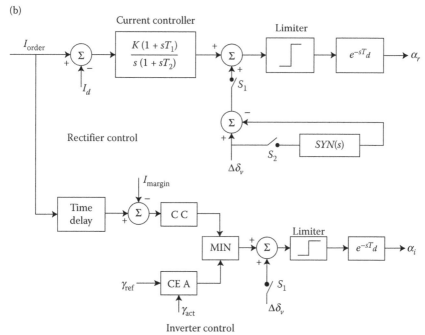

**FIGURE 5.3**
HVDC controllers. (a) Power and auxiliary control. (b) Current and extinction angle control.

$$SYN(s) = \frac{1}{1 + sT_s}, \quad T_s = \frac{1}{2\pi B_w} \tag{5.6}$$

where $B_w$ is the bandwidth of the synchronizing circuit in hertz. As the bandwidth increases, the characteristics of EPC tend toward those of individual phase control (IPC).

The time delay $T_d$ takes into account the discrete nature of the converter control. Typically

$$T_d = \frac{1}{2Pf_0} \tag{5.7}$$

where $P$ is the pulse number and $f_0$ is the operating frequency of the AC system.

For transient stability analysis, any one of the three types of controller models can be used:

1. Detailed models (as shown in Figure 5.3), which represent the dynamics of controllers and the parameters are tuned for particular system requirements. The use of such a model requires knowledge of the actual system conditions for authenticity. The detailed controller model has to be interfaced with the network model considering the transients.

2. A performance model assumes that the controllers are adequately designed to carry out the objectives of the control. For example, it is assumed that the actual DC current in the link faithfully follows the current reference instantaneously or with prespecified time delay. In the former case, it is equivalent to ignoring the controller dynamics and modeling the performance of the controller using steady-state control characteristics. Typical control characteristics in the $V_d - I_d$ plane for a two-terminal system are shown in Figure 5.4. The intersection of the rectifier and inverter characteristics defines the operating point. The operation at A corresponds to current control at the rectifier and constant extinction angle (CEA) control at the inverter. This is the normal operating mode. However, a voltage dip at the rectifier converter bus results in the mode shift when current control is transferred to the inverter and the rectifier operates at minimum $\alpha$ (delay angle). Figure 5.4 also shows the influence of voltage-dependent current order limiter (VDCOL), which is mainly provided for preventing voltage instability and as a back-up protection against commutation failures.

3. The use of a performance model requires the use of mode shift logic to identify the transition from one mode to another. In MTDC systems, the prediction of mode shifts can be cumbersome. Hence, a variation of the performance model has been developed and is termed (Lefebvre et al. 1991) a firing angle-based simplified model. In this model, the identity of the DC network is explicitly retained. The network model can be either purely resistive or include inductances also.

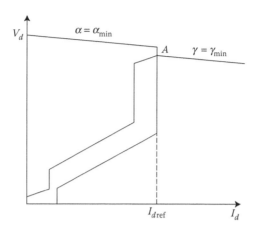

**FIGURE 5.4**
Rectifier and inverter control characteristics.

### 5.2.3 AC System Model

The following assumptions are made in modeling the AC system:

i. The synchronous machine is represented by a classical model.
ii. The AC network is assumed to be lossless.
iii. Machine damping is neglected.

Assumption (i) is made here only for convenience and can be relaxed for detailed analysis including the effects of AVR.

#### 5.2.3.1 Generator Model

The swing equation for the *i*th generator, using COI variables, can be written as

$$M_i \dot{\omega}_i = P_{mi} - P_{ei} - \left( \frac{M_i}{M_T} \right) P_{COI} \tag{5.8}$$

$$\dot{\theta} = \bar{\omega}_i \tag{5.9}$$

where

$$P_{ei} = \frac{E_i V_i \sin(\theta_i - \phi_i)}{x'_{di}}, \quad i = 1, 2, \ldots, m \tag{5.10}$$

$$P_{\text{COI}} = \sum_{i=1}^{m} (P_{mi} - P_{ei}) \tag{5.11}$$

$m$ is the number of machines.

### 5.2.3.2 Load Model

The loads are modeled as arbitrary functions of the respective bus voltages. Thus

$$P_{lj} = f_{pj}(V_j) \tag{5.12}$$

$$Q_{lj} = f_{qj}(V_j) \tag{5.13}$$

### 5.2.3.3 AC Network Equations

The AC network is described by the power flow equations at each bus. At converter buses, $P_d$ and $Q_d$ are treated as dynamic loads.

The power flow equations at bus $i$ are given by

$$P_i + P_{li} + e_i P_{di} = 0 \tag{5.14}$$

$$Q_i + Q_{li} + e_i Q_{di} = 0 \tag{5.15}$$

where $e_i = 1$, if $i \in D$ (where $D$ is the set of converter buses) and $e_i = 0$, otherwise. $P_i$ and $Q_i$ are the injected power and reactive power given by the following expressions:

$$P_i = \frac{m_i E_i V_i \sin(\phi_i - \theta_i)}{x'_{di}} + \sum_{j=1}^{N} B_{ij} V_i V_j \sin \phi_{ij} \tag{5.16}$$

$$Q_i = \frac{m_i \left( V_i^2 - E_i V_i \cos(\theta_i - \phi_i) \right)}{x'_{di}} - \sum_{j=1}^{N} B_{ij} V_i V_j \cos \phi_{ij} \tag{5.17}$$

where $m_i = 1$, if $i \in G$ (where $G$ is the set of generator terminal buses) and $m_i = 0$, otherwise.

### 5.2.4 Structure Preserving Energy Function

Consider the following SPEF defined on the postfault system:

$$W(\bar{\omega}, \theta, V, \phi, t) = W_1(\bar{\omega}) + W_2(\theta, V, \phi, t) \tag{5.18}$$

where

$$W_1(\bar{\omega}) = \sum_{i=1}^{m} \frac{1}{2} M_i \bar{\omega}_i^2 \tag{5.19}$$

$$W_2 = W_{21}(\theta) + W_{22}(t) + W_{23}(V) + W_{24}(\theta,\phi,V) + W_d(V,t) \tag{5.20}$$

$$W_{21}(\theta) = -\sum_{i=1}^{m} P_{mi}(\theta_i - \theta_{i0}) \tag{5.21}$$

$$W_{22}(t) = \sum_{i=1}^{N} \int_{t_0}^{t} P_{li}\left(\frac{d\phi_i}{dt}\right) dt \tag{5.22}$$

$$W_{23}(V) = \sum_{i=1}^{N} \int_{V_{i0}}^{V_i} \frac{f_{qi}(x_i)}{x_i} dx_i \tag{5.23}$$

$$W_{24}(\theta,\phi,V) = \sum_{i=1}^{m} \left[ \frac{\{(V_i^2/2) - E_i V_i \cos(\theta_i - \phi_i)\}}{x'_{di}} - \frac{\{(V_{i0}^2/2) - E_i V_{i0} \cos(\theta_{i0} - \phi_{i0})\}}{x'_{di}} \right]$$

$$- \sum_{i=1}^{N} \sum_{j=1}^{N} \frac{1}{2} B_{ij}(V_i V_j \cos\phi_{ij} - V_{i0} V_{j0} \cos\phi_{ij0}) \tag{5.24}$$

$$W_d(V,t) = \sum_{j \in D} \left[ \int_{t_0}^{t} P_{dj}\left(\frac{d\phi_j}{dt}\right) dt + \int_{V_{j0}}^{V_j} \left(\frac{Q_{dj}}{V_j}\right) dV_j \right] \tag{5.25}$$

$N$ is the number of buses (excluding the generator internal buses). It can be shown that the derivative of $W$ is zero along the system trajectory. For

$$\sum_{i=1}^{m} \left(\frac{\partial W_1}{\partial \bar{\omega}_i}\right)\left(\frac{d\bar{\omega}_i}{dt}\right) + \left(\frac{\partial W_2}{\partial \theta_i}\right)\left(\frac{d\theta_i}{dt}\right) = \sum_{i=1}^{m} (M_i \bar{\omega}_i \dot{\bar{\omega}}_i - P_{mi}\bar{\omega}_i + P_{ei}\bar{\omega}_i) = 0 \tag{5.26}$$

$$\sum_{i=1}^{N} \left[\left(\frac{\partial W_2}{\partial \phi_i}\right)\left(\frac{d\phi_i}{dt}\right) + \left(\frac{\partial W_2}{\partial t}\right)\right] = \sum_{i=1}^{N} (P_i + P_{li} + e_i P_{di})\left(\frac{d\phi_i}{dt}\right) = 0 \tag{5.27}$$

$$\sum_{i=1}^{N} \left(\frac{\partial W_2}{\partial V_i}\right) = \sum_{i=1}^{N} \left(\frac{1}{V_i}\right)(Q_{li} + Q_i + e_i Q_{di}) = 0 \tag{5.28}$$

## Remarks

1. It is assumed that the system model is well defined in the sense that the voltages at the load buses can be solved in a continuous manner at any given time during the transient. This means that the system trajectories are smooth and there are no jumps in the energy function.

2. Consider the terms

$$\int_{t_0}^{t} P_{li}\left(\frac{d\phi_i}{dt}\right)dt \quad \text{or} \quad \int_{t_0}^{t} P_{di}\left(\frac{d\phi_i}{dt}\right)dt$$

These can be expressed as

$$\int_{t_0}^{t} P_{li}\left(\frac{d\phi_i}{dt}\right)dt = P_{li}(\phi_i - \phi_{i0}) - \int_{t_0}^{t}\left(\frac{dP_{li}}{dt}\right)\phi_i\, dt \qquad (5.29)$$

$$\int_{t_0}^{t} P_{di}\left(\frac{d\phi_i}{dt}\right)dt = P_{di}(\phi_i - \phi_{i0}) - \int_{t_0}^{t}\left(\frac{dP_{di}}{dt}\right)\phi_i\, dt \qquad (5.30)$$

If $(dP_{li}/dt)$ and $(dP_{di}/dt)$ are small, the second terms on the RHS of the above equations can be neglected. This approximation has the advantage of making the SPEF path independent.

3. It can be shown that

$$W_{24}(\theta,\phi,V) = \sum_{k=1}^{n_e} \frac{1}{2}(Q_k - Q_{k0}) \qquad (5.31)$$

where $n_e$ is the number of elements of the AC network including machine reactances and $Q_k$ is the reactive power loss in element $k$. From conservation of energy

$$\sum_{k=1}^{n_e} \frac{1}{2}Q_k = \frac{1}{2}\left[\sum_{i=1}^{m} Q_{Gi} - \sum_{j=1}^{N} Q_{lj} - \sum_{k=1}^{n_d} Q_{dk}\right] \qquad (5.32)$$

where $Q_{Gi}$ is the reactive power generation at the internal bus of generator $i$ and $n_d$ is the number of HVDC converters. Equation 5.32 can be used to simplify the computation of SPEF.

4. The component $W_d(V, t)$ of the energy function, which is attributed to the DC system, follows from treating $P_d$ and $Q_d$ at each converter bus as dynamic loads. The effect of the DC link controller is handled indirectly influencing $P_d$ and $Q_d$. Thus, any controller model can be considered.

For a two-terminal DC system, neglecting DC system losses, $W_d$ can be expressed approximately as

$$W_d = W_{d1} + W_{d2} + W_{d3} \tag{5.33}$$

where

$$W_{d1} = P_d[(\phi_r - \phi_i) - (\phi_{r0} - \phi_{i0})] \tag{5.34}$$

$$W_{d2} = k_r a_r I_d \int_{V_{r0}}^{V_r} \sin\zeta_r \, dV_r \tag{5.35}$$

$$W_{d3} = k_i a_i I_d \int_{V_{i0}}^{V_i} \sin\zeta_i \, dV_i \tag{5.36}$$

$\phi_r$ and $\phi_i$ refer to the bus angle at the rectifier and inverter bus, respectively.

Assuming that the inverter controls the power factor angle $\zeta_i$ (by keeping it constant)

$$W_{d2} = k_r a_r I_d \int_{V_{r0}}^{V_r} \left[ \frac{(k_r^2 a_r^2 V_r^2 - V_d^2)^{1/2}}{k_r a_r V_r} \right] dV_r \tag{5.37}$$

$$W_{d3} = k_i a_i I_d \sin\zeta_i (V_i - V_{i0}) \tag{5.38}$$

By noting that

$$\int \left[ \frac{(x^2 - a^2)^{1/2}}{x} \right] dx = (x^2 - a^2)^{1/2} - a\sec^{-1}\left(\frac{x}{a}\right)$$

the integral in Equation 5.37 can be expressed as

$$W_{d2} = k_r a_r I_d [V_r \sin\zeta_r - V_{r0} \sin\zeta_{r0} - (V_r\zeta_r \cos\zeta_r - V_{r0}\zeta_{r0} \cos\zeta_{r0})] \tag{5.39}$$

It is to be noted that during a transient, the DC current $I_d$ is regulated by the rectifier and is assumed to be independent of $V_r$ and $V_i$. $V_d$ is dependent on $V_i$.

If there is a mode shift resulting in current control at the inverter and power factor control at the rectifier, the subscripts $r$ and $i$ are to be interchanged in Equations 5.38 and 5.39.

If $\gamma$ (extinction angle) is to be maintained constant at the inverter, then $W_{d3}$ can be expressed as

$$W_{d3} = I_d \int_{V_{i0}}^{V_i} \left[ \frac{(A + BV_i + CV_i^2)^{1/2}}{V_i} \right] dV_i \qquad (5.40)$$

$$A = -R_{ci}^2 I_d^2, \quad B = 2k_i a_i R_{ci} I_d \cos \gamma_i, \quad C = (1 - \cos \gamma_i^2) k_i a_i^2$$

An explicit expression for the integral of Equation 5.40 can be obtained.

### 5.2.5 Example

For illustration, a three-machine system example is adapted from Anderson and Fouad (1977). The single-line diagram of the system is shown in Figure 5.5. DC link is connected between buses 6 and 9. The power flow over the DC link is 0.3 per unit (p.u.). The controller model is assumed to be a performance model based on DC current and current modulation using an auxiliary controller (see Figure 5.6) or emergency controller (see Figure 5.7) is considered. The control signal for the auxiliary controller is taken from the difference in the phase angles of the two converter buses. A washout circuit is included to eliminate steady-state offset and filter out very-low-frequency components that occur normally. The equations of the auxiliary controller and the emergency controller are given next.

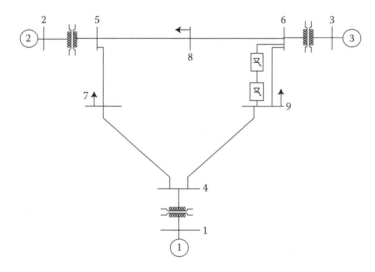

**FIGURE 5.5**
Single-line diagram of a three-machine system.

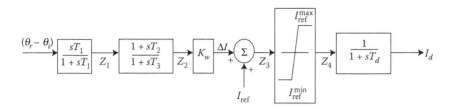

**FIGURE 5.6**
Block diagram of auxiliary controller.

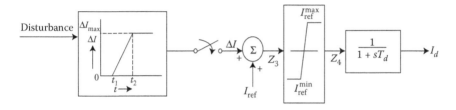

**FIGURE 5.7**
Block diagram of emergency controller.

### 5.2.5.1 Auxiliary Controller

The block diagram of this controller is shown in Figure 5.6.

The dynamic and algebraic equations for this controller are given below:

$$\dot{I}_d = \frac{(Z_4 - I_d)}{T_d} \tag{5.41}$$

$$\dot{y}_1 = \frac{[(\phi_r - \phi_i) - y_1]}{T_1} \tag{5.42}$$

$$\dot{y}_2 = \frac{[(1 - T_2/T_3)Z_1 - y_2]}{T_3} \tag{5.43}$$

$$Z_1 = (\theta_r - \theta_i) - y_1 \tag{5.44}$$

$$Z_2 = Z_1 + y_2 \tag{5.45}$$

$$\Delta I = K_w Z_2 \tag{5.46}$$

$$Z_3 = I_{\text{ref}} + \Delta I \tag{5.47}$$

$$Z_4 = Z_3, \quad \text{if } I_{\text{ref}}^{\min} \leq Z_3 \leq I_{\text{ref}}^{\max}$$

$$= I_{\text{ref}}^{\max}, \quad \text{if } Z_3 > I_{\text{ref}}^{\max}$$

$$= I_{\text{ref}}^{\min}, \quad \text{if } Z_3 < I_{\text{ref}}^{\min} \tag{5.48}$$

### 5.2.5.2 Emergency Controller

This is shown in Figure 5.7. Here, the change in the current reference $\Delta I$ increases from zero to $\Delta I_{max}$ linearly with time, when a disturbance in the AC system is sensed.

The dynamic and algebraic equations are given below:

$$\dot{I}_d = \frac{(Z_4 - I_d)}{T_d} \tag{5.49}$$

$$
\begin{aligned}
\Delta I &= 0, \quad \text{for } 0 \leq t \leq t_1 \\[4pt]
&= \Delta I_{max} \frac{(t - t_1)}{(t_2 - t_1)}, \quad \text{for } t_1 \leq t \leq t_2 \\[4pt]
&= \Delta I_{max}, \quad \text{for } t > t_2
\end{aligned} \tag{5.50}
$$

The expressions for $Z_3$ and $Z_4$ are the same as given earlier.

It is assumed that the disturbance originates at $t = 0^+$ and the time required to sense this is $t_1$. In this type of controller, additional capacitors are switched on at time $t_2$ to meet the increased reactive power requirement at the converter buses due to increased current order.

### 5.2.5.3 Case Study and Results

The disturbance considered is a three-phase fault at $t = 0$ near bus 5 followed by clearing of the fault by switching line 5–7 off. For simplicity, the load characteristics are assumed to be of constant impedance type. The following cases are considered:

*Case 1*: With constant current (CC) reference (no modulation)

*Case 2*: With auxiliary controller $K_w = 3$

*Case 3*: With emergency controller $I_{ref}^{max} = 1.0$, $I_{ref}^{min} = 0.1$, $t_1 = 0.02$, and $t_2 = 0.12$

In case 3, it is assumed that an additional capacitor bank (of 0.25 p.u.) is added at both converter buses after time $t_2$ to account for the increased reactive power requirements.

The CCT ($T_{cr}$) obtained by digital simulation and prediction is shown in Table 5.1 along with the critical energies. The predicted results are based on the PEBS method. It is observed that $T_{cr}$ obtained by prediction agrees well with that obtained by digital simulation. $T_{cr}$ is slightly reduced for case 2 compared to that for case 1; it is to be noted that the auxiliary controller is provided primarily for damping oscillations in the system. The effectiveness of this controller also depends on the choice of control parameters and the reactive power constraints. The latter can be explained as follows: as DC

**TABLE 5.1**

Critical Clearing Times ($T_{cr}$) (HVDC Link)

| Case Number | Simulation $T_{cr}$ | Prediction $T_{cr}$ | Critical Energy |
| --- | --- | --- | --- |
| 1 | 0.164–0.165 | 0.159–0.160 | 0.594 |
| 2 | 0.161–0.162 | 0.158–0.159 | 0.569 |
| 3 | 0.173–0.174 | 0.175–0.176 | 0.813 |

current is increased, the reactive power requirements are also increased. If adequate reactive power is not available, the AC voltage will drop, thus (partially) nullifying the effect of the controller.

As expected, the emergency controller helps in improving the transient stability. The results obtained by TEF are accurate enough to predict the effect of controllers.

## 5.3 Static Var Compensator

### 5.3.1 Description

SVC is a first-generation FACTS controller based on thyristor valve technology, which is applied in HVDC converters. SVC is a shunt-connected FACTS controller that is applied in transmission networks for the following reasons (Hingorani and Gyugyi 2000, Padiyar 2007):

1. Increase power transfer in long AC lines
2. Improve stability (synchronous and voltage) with fast-acting voltage regulation
3. Damp low-frequency oscillations due to (rotor) swing modes
4. Damp subsynchronous frequency oscillations due to torsional modes
5. Control dynamic overvoltages
6. Improve HVDC converter operation at low short circuit ratios (SCR) by providing for voltage regulation at the converter bus by rapid control of the reactive power

An SVC has no inertia compared to synchronous condensers and can be extremely fast in response (2–3 cycles). There are two types of SVC—(a) fixed capacitor–thyristor-controlled reactor (FC–TCR) and (b) thyristor-switched capacitor–thyristor-controlled reactor (TSC–TCR). The second type is more flexible than the first one and requires a smaller rating of the reactor and consequently generates less harmonics. The schematic diagram of a TSC–TCR-type SVC is shown in Figure 5.8.

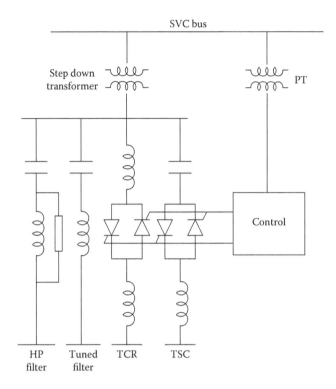

**FIGURE 5.8**
A TSC–TCR-type SVC.

### 5.3.2 Control Characteristics and Modeling of SVC Controller

The steady-state control characteristics of an SVC are shown in Figure 5.9. Here, the line AB is the control range and OA represents the characteristic when SVC hits the capacitive limit. BC represents the inductive limit. Note that the SVC current is considered positive when the SVC susceptance is inductive. Thus

$$I_{SVC} = -B_{SVC}\, I_{SVC} \tag{5.51}$$

The slope of OA is $-B_C$, where $B_C$ is the susceptance of the capacitor and the slope of OBC is the magnitude of $B_L$, the susceptance of the reactor, assuming the capacitor is switched off. A positive slope (in the range of 1–5%) is given in the control range to

a. Enable parallel operation of more than one SVC connected at the same or neighboring buses
b. Prevent SVC hitting the limits frequently

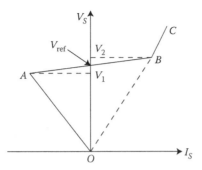

**FIGURE 5.9**
Control characteristics of an SVC.

Assuming that the SVC operates in its control range, the bus voltage $V_{SVC}$ is determined by the intersection of the system characteristic and the control characteristic. The system characteristic is a straight line with a negative slope and is defined by

$$V_{SVC} = V_{Th} - X_{Th}\, I_{SVC} \tag{5.52}$$

The equation of the line AB is given by

$$V_{SVC} = V_{ref} + X_s\, I_{SVC} \tag{5.53}$$

where $X_s$ is the slope of the line. From Equations 5.52 and 5.53, we obtain

$$I_{SVC} = \frac{V_{Th} - V_{ref}}{(X_s + X_{Th})} \tag{5.54}$$

Note that Equation 5.52 assumes that the system Thevenin impedance is reactive. This is valid if the network losses are neglected. If this is not the case (as the system impedance has a resistive component), the solution is obtained by applying the compensation theorem. This will be discussed in the next subsection.

The SVC controller block diagram is shown in Figure 5.10. Here, $H_m(s)$ represents a low-pass filter. The first-order low-pass filter is represented as

$$H_m(s) = \frac{1}{1 + sT_m}$$

The voltage regulator is typically a *P–I* controller. $T_m$ is typically 2–3 ms and the transport delay $T_d = T/12$, where $T$ is the period of the supply voltage. $T_b$ represents the average delay in responding to a control signal caused by

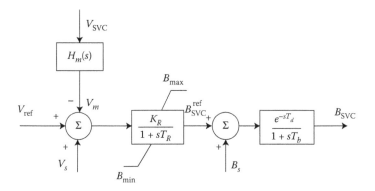

**FIGURE 5.10**
Block diagram of an SVC controller.

the discrete nature of the firing pulse. $V_s$ represents an auxiliary signal used for damping low-frequency or subsynchronous-frequency oscillations. $B_s$ is also an alternative auxiliary signal instead of $V_s$.

### 5.3.3 Network Solution with SVC: Application of Compensation Theorem

The model of SVC is represented by the equivalent circuit shown in Figure 5.11. This shows a voltage source $\hat{E}_{SVC}$ in series with a reactance $X_{SVC}$. The values of $\hat{E}$ and $X_{SVC}$ depend on the region of operation

   i. Control region (line AB): $\hat{E}_{SVC} = V_{ref} \angle \phi_{SVC}, X_{SVC} = X_s$
   ii. Capacitive limit (line OA): $\hat{E}_{SVC} = 0, X_{SVC} = -(1/B_C)$
   iii. Inductive limit (line BC): $\hat{E}_{SVC} = 0, X_{SVC} = (1/B_L)$

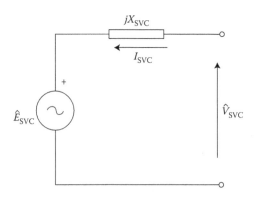

**FIGURE 5.11**
Equivalent circuit of an SVC.

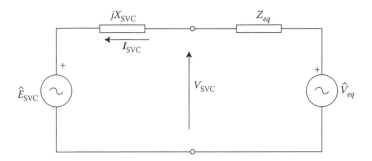

**FIGURE 5.12**
Combined equivalent circuit of an SVC connected to the network.

where $B_C$ is the maximum capacitive susceptance of the SVC and $B_L$ is the maximum value of the magnitude of inductive susceptance. $\phi_{SVC}$ is the phase angle of the voltage at the SVC bus.

The SVC model is both nonlinear and time varying. Assuming that the SVC is connected to a linear network excited by time-varying voltage sources at the internal nodes of the synchronous generators, we can represent the network by a (Thevenin) equivalent at the SVC bus. The combined equivalent circuit is shown in Figure 5.12. If the network impedances are constant, $Z_{eq}$ remains constant. $V_{eq}$ is obtained from the network solution as the SVC terminal voltage when SVC current is set to zero.

From Figure 5.12, the SVC current is computed as

$$\hat{I}_{SVC} = \frac{\hat{V}_{eq} - \hat{E}_{SVC}}{Z_{eq} + jX_{SVC}} \tag{5.55}$$

The magnitude of the SVC bus voltage is

$$V_{SVC} = \left| \hat{E}_{SVC} + jX_{SVC}\hat{I}_{SVC} \right| \tag{5.56}$$

If $\hat{E}_{SVC} = 0$, the solution for $\hat{I}_{SVC}$ is straightforward as both $V_{eq}$ and $Z_{eq}$ are known at any given time instant. The magnitude of $\hat{E}_{SVC}$ is known for the control region, but $\phi_{SVC}$ depends on $\hat{V}_{SVC}$.

### 5.3.3.1 Calculation of $\phi_{SVC}$ in Control Region

It can be shown that $\phi_{SVC}$ is obtained from the solution of a quadratic equation given by

$$a \tan^2 \phi_{SVC} + b \tan \phi_{SVC} + c = 0 \tag{5.57}$$

*Proof*

From Equations 5.55 and 5.56, we can express

$$\hat{V}_{SVC} = V_S \angle \phi_{SVC} = (1 - \hat{A})\hat{V}_{eq} + \hat{A}\hat{E}_{SVC} \qquad (5.58)$$

where

$$\hat{A} = \frac{Z_{eq}}{Z_{eq} + jX_{SVC}} = A \angle \alpha$$

Separating the real and imaginary components in Equation 5.58, we get

$$V_S \cos \phi_{SVC} = x + z \cos (\phi_{SVC} + \alpha) \qquad (5.59)$$

$$V_S \sin \phi_{SVC} = y + z \sin (\phi_{SVC} + \alpha) \qquad (5.60)$$

where

$$x = \mathrm{Re}[(1 - \hat{A})\hat{V}_{eq}], \quad y = \mathrm{Im}[(1 - \hat{A})\hat{V}_{eq}], \quad z = |\hat{A}\hat{E}_{SVC}|, \quad V_S = |\hat{V}_{SVC}|$$

From Equations 5.59 and 5.60, we get

$$
\begin{aligned}
\tan \phi_{SVC} &= \frac{y + z \cos \alpha \sin \phi_{SVC} + z \sin \alpha \cos \phi_{SVC}}{x + z \cos \alpha \cos \phi_{SVC} - z \sin \alpha \sin \phi_{SVC}} \\
&= \frac{y \sec \phi_{SVC} + z \cos \alpha \tan \phi_{SVC} + z \sin \alpha}{x \sec \phi_{SVC} + z \cos \alpha - z \sin \alpha \tan \phi_{SVC}}
\end{aligned}
\qquad (5.61)
$$

After some manipulations and using the identity

$$\sec^2 \phi = 1 + \tan^2 \phi$$

we can finally derive Equation 5.57, where

$$a = x^2 - z^2 \sin^2 \alpha, \quad b = -2xy, \quad c = y^2 - z^2 \sin^2 \alpha$$

Once $\phi_{SVC}$ is found, $I_{SVC}$ is readily calculated.

### 5.3.3.2 Network Solution

The network solution is carried out in two steps. In the first step, the voltage solution is obtained by putting $\hat{I}_{SVC} = 0$. The voltage calculated at the SVC bus at the end of the first step is the same as $\hat{V}_{Th}$ or $\hat{V}_{eq}$. The knowledge of $\hat{V}_{eq}$ and $Z_{eq}$ (which has been calculated in advance and stored) enables the computation of $\hat{I}_{SVC}$ as described earlier.

The network is solved again with the injection of $\hat{I}_{SVC}$ at the SVC bus (all other sources put equal to zero). The second solution does not require much computations as the current vector is sparse.

The voltages at all the buses are obtained from the addition (superposition) of the voltages calculated in the two network solutions.

**Remarks**

1. It is assumed that the generator stator is represented by a constant impedance (neglecting dynamic saliency) behind a voltage source (see Appendix A). Dynamic saliency results in time-varying impedance (with respect to network or common reference frame). Also, the impedance needs to be expressed (as a $2 \times 2$ matrix) in $D$–$Q$ axes. In this case, it can be shown that $\phi_{SVC}$ can be obtained by solving a quartic equation. However, dynamic saliency can also be handled by introducing a dummy rotor coil in the quadrature axis with arbitrarily chosen small time constant (see Appendix C).

2. $Z_{eq}$ changes whenever there is a change in the network configuration.

3. If $Z_{eq}$ is a purely reactive (inductive) impedance, then $\phi_{SVC}$ is identical to the phase angle of $\hat{V}_{eq}$, which is known. This eliminates the need for the solution of the quadratic Equation 5.57.

Even if a quadratic solution is to be solved, the correct value of $\phi_{SVC}$ is obtained as that solution that is closer to the phase angle of $\hat{V}_{eq}$.

### 5.3.4 Potential Energy Function for SVC

The energy function for an SVC can be derived from its steady-state control characteristics shown in Figure 5.9 (Padiyar and Immanuel 1994). There are three modes of operation of SVC based on the operating value of $V_{SVC}$. These are given below:

*Mode 1:* $V_{SVC} < V_1$
*Mode 2:* $V_1 \leq V_{SVC} \leq V_2$
*Mode 3:* $V_{SVC} > V_2$

The SVC can be modeled as a voltage-dependent reactive load. Assuming that an SVC is connected at any bus $j$, we can express the contribution to the PE from SVC as

$$W_{SVC} = \int_{V_{S0}}^{V_S} I_S \, dV_S \tag{5.62}$$

where $I_S = I_{SVC}$ and $V_S = V_{SVC}$ (to simplify the notation).

In the above expression, we are considering the SVC current $I_s$ as the reactive current flowing from the bus $j$ to the SVC. The losses in the SVC are neglected. For the three modes of operation, $I_s$ and the energy functions are as follows:

*Mode 1:*

$$I_S = -B_C V_S$$

$$W_{SVC1} = \int_{V_{S0}}^{V_S} I_S \, dV_S = -\frac{B_C}{2}(V_S^2 - V_{S0}^2)$$

*Mode 2:*

$$I_S = \frac{V_S - V_{ref}}{X_s}$$

where $X_s$ is the slope of the line AB.

$$W_{SVC2} = \int_{V_1}^{V_S} I_S \, dV_S + W_{SVC1}(V_1) = \frac{1}{2X_s}[(V_S - V_{ref})^2 - (V_1 - V_{ref})^2] + W_{SVC1}(V_1)$$

*Mode 3:*

$$I_S = B_L V_S$$

$$W_{SVC3} = \int_{V_2}^{V_S} I_S \, dV_S + W_{SVC2}(V_2) = \frac{1}{2}B_L(V_S^2 - V_2^2) + W_{SVS2}(V_2)$$

## Remarks

1. The PE contribution from an SVC varies depending on the mode of operation of the SVC.

2. The extension of the energy function to multiple SVCs in a network is straightforward. The PE due to multiple SVCs is the sum of PEs contributed by individual SVCs.

3. Hiskens and Hill (1992) present a method to incorporate SVCs into energy function computation. Their formulation considers the SVC susceptance as a part of the network admittance. At a limit point, the network description undergoes a change and consequently, the system behavior is not smooth. They have investigated the lack of smoothness and the effect of SVC limits.

## 5.3.5 Example

The example is based on an SMIB system described in Padiyar and Varma (1991). The single-line diagram is similar to that shown in Figure 4.3. The SVC is connected at the midpoint of the lines that are bussed together. The disturbance considered is a three-phase fault at bus 1. Two types of fault clearing are considered: (a) without line outage and (b) with line outage.

The generator is represented by a two-axis model. The SVC control is modeled (i) without dynamics, considering only steady-state control characteristic, and (ii) with dynamics (based on controller block diagram shown in Figure 5.10).

The CCTs ($t_{cr}$) obtained by prediction and simulation are shown in Table 5.2. The prediction is based on the PEBS method using sustained fault trajectory.

## Discussion

A very significant observation is that the exclusion of controller dynamics in the model does not affect the CCT. This implies that steady-state control characteristics are adequate to predict the system performance with SVC.

**TABLE 5.2**

Critical Clearing Times ($t_{cr}$) from Prediction and Simulation (SVC in SMIB System)

|  |  | Simulation $t_{cr}$(s) | Prediction $t_{cr}$(s) |
| --- | --- | --- | --- |
| Without AVR | Without line outage |  |  |
|  | Without SVC control | 0.216–0.217 | 0.214–0.215 (0.208–0.209) |
|  | With SVC control |  |  |
|  |   With dynamics | 0.258–0.259 | 0.256–0.257 (0.252–0.253) |
|  |   Without dynamics | 0.258–0.259 | 0.256–0.257 (0.252–0.253) |
|  | With line outage |  |  |
|  | Without SVC control | 0.128–0.129 | 0.131–0.132 (0.108–0.109) |
|  | With SVC control |  |  |
|  |   With dynamics | 0.199–0.200 | 0.202–0.203 (0.189–0.190) |
|  |   Without dynamics | 0.199–0.200 | 0.197–0.198 (0.188–0.189) |
| With AVR | Without line outage |  |  |
|  | Without SVC control | 0.251–0.252 | 0.246–0.247 (0.242–0.243) |
|  | With SVC control |  |  |
|  |   With dynamics | 0.282–0.283 | 0.279–0.280 (0.276–0.277) |
|  |   Without dynamics | 0.282–0.283 | 0.278–0.279 (0.276–0.277) |
|  | With line outage |  |  |
|  | Without SVC control | 0.198–0.199 | 0.194–0.195 (0.178–0.179) |
|  | With SVC control |  |  |
|  |   With dynamics | 0.242–0.243 | 0.236–0.237 (0.231–0.232) |
|  |   Without dynamics | 0.242–0.243 | 0.237–0.238 (0.230–0.231) |

Note that the PE for SVC presented in the previous section is accurate and is not path dependent.

In all the cases, the predicted values of $t_{cr}$ are quite close to the values obtained from simulation. In applying the PEBS method, it is necessary to approximate the PE component ($W_{26}$) by the following relation (as explained in the previous chapter):

$$W_{26} = -\int i_q \, dE'_d = \frac{1}{2}(x_q - x'_q)(i_q^2 - i_{q0}^2)$$

The predicted values of $t_{cr}$ (values given in parentheses in the last column of Table 5.2) are pessimistic when the PEBS method is not corrected to account for the fast dynamics of the damper winding. For the case with line outage, there is considerable error in the predicted values of $t_{cr}$. The corrected PEBS method gives accurate results.

It is interesting to observe from the results shown in Table 5.2 that the provision of SVC controller has a similar effect to that of the AVR. The simultaneous action of AVR and SVC control has further improvement in the $t_{cr}$.

### 5.3.6 Case Study of New England Test System

The 10-machine, 39-bus New England test system is taken up for study. Byerly et al. (1982) proposed voltage sensitivity analysis to identify buses with poor voltage regulation that may contribute to system instability. They identified buses 26, 27, 28, and 29 in the New England test system that have relatively large voltage drops as the generation at bus 9 is increased. They reported studies to determine the increase in transient limits by providing SVCs at the four buses 26, 27, 28, and 29. The study also determines the requirements for controlled reactive power (essentially the ratings of SVC) at these buses.

The transient stability limit of generator 9 (connected at bus 38) for a three-phase fault at bus 26 (that is cleared by tripping the line to bus 28 after 0.05 s (three cycles)) can be determined both from digital simulation and application of SPEF (using the PEBS method to determine the critical value of the energy).

The following assumptions are made in the study:

1. The generators are represented by a classical model (for simplicity).
2. The limits on the susceptances of SVC are deliberately chosen large such that each SVC always operates within its regulating range.
3. The initial value of the reactive power for each SVC is zero.
4. The loads are modeled as constant impedances (for convenience).
5. The SVC is modeled by its steady-state control characteristics (i.e., a straight line with positive slope and the operating point lying on the line).

**TABLE 5.3**

Regulating Slopes for SVCs (10-Generator System)

| Bus Number | Regulation Slope $S_k$(p.u.) |
|---|---|
| 26 | 0.0169 |
| 27 | 0.0171 |
| 28 | 0.0282 |
| 29 | 0.0247 |

The values of the regulating slopes are taken from Byerly et al. (1982) and are shown in Table 5.3.

### 5.3.6.1 Network Calculation with Multiple SVCs

Consider the network equations (for the linear transmission network) given by

$$Y\hat{V} = \hat{I} \qquad (5.63)$$

where

$$I^t = \begin{bmatrix} \hat{I}_G & \hat{I}_L & \hat{I}_{SVC} \end{bmatrix}$$

Applying compensation theorem to multiple SVCs, we get

$$\hat{I}_{SVC} = [Z_{eq} + jX_{SVC}]^{-1}(\hat{V}_{eq} - \hat{E}_{SVC}) \qquad (5.64)$$

where $Z_{eq}$ is a square matrix of size $n_S$ where $n_S$ is the number of SVCs. $X_{SVC}$ is a diagonal matrix with entries of regulating slopes of individual SVCs. The solution of Equation 5.64 is complicated by the fact that $\hat{E}_{SVC}$ is not known in advance unless $Z_{eq}$ is purely reactive.

In the special case, $\hat{E}_{SVCi} = V_{refi}\angle\phi_{eqi}$, where $\phi_{eqi}$ is the phase angle of $V_{eqi}$ ($i = 1,2,...,n_S$). In general, an iterative solution is required, which complicates the overall network solution.

We note the fact that the SVC model in the control range is nonlinear. Since the network also has nonlinear loads (that require iterative solution), it is convenient to represent SVC as a nonlinear reactive load (inductive or capacitive). Dommel and Sato (1972) model a nonlinear load by an equivalent circuit of a constant admittance in parallel with a nonlinear current source. Along similar lines, we can model an SVC (at bus $i$) by the equivalent circuit shown in Figure 5.13, where

$$y_{SVCi} = -\frac{j}{X_{SVCi}}, \quad i = 1,2,...,n_S$$

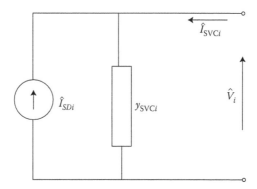

**FIGURE 5.13**
Equivalent circuit of an SVC connected at bus *i* (with multiple SVCs).

$$\hat{I}_{SDi} = y_{SVCi}\hat{E}_{SVCi}$$

$$\hat{E}_{SVCi} = V_{refi}\angle\phi_{refi} = V_{refi}\angle\phi_{SVCi}$$

Note that $V_{refi}$ is known but $\phi_{SVCi}$ is unknown. We can solve the network iteratively by assuming $\phi_{SVC}(t_n) = \phi_{SVC}(t_{n-1})$ at the start of the iteration (where $t_n$ and $t_{n-1}$ are the current and previous time instants, respectively). The iteration can be stopped using the convergence criteria

$$\left| V_i^K - V_i^{(K-1)} \right| < \varepsilon, \quad i = 1, 2, \dots, N$$

where $K$ is the iteration number and $\varepsilon$ is an appropriately selected value (say $10^{-4}$).

### 5.3.6.2 Structure Preserving Energy Function

The SPEF for the classical model of generators and using COI variables is given by

$$W(\theta, \bar{\omega}, V, \phi) = W_1(\bar{\omega}) + W_{21}(\theta) + W_{22}(V, \phi) + W_{23}(V) + W_{24}(\theta, V, \phi) + W_{SVC} \quad (5.65)$$

The expression for $W_1$, $W_{21}$, $W_{22}$, and $W_{23}$ is the same as that given in Chapters 3 and 4. The term $W_{24}$ represents the sum of energy stored in the machine transient reactances and transmission lines. Since this is equal to half of the sum of reactive losses, we can also express $W_{24}$ as

$$W_{24} = \frac{1}{2}\sum_{k=1}^{nb} Q_k = \frac{1}{2}\left\{\sum_{i=1}^{m}(Q_{Gi} - Q_{Gi0}) - \sum_{j=1}^{n}(Q_{Lj} - Q_{Lj0}) - \sum_{i=1}^{ns}(Q_{SVCi} - Q_{SVCi0})\right\}$$

$$(5.66)$$

where *nb* is the total number of series-connected branches, including machine reactances.

The PE component due to SVC is

$$W_{SVC} = \sum_{i=1}^{n_S} \int_{V_{Si0}}^{V_{Si}} I_{Si} \, dV_{Si} = \sum_{i=1}^{n_S} W_{SVCi} \tag{5.67}$$

In the control region of SVC at bus *i*, the SVC current is given by

$$I_{Si} = \frac{V_{Si} - V_{refi}}{X_{SVCi}}$$

Hence

$$W_{SVCi} = \frac{1}{2X_{SVCi}}[(V_{Si} - V_{refi})^2 - (V_{Si0} - V_{refi0})^2]$$

### 5.3.6.3 Results and Discussion

Table 5.4 shows the effects of providing 1–4 SVCs at any or all locations among buses 26, 27, 28, and 29. It is observed that from the disturbance considered, no SVC is required to improve stability when the output of generator 9 is 830 MW (the base case). As the generator output is increased up to 1390 MW, 1–4 SVCs are required. At 1400 MW, the system is unstable. Similarly, when the generator output is 1330 MW, two SVCs are inadequate to maintain stability. At least three SVCs are required. When the generator output is 1080 MW, providing a single SVC at bus 28 or 29 can stabilize the system.

It was observed that generator 9 separates from the rest when the system is unstable. The load model has an influence on the peak values of $\theta_9$ (the rotor angle of generator 9 referred to COI). The load model of CC type has the effect of reducing the acceleration and peak of the rotor angle. The required dynamic range of SVC output is also less with CC-type loads.

## 5.4 Static Synchronous Compensator

### 5.4.1 General

This was initially called as advanced SVC based on VSC and is a part of the emerging FACTS controllers. It is shunt connected (similar to an SVC)

**TABLE 5.4**

Effect of SVCs on Transient Stability Limit (10-Generator System)

| Generation at Generator #9 (MW) | Number of SVCS | Location of SVCs (Bus Numbers) | Transient Stability |
|---|---|---|---|
| 830 | 0 | — | Stable |
| 1080 | 0 | — | Unstable |
| | 1 | 26 | Unstable |
| | | 27 | Unstable |
| | | 28 | Stable |
| | | 29 | Stable |
| | 2 | 26 & 27 | Unstable |
| 1300 | 2 | 26 & 29 | Unstable |
| | | 28 & 29 | Stable |
| 1330 | 1 | 26/27/28/29 | Unstable |
| | 2 | All six combinations | Unstable |
| | 3 | 26, 27, & 28 | Unstable |
| | | 26, 27, & 29 | Unstable |
| | | 26, 28, & 29 | Stable |
| | | 27, 28, & 29 | Stable |
| | 4 | 26, 27, 28, & 29 | Stable |
| 1390 | 3 | 26, 28, & 29 | Unstable |
| | | 27, 28, & 29 | Unstable |
| | 4 | 26, 27, 28, & 29 | Stable |
| 1400 | 4 | 26, 27, 28, & 29 | Unstable |

as shown in Figure 5.14. A STATCOM has several advantages over SVC, some of which are as follows:

1. It has faster response.
2. It has a smaller footprint as bulky passive components are not present.
3. It can be interfaced with real power sources.
4. It has superior performance during low-voltage condition as the reactive current can be maintained constant. The steady-state control characteristics of a STATCOM are shown in Figure 5.15. It is even possible to increase the reactive current under transient conditions by rating the power semiconductor devices (gate turn-off (GTO) or IGBT) to provide the transient overload.

### 5.4.2 Modeling of a STATCOM

A STATCOM can be modeled from its steady-state control characteristics (shown in Figure 5.15) in a similar manner as done for an SVC. The control

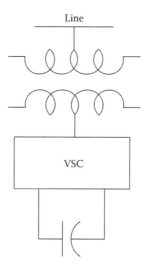

**FIGURE 5.14**
Schematic of a STATCOM.

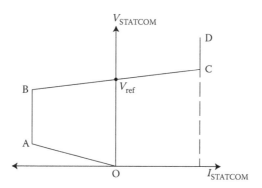

**FIGURE 5.15**
Control characteristics of a STATCOM.

range BC can be represented by an equivalent circuit similar to what is shown in Figure 5.11. The slope of the characteristics in the control range is represented by an equivalent inductive reactance.

The VSC shown in Figure 5.14 can be a six-pulse (two-level converter) with or without pulse width modulation (PWM). If PWM is not feasible (as with VSC utilizing GTO devices), it is necessary to utilize a higher pulse number or multilevel converters to eliminate the voltage harmonies injected by the VSC.

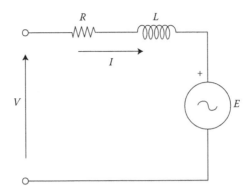

**FIGURE 5.16**
Equivalent circuit of a STATCOM.

Neglecting harmonics, it is possible to obtain a simple yet accurate equivalent circuit of a STATCOM shown in Figure 5.16, where

$$\hat{I} = I_Q + jI_D, \quad \hat{V} = V_Q + jV_D = V\angle\theta \quad \text{and} \quad \hat{E} = kV_{dc}|(\theta + \alpha)$$

$$C\frac{dV_{dc}}{dt} + GV_{dc} = k[\sin(\alpha + \theta)i_D + \cos(\alpha + \theta)i_Q] \tag{5.68}$$

$R$ and $L$ are resistance and leakage reactance of the interconnecting transformer (see Figure 5.14). Sometimes, a separate reactor is used to connect the VSC to the system. This has the advantage of applying a suitable harmonic filter across the transformer winding connected to the converter through a reactor (the transformer does not have to carry harmonic currents).

The equivalent circuit shown in Figure 5.16 is represented by the following complex equation:

$$L\frac{d\hat{I}}{dt} + (R + j\omega_0 L)\hat{I} = \hat{V} - \hat{E} \tag{5.69}$$

If network transients are neglected (as normally assumed), the derivative of $\hat{I}$ can be set to zero.

### 5.4.3 STATCOM Controller

There are two control variables in a STATCOM, $k$ and $\alpha$. The control of $k$ enables variation of the voltage magnitude ($E$) injected by the VSC. This is normally done by PWM. The converters where this is feasible are called as Type 1 converters and converters with only control over $\alpha$ (with fixed value of $k$) are called Type 2 converters (Schauder and Mehta 1993). The fixed value $k$ for Type 2 converters in a multipulse converter is given by

$$k_{fix} = \left(\frac{p}{6}\right)\frac{\sqrt{6}}{\pi} \tag{5.70}$$

where $p$ is the pulse number.

Since PWM using IGBT devices is an emerging technology that is expected to replace Type 2 converters using GTO devices, we consider here only Type 1 converter with associated control. In Type 1 controller, the active component ($i_p$) and reactive component ($i_r$) of the currents drawn by the STATCOM can be regulated by vector control, which acts in a decoupled manner. The current controllers are shown in Figure 5.17. The outputs are $e_p$ and $e_r$ from which $k$ and $\alpha$ can be computed as

$$k = \frac{\sqrt{e_p^2 + e_r^2}}{V_{dc}}, \quad \alpha = -\tan^{-1}\frac{e_r}{e_p}$$

It is to be noted that $e_p$ and $e_r$ are in phase and quadrature (lagging) components of $\hat{E}$ with the bus voltage $\hat{V}$. Thus

$$E_Q + jE_D = (e_p - je_r)e^{j\theta} \tag{5.71}$$

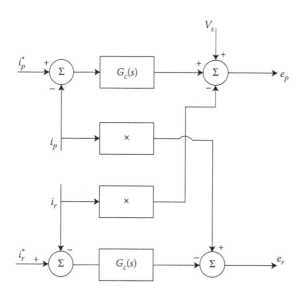

**FIGURE 5.17**
Current controllers of a STATCOM.

Similarly

$$i_Q + ji_D = \left(i_p - ji_r\right)e^{j\theta} \tag{5.72}$$

The active current reference $i_p^*$ is obtained as the output of the DC voltage controller, whereas $i_r^*$ is either set by the reactive power control requirements or as the output from the AC voltage controller.

The current controller is a *P–I* controller whose transfer function is given by

$$G_c(s) = \frac{K_c(1 + sT_c)}{sT_c} \tag{5.73}$$

By selecting $T_c = L/R$ and $K_c = L/T_i$, we obtain

$$i_p = \frac{1}{(1 + sT_i)}i_p^*, \quad i_r = \frac{1}{(1 + sT_i)}i_r^* \tag{5.74}$$

The time constant $T_i$ is generally chosen to provide a specified response time, which can be less than a cycle.

**Remarks**

1. If losses in the STATCOM are neglected, $i_p = 0$. The STATCOM model can be simplified by considering only the reactive current control (see Equation 5.74). Even here, since $T_i$ is small, it may be adequate to assume $i_r = i_r^*$ for transient stability studies. This approximation enables the formulation of the PE contribution of a STATCOM in a manner similar to that of an SVC.

2. If an energy source is connected to the DC side of a STATCOM, $i_p$ can be controlled to supply or absorb active power. In this case, the output of the power controller determines $i_p^*$.

3. With PWM, it is usual to define a modulation index $m$ $(0 < m < 1)$, such that $k = mk_{fix}$. Here, $m$ is viewed as a control variable.

### 5.4.4 Potential Energy Function for a STATCOM

As mentioned earlier, we can define a path-independent energy function from the steady-state control characteristics shown in Figure 5.15. Since the voltage corresponding to point A is quite low, we can neglect the operation along OA and consider only three operating modes as given below:

*Mode 1: $(V_S < V_1)$*

$$I_S = -I_{rmax}$$

$$W_{ST1} = \int_{V_{S0}}^{V_S} I_S \, dV_S = I_S(V_S - V_{S0})$$

*Mode 2: $(V_1 < V_S < V_2)$*

$$I_S = \frac{V_S - V_{ref}}{X_s}$$

where $X_s$ is the slope of the line BC.

$$W_{ST2} = \int_{V_1}^{V_S} I_S \, dV_S + W_{ST1}(V_1)$$

$$= \frac{1}{2X_s}[(V_S - V_{ref})^2 - (V_1 - V_{ref})^2] + W_{ST1}(V_1)$$

*Mode 3: $(V_S > V_2)$*

$$I_S = I_{rmax}$$

$$W_{ST3} = I_{rmax}(V_S - V_2) + W_{ST2}(V_2)$$

where $I_{rmax}$ is the magnitude of the (symmetrical) limits of the current drawn by the STATCOM.

## 5.5 Series-Connected FACTS Controllers

Series capacitors have been used in long-distance transmission lines to enhance the power flow. However, fixed series capacitors can introduce the problem of torsional interactions with turbo-generators due to the phenomenon of subsynchronous resonance (SSR). Also, control of power flow in a line to improve system security requires fast response, which can be provided by series-connected FACTS controllers.

There are two types of series FACTS controllers, namely

1. TCSC, which is essentially a TCR connected across the series capacitor.
2. SSSC, which is based on VSC technology using self-commutated devices (such as GTO or IGBT). The VSC is connected to the transmission line through a series-connected transformer.

### 5.5.1 Thyristor-Controlled Series Capacitor

For details of principles and operation of a TCSC, the reader is referred to Hingorani and Gyugyi (2000), Mathur and Varma (2002), and Padiyar (2007). Here, we will mainly consider the modeling of TCSC for stability studies.

It is not necessary to model the gate pulse unit and the generation of gate pulses. It is adequate to assume that the desired value of TCSC reactance is implemented within a well-defined time frame. This delay can be modeled by a first-order lag as shown in Figure 5.18. The value of $T_{TCSC}$ is typically 15–20 ms. $X_{ref}$ is determined by the power scheduling controller or in its absence by manual control based on order from load dispatch.

#### *5.5.1.1 Power Scheduling Control*

The simplest type of power scheduling control adjusts the reactance order (or setpoint) slowly to meet the required steady-state power flow requirements of the transmission network. The adjustment may be done manually or by a slow-acting feedback loop.

An alternative approach is to use a closed-loop current control in which the measured line current is compared to a reference current (which may be derived from the knowledge of the required power level).

An interesting approach to power scheduling is one, where during a transient, the line in which TCSC is situated carries the required power so that the power flow in parallel paths is kept constant. This is equivalent to maintaining the angular difference across the line a constant and has been termed as constant angle (CA) control (Padiyar et al. 1996). Assuming that the voltage magnitudes at the two ends of the line are regulated, maintaining CA is equivalent to maintaining constant voltage difference between two ends of the line. Hence, we can also call this constant voltage drop (CVD) control.

Both CC and CA controllers can be of P-I type with dynamic compensation for improving the response. The steady-state control characteristics of both CC

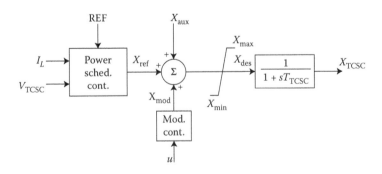

**FIGURE 5.18**
Block diagram of a TCSC controller.

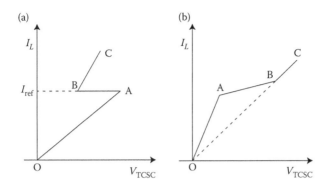

**FIGURE 5.19**
Control characteristics. (a) Constant current (CC). (b) Constant angle (CA).

and CA control are shown in Figure 5.19a and b, respectively. Assuming $V_{TCSC}$ to be positive in the capacitive region, the characteristics have three segments OA, AB, and BC. The control range is AB. OA and BC correspond to the limits on $X_{TCSC}$. In Figure 5.19b, the control range AB is described by the equation

$$V_{TCSC} = I_L X - V_{Lref} \qquad (5.75)$$

where $I_L$ is the magnitude of the line current, $X$ is the net line reactance (taking into account the fixed series compensation, if any), and $V_{Lref}$ is the constant (regulated) voltage drop across the line (including TCSC). Thus, the slope of the line AB is the reciprocal of $X$. OA in Figure 5.19b corresponds to the lower limit on TCSC reactance while BC corresponds to the higher limit on TCSC reactance. On the other hand, OA in Figure 5.19a represents higher TCSC reactance. $I_{ref}$ is the operating value of the line current.

### 5.5.1.2 Power Swing Damping Control

This is designed to modulate the TCSC reactance in response to an appropriately chosen control signal derived from local measurements. The objective is to damp low-frequency swing modes (corresponding to oscillation of generator rotors) of frequencies in the range of 0.2–2.0 Hz. One of the signals that is easily accessible is the line current magnitude. Alternatively, the signal corresponding to the frequency of Thevenin (equivalent) voltage of the system across the TCSC can be used. This signal can be synthesized from the knowledge of voltage and current measurements. In Chapter 7, the design of damping controllers is presented.

### 5.5.1.3 Transient Stability Control

This is generally a discrete control in response to the detection of a major system disturbance.

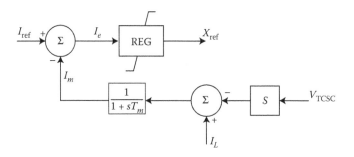

**FIGURE 5.20**
Block diagram of CC or CA controller.

The discrete or bang-bang control of TCSC in response to signals from locally measured variables is described in Chapters 8 and 9. The controller is activated immediately after detecting a major disturbance such as clearing of a fault and is deactivated when the magnitude of frequency deviation is below threshold. This type of control is beneficial not only in reducing the first swing but also for damping subsequent swings.

The block diagram of CC or CA controller is shown in Figure 5.20. $T_m$ is the time constant of first-order low-pass filter associated with the measurement of line current $I_L$ and the TCSC voltage. $S = 0$ for CC control and $S = 1/X$ for CA control. $X$ is the net reactance of the line given by

$$X = X_{\text{Line}} - X_{FC}$$

where $X_{\text{Line}}$ is the line reactance and $X_{FC}$ is the reactance of the fixed series capacitor, if any. Generally, TCSC will be used in conjunction with a fixed series capacitor to minimize the overall cost of compensation while providing effective control for stability improvement.

The simplest regulator is an integrator with gain, which is positive in the case of current control and negative in the case of CA control. In the latter case, $I_{\text{ref}}$ is actually the voltage reference divided by $X$. Hence, positive error signal implies that the net voltage drop in the line is less than the reference and $X_{\text{TCSC}}$ (assumed to be positive in the capacitive region) is to be reduced. On the other hand, for current control, if the error is positive, the controller has to increase $X_{\text{TCSC}}$ to raise the line current to reduce the error.

### 5.5.2 Static Synchronous Series Compensator

An SSSC injects a reactive voltage in the transmission line in which it is connected. To simplify the analysis, we can ignore the losses. An SSSC is faster in response compared to the TCSC.

SSSC has two major advantages over TCSC in terms of performance:

a. Unlike TCSC, an SSSC injects both inductive and capacitive voltages in a symmetric fashion. The rating of SSSC depends on the magnitude of the injected reactive voltage.
b. If an energy source (such as battery) is connected on the DC side of the VSC, an SSSC can inject both active and reactive voltages. It can supply or absorb active power depending on the control mode.

The power scheduling control can be similar to that of a TCSC. We can either regulate the current (and hence the power) in the line having an SSSC or regulate the current (and power) in the parallel paths (with CVD or CA control). However, the range of $V_{SSSC}$ extends to both first and second quadrants in the $I_L - V_{SSSC}$ plane as $V_{SSSC}$ can be negative or positive.

## 5.6 Potential Energy in a Line with Series FACTS Controllers

Consider a transmission line $k$ connected between buses $i$ and $j$. The power flow $P_k$ in the line can be expressed as

$$P_k = P_{k0} + \Delta P_k = P_k (\phi_k, u)$$

where

$$P_{k0} = \frac{V_i V_j}{x_k} \sin \phi_k, \quad \phi_k = \phi_i - \phi_j$$

$u$ is the control variable associated with the series FACTS controller. For a TCSC, it is $x_{TCSC}$ and for an SSSC, it is $V_{SSSC}$, a reactive voltage. Neglecting the controller dynamics, the PE due to the controlled series compensator (CSC) is given by

$$W_{CSC} = \int_{\phi_{k0}}^{\phi_k} \Delta P_k \, d\phi_k \tag{5.76}$$

In the absence of CSC, the PE contribution of a line is

$$W_k = \frac{1}{2} I_k^2 x_k = \frac{V_k^2}{2 x_k}$$

where $V_k$ is the voltage across the line.

There is more than one way of obtaining the expression for $W_{CSC}$. This will be illustrated as we consider individual cases.

### 5.6.1 Thyristor-Controlled Series Capacitor

If we consider the controllable portion of the TCSC reactance (capacitive) as $x_c$ and the net reactance of the line as $x_k$ (including the FC), we get

$$P_k = \frac{V_i V_j}{(x_k - x_c)} \sin \phi_k = \frac{V_i V_j \sin \phi_k}{x_k} + \Delta P_k$$

From the above, we can derive

$$\Delta P_k = \frac{x_c}{x_k(x_k - x_c)} V_i V_j \sin \phi_k$$

Since $\int P_{k0}\, d\phi = (V_k^2/2x_k)$, we can obtain

$$W_{TCSC} = \int \Delta P_k\, d\phi = \frac{V_k^2}{2x_k}\left(\frac{x_c}{x_k - x_c}\right) \tag{5.77}$$

### *Alternative Derivation*

Since $V_k$ (the voltage drop across the line $k$) is leading the current $I_k$ by $90°$ (in a lossless line), we can express (see Section 4.6 in Chapter 4) the PE in line $k$ as

$$\int I_k\, dV_k = \int P_k\, d\phi_k \tag{5.78}$$

The effect of introducing a TCSC in line $k$ can be expressed as injecting a reactive voltage (capacitive) in the line (see Figure 5.21).
Since

$$V_{rk} = I_k x_c = \frac{V_k}{(x_k - x_c)} x_c$$

$$\Delta I_k = \frac{V_{rk}}{x_k} = \frac{V_k x_c}{x_k(x_k - x_c)}$$

$$W_{TCSC} = \int \Delta I_k\, dV_k = \frac{V_k^2 x_c}{2x_k(x_k - x_c)} = \frac{\Delta I_k V_k}{2} = \frac{\Delta Q_k}{2}$$

where $\Delta Q_k$ is the increment in the reactive power loss in the line due to the TCSC.

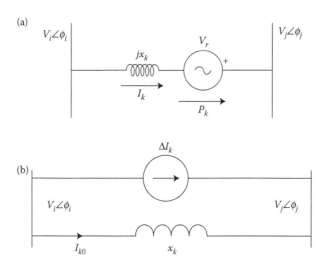

**FIGURE 5.21**
A reactive voltage injected in line $k$. (a) TCSC or SSSC. (b) Current injection model.

### 5.6.2 Static Synchronous Series Compensator

To compute $W_{SSSC}$, we will use the alternative derivation. Referring to Figure 5.21a, we note that this also applies to an SSSC, except that $V_{rk}$ is a constant reactive voltage injected by SSSC.

Thus,

$$\Delta I_k = \frac{V_{rk}}{x_k}$$

and

$$W_{SSSC} = \int \Delta I_k \, dV_k = \frac{V_{rk}}{x_k} V_k = \frac{V_{rk}\sqrt{V_i^2 + V_j^2 - 2V_i V_j \cos \phi_k}}{x_k} \qquad (5.79)$$

Since $\Delta Q_k = V_k \Delta I_k$, we also obtain

$$W_{SSSC} = \Delta Q_k \qquad (5.80)$$

In Mihalic and Gabrijel (2004), the expression for the energy function has a negative sign as $V_{rk}$ is assumed to be positive when it is inductive.

### Remarks

1. It is to be noted that the introduction of a CSC (TCSC or SSSC) will affect the bus voltages (both magnitudes and phase angles). The

expression for the PE associated with the controller is essentially an additional term for the PE in the presence of CSC that introduces a reactive voltage.

With more than one CSC connected in different lines, we get the total energy due to the controllers as the sum of energies in individual lines where controllers are present. Thus

$$W_{CSC} = \sum_{k \in L_c} W_{CSCk}$$

where $L_c$ is a set of lines with CSC. $W_{CSCk}$ represents the energy in line $k$, associated with the CSC in that line.

2. In deriving the expression for the energy, we have assumed that $x_c$ or $V_r$ are constants that are permissible if we consider that $x_c$ and $V_r$ remain at their maximum values during the first swing following any major disturbance such as a fault followed by clearing.

3. The controller dynamics is neglected in deriving the expressions for the energy associated with CSC. This approximation is acceptable as we have observed from the results of the case study on SVC.

### 5.6.3  Potential Energy in the Presence of CC and CA Controllers

The CC controller tries to maintain the current magnitude in the line with the CSC (TCSC or SSSC) constant. This control action is useful in ensuring that the line is not overloaded following a contingency. In contrast, the CA controller attempts to maintain the voltage magnitude (across the line) constant. The effect of this controller is to ensure that parallel lines are not overloaded following a contingency. This implies that the CA controller is to be adopted in lines with a large spare capacity. On the other hand, the CC control is to be used in lines with a little spare capacity.

### 5.6.3.1  Potential Energy with CC Control

From Figure 5.19a, we observe that the control characteristics OA and BC are associated with upper and lower limits on the control variable ($x_c$ or $V_r$). Since energy functions are already derived with constant $x_c$ or $V_r$, we consider only the control range AB.

In this control range, the line current $I_k$ is regulated at the value of $I_{ref}$. The voltage across the line $k$ is

$$V_k = I_k x_k - V_{rk}$$

Since $I_k$ is a constant, $dV_k = -dV_{rk}$. Hence

$$W_k = -\int I_{ref} \, dV_{rk} = -I_{ref} V_{rk}$$

**TABLE 5.5**

Transient Stability Limits (with TCSC)

| Controller | Without AVR (MW) | With AVR (MW) |
|---|---|---|
| CC | 700 | 730 |
| CR | 760 | 800 |
| CA | 840 | 870 |

Note that when the line current tends to increase following a major disturbance, $V_{rk}$ decreases to regulate the line current. However, it is to be noted that the current cannot be regulated in the first swing due to the limits on the FACTS compensation. Hence, the CC control contributes negative energy. Thus, it can be expected that a CC controller will reduce the transient stability limit compared to a constant reactance (CR) control.

### 5.6.3.2 Potential Energy with CA Control

From Figure 5.19b, it is observed that the control characteristics OA and BC are associated with the lower and upper limits on the control variable. Hence, we will consider here the control range AB.

In the control range, the voltage $V_k$ across the line is regulated at the value $V_{ref}$. This implies $dV_k = 0$. Hence, the operation in the control range does not contribute to either increase or decrease in the energy. However, following a major disturbance, the controller will insert a maximum capacitive reactance or voltage in the first swing to regulate the voltage across the line. In contrast, the CC control inserts a minimum capacitive reactance or voltage (zero in the case of TCSC and inductive in the case of SSSC), thereby increasing the peak value of the rotor swing.

The results of a case study on IEEE Second Benchmark System (Padiyar 1999), (IEEE SSR Working Group 1985) is reported in Padiyar et al. (1996) and Uma Rao (1996). The system considered is an SMIB system with two transmission lines, one of which is series compensated with 25% of fixed compensation and a TCSC with variable compensation from 10% to 45%. The operating value of the series compensation is 50%, including TCSC. The transient stability limits for a three-phase fault at the generator terminals cleared in three cycles is shown in Table 5.5 for (a) CR, (b) CC, and (c) CA control of TCSC. The results are given for (i) with and (ii) without AVR. The AVR is also provided with PSS.

## 5.7 Unified Power Flow Controller

### 5.7.1 Description

The UPFC (shown in Figure 5.22) is the most versatile FACTS controller with three control variables (the magnitude and phase angle of the series-injected

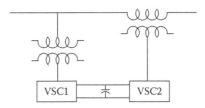

**FIGURE 5.22**
Schematic of a UPFC.

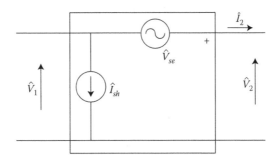

**FIGURE 5.23**
Equivalent circuit of a UPFC.

voltage in addition to the reactive current drawn by shunt-connected VSC). The equivalent circuit of a UPFC on a single-phase basis is shown in Figure 5.23. The current $i_{sh}$ is drawn by the shunt-connected VSC while the voltage $v_{se}$ is injected by the series-connected VSC. Neglecting harmonics, both the quantities can be represented by phasors $\hat{I}_{sh}$ and $\hat{V}_{se}$.

Neglecting power losses in the UPFC, the following constraints equation applies:

$$\mathrm{Re}[\hat{V}_1 \hat{I}_{sh}^*] = \mathrm{Re}[\hat{V}_{se} \hat{I}_2^*] \tag{5.81}$$

Assuming that $\hat{V}_1 = V_1 e^{j\theta_1}$ and $\hat{I}_2 = I_2 e^{j\phi_2}$, $\hat{I}_{sh}$ and $\hat{V}_{se}$ can be expressed as

$$\hat{I}_{sh} = (I_p - jI_r)e^{j\theta_1} \tag{5.82}$$

$$\hat{V}_{se} = (V_p + jV_r)e^{j\phi_2} \tag{5.83}$$

where $I_p$ and $I_r$ are "real" and "reactive" components of the current drawn by the shunt-connected VSC. Similarly, $V_p$ and $V_r$ are the "real" and "reactive" voltages injected by the series-connected VSC. Positive $I_p$ and $V_p$ indicate positive "real" (active) power flowing into the shunt-connected VSC and flowing out of the series-connected VSC. The positive $I_r$ and $V_r$ indicate reactive power drawn by the shunt convertor and supplied by the series converter.

Using Equations 5.82 and 5.83, Equation 5.81 can be expressed as

$$V_1 I_p = I_2 V_p \tag{5.84}$$

### 5.7.2 Energy Function with Unified Power Flow Controller

Consider a UPFC connected in line $k$ near the sending end as shown in Figure 5.24a and the injection model of the UPFC as shown in Figure 5.24b.

It is assumed that the three independent parameters of the UPFC, $V_C$, $\beta$, and $I_r$ are assumed to be constants. The UPFC is assumed to be lossless.

The power flow ($P_k$) in the line $k$ can be obtained as

$$P_k = \text{Re}[\hat{I}_j^* \hat{V}_j]$$

where $\hat{I}_j$ is given by

$$\hat{I}_j = \frac{V_i \angle \phi_i + V_C \angle (\phi_i + \beta) - V_j \angle \phi_j}{jx_k} \tag{5.85}$$

The incremental power flow in the line ($\Delta P_k$) due to UPFC is given by

$$\Delta P_k = \frac{V_j V_C \sin(\phi_i + \beta - \phi_j)}{x_k} \tag{5.86}$$

Since $\text{Re}[\hat{V}_C \hat{I}_j^*] = V_i I_p$, we have the constraint

$$V_i I_p + \frac{V_j V_C \sin \beta}{x_k} = \frac{V_j V_C \sin(\phi_i + \beta - \phi_j)}{x_k} = \Delta P_k \tag{5.87}$$

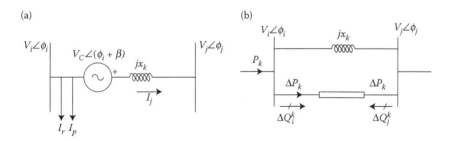

**FIGURE 5.24**
A UPFC connected in line $k$. (a) Basic model. (b) Injection model.

The incremental injections of the reactive power at the buses $i$ and $j$ are given by

$$\Delta Q_i^k = V_i I_r + \frac{V_j V_C \cos \beta}{x_k} \tag{5.88}$$

$$\Delta Q_j^k = \frac{V_j V_C \cos(\phi_i + \beta - \phi_j)}{x_k} \tag{5.89}$$

The energy function due to the UPFC is given by

$$W_{\text{UPFC}} = \int \Delta P_K \, d\phi_k = \int \frac{V_j V_C \sin(\phi_i + \beta - \phi_j)}{x_k} \, d\phi_{ij} \tag{5.90}$$

Integrating the above expression by parts, we can obtain

$$W_{\text{UPFC}} = \Delta Q_i^k + \Delta Q_j^k - \int \frac{\Delta Q_i^k}{V_i} \, dV_i - \int \frac{\Delta Q_j^k}{V_j} \, dV_j \tag{5.91}$$

Since the last two terms in the RHS of the above expression can be clubbed together with $\int (Q_{Li}/V_i) dV_i$, $\int (Q_{Lj}/V_j) dV_j$, we get the additional term to be added in the energy function due to the UPFC as

$$W'_{\text{UPFC}} = \frac{V_i V_C}{x_k} \cos \beta - \frac{V_j V_C \cos(\phi_i + \beta - \phi_j)}{x_k} + V_i I_r \tag{5.92}$$

Note that we have assumed $V_C$, $\beta$, and $I_r$ as constants in deriving the above expression.

A similar approach is used by Azbe et al. (2005) in deriving the energy function for a UPFC. They have also attempted to relax the restriction on the controllable parameters (of being constants) by considering their piecewise variation over a time interval.

UPFC is a versatile two-converter (that can exchange power between the converters through the DC link) FACTS controller that can be used to regulate power and reactive power in a line in addition to controlling the reactive power injected at the bus where the shunt converter is connected. There are two other configurations of the two VSCs: (i) connected in series in two lines (to form an interline power flow controller (IPFC)) (Gyugyi et al. 1999) and (ii) connected in shunt to form a BTB VSC HVDC link. Padiyar (2007) describes the convergence (or equivalence) between the three configurations—UPFC, IPFC, and VSC–HVDC. It is possible to extend the multiconverter concept to more than two VSCs interlinked on the DC side. Azbe and Mihalic (2008) present an energy function for IPFC to derive an optimum control strategy based on minimizing its derivative.

The application of SPEF for the design of linear damping controllers (for small oscillations) based on FACTS is presented in Chapter 7. The application of SPEF for emergency control that involves stabilizing the system subjected to a large disturbance and steering it close to the postfault equilibrium state is presented in Chapters 8 and 9.

# 6

## Detection of Instability Based on Identification of Critical Cutsets

## 6.1 Introduction

There have been several attempts in the 1980s to develop on-line DSA techniques to ensure that the system will survive any credible contingency involving a large disturbance. In a stressed power system when the line loading is close to the stability limits, a large disturbance such as the fault followed by clearing can lead to uncontrolled tripping of generators and cascading outages that may finally result in a blackout if proper remedial actions are not taken. Since on-line time domain simulation is not feasible, "the common practice in industry is to perform off-line studies using a manually identified set of critical operating conditions. The results of these studies are provided to the operator as an instruction and/or in a computerized form for the on-line monitoring of the power system, but will not encompass many situations which the operators may face. On-line computation of approximate limits (on power flow in critical lines) is currently limited to the use of various forms of the equal area criterion. Advanced methods should be developed to provide both accuracy and speed before on-line DSA becomes a reality" (Findlay et al. 1988, Pai 1989).

The work reported in this chapter is aimed at developing a reliable and fast method for on-line detection of loss of synchronism in power systems based on the identification of the critical cutsets. No assumptions are made about the nature of the system model. Wide area measurement system (WAMS) is an emerging technology that can assist in the development of real-time control strategies that will enable robust system operation utilizing high-power electronic controllers based on HVDC converters and FACTS.

In developed countries, with adequate margins of generation and transmission facilities, it has been the practice to employ preventive control to ensure system security and to avoid blackouts. However, the major blackout in the United States and Canada (in August 2003) has shown that preventive control is not always successful. The preventive control requires fast and reliable state estimation. Further, it requires a list of credible contingencies that may not be exhaustive. A viable option to ensure security in systems (particularly beset with shortages) is to detect on-line the loss of stability and take emergency control actions to minimize load shedding and loss of system integrity.

The adaptive system protection, based on the identification of the critical cutset, will also be discussed in this chapter. The conventional out-of-step relaying based on the impedance measurement in a transmission line has several limitations. To overcome these limitations, an adaptive out-of-step relaying scheme was proposed by Centeno et al. (1997). However, this scheme requires the knowledge of the equivalent system parameters with reasonable accuracy.

Some techniques of on-line detection of instability have been discussed in the literature—application of global phasor measurements (Rovnyak et al. 1995), heuristic algorithms (Wang and Girgis 1997), application of intelligent techniques such as decision trees (Rovnyak et al. 1994), and artificial neural network (ANN) (Liu et al. 1999).

We present here an algorithm for on-line detection of synchronous instability. The algorithm has two steps. In the first step, the candidate lines are checked for instability using transient energy function (Padiyar and Krishna 2006), which requires only the knowledge of variations in the power flow in the line and angle across the line. In the second step, the logical data obtained from applying the instability criterion (in candidate lines) are processed to check for system stability, using a graph theoretic algorithm to determine whether the lines, which satisfy the instability criterion, form a cutset.

The results of the application of the algorithm to three systems (10 machine, 17 machine, and a large, practical system of 493 buses and 193 generators) are presented with discussions.

## 6.2 Basic Concepts

In this section, we present the basic concepts underlying the development of the algorithm.

Basically, there are two major propositions that are applicable to the process of loss of synchronism in a power system.

> *Proposition 1* (Pavella et al. 2000): However complex, the mechanism of loss of synchronism in a power system originates from the irrevocable separation of its machines into two groups.
>
> *Proposition 2* (Krishna and Padiyar 2010): There is a unique cutset (called as the critical cutset) consisting of transmission lines, transformers (or series elements), connecting the two groups of

machines that separate. The cutset angles (difference between the angles of the terminal buses of each member of the cutset) become unbounded.

Proposition 1 has been utilized extensively to determine the MOI that characterizes the loss of synchronous stability when a disturbance occurs. The knowledge of MOI has been used to compute the controlling UEP and the correct value of KE that leads to system separation. However, the determination of MOI is not straightforward and several empirical approaches (Pavella and Murthy 1994) have been suggested and tried.

Xue et al. (1989) proposed the EEAC based on Proposition 1. In this method, the system is reduced to the OMIB system and EAC is applied to infer the stability properties of the original system. Further developments have led to SIME, which is a transient stability method based on generalized OMIB. It is a hybrid method that also uses time domain simulation.

Unlike Proposition 1, which is justified based on observations of a large number of system stability studies, it is possible to prove Proposition 2. Before we present the proof, it is interesting to note that there is only one and unique outset that separates the system into two areas (see Figure 6.1). As an example, consider the 17-generator, 162-bus IEEE test system. The swing curves for the critically unstable three-phase fault at bus 129, which is cleared by tripping the line 5–129, are shown in Figure 6.2. (The clearing time is 0.31 s.) The system separates into two groups; group A consists of seven generators connected at buses 6, 73, 76, 114, 121, 130, and 131. There are 19 lines in the critical cutset. The angles across these lines are plotted in Figure 6.3. For the 10-generator New England test system, for several faults at different buses, generator 2 (which is the largest) separates from the rest and the critical cutset consists of two lines 11–12 and 18–19 (Padiyar and Krishna 2006). It is obvious that even for the 10-generator system involving the separation of one generator from the rest, there can be several possible cutsets. However, there is only one unique critical cutset, where the angles across the members (made of series elements) become unbounded.

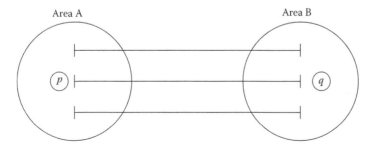

**FIGURE 6.1**
A system with two areas each having coherent generators.

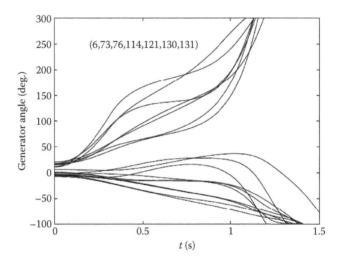

**FIGURE 6.2**

Swing curves for critically unstable case (fault at bus 129, 17-generator system).

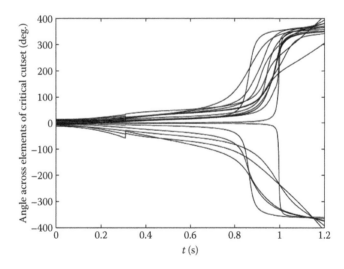

**FIGURE 6.3**

Angles across lines of the critical cutset for critically unstable case.

## 6.3 Prediction of the Critical Cutset

### 6.3.1 Analysis

It is possible to predict the critical cutset if we know the MOI. If we know that the set of advanced generators lie in area A (see Figure 6.1) and the rest of the generators lie in area B, we can identify the critical cutset that connects the two areas, A and B. Note that even if we know the MOI, we do not know in advance the lines and nodes that belong to each area. Only after the identification of the critical cutset we know the exact composition of areas A and B.

Consider a candidate line (or a series element) that is connected between buses $p$ and $q$ as a member of the critical cutset. The bus voltages can be written as the sum of the two components—one due to the current sources in the area having advanced generators (area A) and the other due to the current sources in the remaining area (area B). Thus

$$V_p \angle \phi_p = \hat{V}_{pA} + \hat{V}_{pB} \tag{6.1}$$

$$V_q \angle \phi_q = \hat{V}_{qA} + \hat{V}_{qB} \tag{6.2}$$

We can obtain $\hat{V}_{pA}$ and $\hat{V}_{qA}$ as solutions of

$$[Y]\hat{V}_A = \hat{I}_A \tag{6.3}$$

where $\hat{I}_A$ is the vector of current sources in area A. $\hat{V}_A$ is the vector of bus voltages (including $\hat{V}_{pA}$ and $\hat{V}_{qA}$). $[Y]$ is the postfault bus admittance matrix, which includes shunt admittances of all the generators (represented by Norton equivalents) and constant impedance loads. The source current $\hat{I}_{kA}$ of generator $k$ belonging to A is given by

$$\hat{I}_{kA} = \frac{E'_{qk} + j(E'_{dk} + E'_{dck})}{R_{ak} + jx'_{dk}} e^{j\delta_k} \tag{6.4}$$

where $E'_{qk}$ and $E'_{dk}$ are the $q$-axis and $d$-axis voltages of the generator $k$, respectively, $\delta_k$ is the rotor angle, $R_{ak}$ and $x'_{dk}$ are the stator resistance and $d$-axis transient reactance, respectively, and $E'_{dck}$ is the voltage due to the dummy coil in the $q$-axis (see Appendix C).

The differential equations governing $E'_{qk}$, $E'_{dk}$, and $E'_{dck}$ are

$$\frac{dE'_{qk}}{dt} = \frac{1}{T'_{dok}}[-E'_{qk} + (x_{dk} - x'_{dk})i_{dk} + E_{fdk}] \tag{6.5}$$

$$\frac{dE'_{dk}}{dt} = \frac{1}{T'_{dok}}[-E'_{dk} - (x_{qk} - x'_{qk})i_{qk}] \tag{6.6}$$

$$\frac{dE'_{dck}}{dt} = \frac{1}{T_{ck}}[-E'_{dck} - (x'_{qk} - x'_{dk})i_{qk}] \tag{6.7}$$

$T_{ck}$ is the arbitrarily chosen time constant of the dummy coil introduced to account for the transient saliency.

The following equation (similar to Equation 6.3) is used to solve for the bus voltage vector $\hat{V}_B$:

$$[Y]\hat{V}_B = \hat{I}_B \tag{6.8}$$

$\hat{I}_B$ is the vector of current sources in area B, and $\hat{V}_{pB}$ and $\hat{V}_{qB}$ are elements of the vector $\hat{V}_B$.

The candidate line connected across buses $p$ and $q$ belongs to the critical cutset if and only if $\phi_{pq} = \phi_p - \phi_q$ becomes unbounded. The following theorem gives the necessary and sufficient conditions for this to happen.

**Theorem 1**

$\phi_{pq}$ becomes unbounded if and only if $t \geq t_{cl}$

$$\left|\hat{V}_{pA}\right| > \left|\hat{V}_{pB}\right| \quad \text{and} \quad \left|\hat{V}_{qA}\right| < \left|\hat{V}_{qB}\right| \tag{6.9}$$

$t_{cl}$ is the fault clearing time.

*Proof.*
Multiplying both sides of Equation 6.1 by $e^{-j\delta_{0B}}$ (where $\delta_{0B}$ is the angle of the COI of the generators in area B) gives

$$V_p\angle(\phi_p - \delta_{0B}) = \hat{V}'_p = \hat{V}'_{pA}e^{j\delta} + \hat{V}'_{pB} \tag{6.10}$$

where $\hat{V}'_{pA} = \hat{V}_{pA}e^{-j\delta_{0A}}$, $\hat{V}'_{pB} = \hat{V}_{pB}e^{-j\delta_{0B}}$, and $\delta = \delta_{0A} - \delta_{0B}$. $\delta_{0A}$ is the angle of COI of the generators in area A. Note that $\left|\hat{V}'_{pA}\right| = \left|\hat{V}_{pA}\right|$ and $\left|\hat{V}'_{pB}\right| = \left|\hat{V}_{pB}\right|$. Similarly, Equation 6.2 can be transformed to

$$V_q\angle(\phi_q - \delta_{0B}) = \hat{V}'_q = \hat{V}'_{qA}e^{j\delta} + \hat{V}'_{qB} \tag{6.11}$$

Note also that $\left|\hat{V}'_{qA}\right| = \left|\hat{V}_{qA}\right|$ and $\left|\hat{V}'_{qB}\right| = \left|\hat{V}_{qB}\right|$. If the generators in each group remain in synchronism, then the average variations in the angles of $\hat{V}'_{pA}$ and $\hat{V}'_{pB}$ are zero, even when area A separates from area B. Similar comments apply to the average variations in the angles of $\hat{V}'_{qA}$ and $\hat{V}'_{qB}$.

To prove the theorem, it must be shown that the angle across the candidate line (connected between buses $p$ and $q$) ($\phi_p - \phi_q$) has an average increase in magnitude of 360° when $\delta$ increases by 360°, if and only if the inequality relations in Equation 6.9 are applicable. In such a case, the candidate line is a member of the critical cutset that separates area A from area B as the system becomes transiently unstable.

For proving the sufficiency condition, assume that the inequality conditions of Equation 6.9 are applicable. The phasor relationships of Equations 6.10 and 6.11 are shown in Figures 6.4a and 6.4b, respectively. It is obvious that the average variation in $\phi_{pq}$ ($\phi_p - \phi_q$) is 360° as $\delta$ increases by 360°.

If the inequality conditions of Equation 6.9 do not apply, then it can be shown that the average variation of $\phi_{pq}$ is zero as $\delta$ increases by 360°. This implies that Equation 6.9 gives the necessary conditions for the line $p$–$q$ to be a member of the critical cutset.

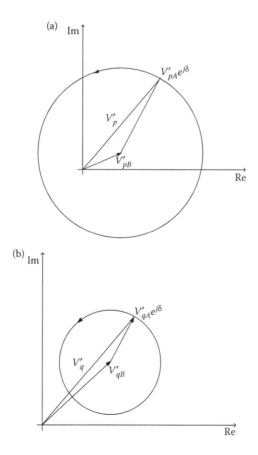

**FIGURE 6.4**
Locus of voltage magnitudes at the two ends of a line connecting areas A and B. (a) Locus of magnitude of $V_p$. (b) Locus of magnitude of $V_q$.

**Remarks**

1. A nonlinear load at bus $k$ given by $P_{Lk}(V_k) + jQ_{LK}(V_k)$ can be represented by an admittance $Y_{lk}$ in parallel with a current source $\hat{I}_{lk}$ given by

$$Y_{lk} = \frac{P_{Lk0} - jQ_{Lk0}}{V_{k0}^2} \tag{6.12}$$

$$-\hat{I}_{lk} = \frac{P_{Lk} - jQ_{Lk}}{\hat{V}_k^*} - Y_{lk}\hat{V}_k \tag{6.13}$$

where the subscript 0 indicates operating values. $Y_{lk}$ is absorbed into the bus admittance matrix. The nonlinear equation (Equation 6.13) can be approximated by the following differential equation:

$$\varepsilon\dot{\hat{I}}_{lk} = \frac{P_{Lk} - jQ_{Lk}}{\hat{V}_k^*} + Y_{lk}\hat{V}_k - \hat{I}_{lk} \tag{6.14}$$

where $\varepsilon$ is small. With this approximation, the network equation becomes linear even in the presence of nonlinear loads. Similarly, the current due to nonlinear FACTS controllers can be obtained by approximating the algebraic equations by differential equations (if FACTS dynamics are neglected in the system model).

2. The critical cutset can be determined by monitoring the values of $|\hat{V}_{pA}|$, $|\hat{V}_{pB}|$, $|\hat{V}_{qA}|$, and $|\hat{V}_{qB}|$ for all eligible lines and transformers in the postfault period, if the MOI is known.

3. If the generators are represented by the classical model and there are no nonlinear elements (loads and FACTS controllers), and the two areas are strictly coherent, then $|\hat{V}_{pA}|$, $|\hat{V}_{pB}|$, $|\hat{V}_{qA}|$, and $|\hat{V}_{qB}|$ are constants in the postfault period. This fact can be used to determine the critical cutset based on the values of $|\hat{V}_{pA}|$, $|\hat{V}_{pB}|$, $|\hat{V}_{qA}|$, and $|\hat{V}_{qB}|$ at the instant of fault clearing; this reduces the amount of computation to a large extent. Note that we consider an area as coherent if the frequencies of all the buses in that area, following a major disturbance, tend to be equal (after the intermachine oscillations decay).

4. It was assumed that the generators in area A accelerate with respect to the generators in area B. If the reverse is true, then it can be shown that the following conditions apply:

$$|\hat{V}_{qA}| > |\hat{V}_{qB}| \quad \text{and} \quad |\hat{V}_{pA}| < |\hat{V}_{pB}| \tag{6.15}$$

### 6.3.2 Case Study

Krishna and Padiyar (2010) report studies conducted on the 10-generator, 39-bus New England system. The loads are assumed to be constant impedances. The MOI and critical cutset do not depend on the generator model. For the classical generator model, $|\hat{V}_{pA}|$ and $|\hat{V}_{qA}|$ are constants when area A consists of only one generator. The critical cutset (based on the calculation of $|\hat{V}_{pA}|$, $|\hat{V}_{pB}|$, $|\hat{V}_{qA}|$, and $|\hat{V}_{qB}|$ at the instant of fault clearing) is in agreement with the actual critical cutset observed in simulation in practically all the cases. For all the cases, the predicted cutsets are determined at the critically unstable fault clearing time. For faults at buses 27, 28, 29, and 26, generator 9 separates from the rest of the system; for other faults, generator 2 separates from the rest of the system.

### 6.3.3 Discussion

The analytical criterion for identifying a candidate line as a member of the critical cutset is simple yet quite general. The bus voltage vector and the voltages across critical lines are computed during system simulation, as the sum of two components, one due to injected currents in area A and the other due to injected currents in area B. The magnitudes of these components are used to formulate the criterion. In most cases, the system simulation is required only up to the instant of fault clearing. This is because of the fact that, with the classical model of the generator and the assumption of strict coherency among the generators belonging to a group, the magnitudes of the two voltage components remain constant for $t > t_{c1}$. In most cases, the prediction based on the classical model is valid for the detailed model unless $|\hat{V}_{pA}|$, $|\hat{V}_{pB}|$, $|\hat{V}_{qA}|$, and $|\hat{V}_{qB}|$ are quite close at the fault clearing time. In any case, the system simulation, only for a short term, is required to accurately determine whether the candidate line is a member of the critical cutset.

It is to be noted that it is difficult to accurately predict the MOI (Pavella and Murthy 1994). It is easier to monitor the angles across series elements that are members of the critical cutset. From graph-theoretic analysis, it is feasible to predict accurately MOI from the knowledge of the critical cutset.

## 6.4 Detection of Instability by Monitoring Critical Cutset

If we know the critical cutset, it is quite straightforward to monitor the angles across each members of the cutset using PMU and wide area measurements (WAM). Strictly speaking, there is no need for using GPS technology to measure the angle across a transmission line. The phasor measurements of line current and the bus voltage at either end of the line coupled with knowledge of series line impedance is adequate to compute on-line the angle across a line. As a matter of fact, it will be shown in the next section that the criterion for instability requires measurements of bus frequencies and line power flows. The frequency monitoring network (FNET) (Zhong et al. 2005) can be employed for the purpose of wide-area frequency monitoring.

Since we do not have prior knowledge of the critical cutset, which is also a function of the system characteristics, and the disturbance, it would appear that we need to monitor all the series elements. Fortunately, this is not required in practice. Figure 6.1 shows two areas connected by the critical cutset when each area is coherent in the sense that generators within the area swing together and all the buses have the same frequency following a disturbance. In general, there could be several coherent areas in a power system connected by tie lines (Khorasani et al. 1986). This implies that we need to consider only the tie lines connecting the coherent areas as candidate lines.

We develop next a criterion for instability, based on the transient energy function.

### 6.4.1 Criterion for Instability

The transient energy for the system shown in Figure 6.1 can be expressed as

$$W = W_1 + W_2$$

where $W_1$ is the effective KE that leads to system separation and $W_2$ is the PE. The PE can be expressed as

$$W_2 = \sum_{k=1}^{ns} \int_{t_0}^{t} (P_k - P_{ks}) \frac{d\delta_k}{dt} dt \tag{6.16}$$

where $P_k$ is the transient power flow in the series element, $P_{ks}$ is the steady-state value of $P_k$, $\delta_k$ is the angle across $k$, and $ns$ is the total number of series elements.

The PE can be decomposed into the energy within the two areas and the energy along the critical cutset (Padiyar and Uma Rao 1997). Assuming coherent areas, the PE within an area is zero as all the buses in that area have the same frequency ($d\delta_k/dt$ is zero for all series elements within an area). Hence the PE given by Equation 6.16 can be written as follows:

$$W_2 = \sum_{k=1}^{nc} \int_{t_0}^{t} (P_k - P_{ks}) \frac{d\delta_k}{dt} dt \tag{6.17}$$

where $nc$ is the number of elements in the critical cutset. It can be shown that the variation of PE in all the lines in the critical cutset is similar. If a series element (line or transformer) $k$ in the critical cutset connects buses $i$ and $j$

$$P_k = V_i V_j b_k \sin \delta_k \qquad (6.18)$$

$$\delta_k = \phi_i - \phi_j \qquad (6.19)$$

where $V_i$ and $V_j$ are the voltage magnitudes at buses $i$ (in area A) and $j$ (in area B), respectively, and $b_k$ is the magnitude of the susceptance of the series element. Owing to the assumption of coherency, the variations of $V_i$ and $\phi_i$ are similar for all the elements in the critical cutset. This is also true of the variations of $V_j$ and $\phi_j$. Hence, the variation of PE can be monitored from the energy in the individual lines in the cutset. Thus

$$W_2 = A_k W_{2k} \qquad (6.20)$$

where $A_k$ is a constant and subscript $k$ refers to any element in the cutset, and

$$W_{2k} = \int_{t_0}^{t} (P_k - P_{ks}) \frac{d\delta_k}{dt} dt \qquad (6.21)$$

The corrected KE $W_1'$ to properly account for the portion of the KE that contributes to system separation (Fouad and Vittal 1992) is given by

$$W_1' = \frac{1}{2} M_{eq} \omega_{eq}^2 \qquad (6.22)$$

where

$$M_{eq} = \frac{M_A M_B}{M_A + M_B}, \quad \omega_{eq} = \omega_A - \omega_B, \quad M_A = \sum_{i \in area\, A} M_i, \quad M_B = \sum_{i \in area B} M_i,$$

$$\omega_A = \frac{1}{M_A} \sum_{i \in area\, A} M_i \omega_i, \quad \omega_B = \frac{1}{M_B} \sum_{i \in area\, B} M_i \omega_i$$

By assumption of strict coherency, the rotor speeds of all the generators in an area are equal and the derivatives of the angles across all the elements in the critical cutset are identical. Hence

$$\omega_{eq} = \frac{d\delta_k}{dt} \qquad (6.23)$$

where $\delta_k$ is the angle across any line in the critical cutset. The corrected KE is given by

$$W_1' = \frac{1}{2} M_{eq} \left( \frac{d\delta_k}{dt} \right)^2 \tag{6.24}$$

The criterion derived for the detection of instability is based on energy function analysis. The power system gains kinetic and PE due to a disturbance. For transient stability, the system must be capable of absorbing the KE (responsible for system separation) completely. If the KE is not completely converted to PE, the system becomes unstable. Therefore, for a stable swing, KE is zero when PE attains a maximum, and for an unstable swing, KE is not zero (positive) when PE attains a maximum. This criterion is used for the detection of instability, with KE and PE given by Equations 6.24 and 6.17, respectively. Since the criterion checks whether KE is zero or positive when PE is maximum, it is adequate to monitor $d\delta_k/dt$ instead of the KE given by Equation 6.24, and the PE given by Equation 6.20 can be used instead of Equation 6.17.

The PE attains a maximum value when $P_k = P_{ks}$ or $d\delta_k/dt = 0$. For stable cases, $d\delta_k/dt = 0$ ($\delta_k$ reaches a maximum value) when PE attains the first maximum; for unstable cases, $P_k = P_{ks}$ when PE attains the first maximum. Hence, the system is unstable if $P_k$ decreases to $P_{ks}$ before $d\delta_k/dt$ becomes zero. The detection criterion requires $P_k$ and $\delta_k$. These two quantities can be obtained by local measurements at one end of a line. $\delta_k$ is obtained from the measurement of voltage and current at one end of a line with the knowledge of line impedance.

## Remarks

1. Unlike other techniques such as extended EAC and SIME, the proposed stability criterion does not require the knowledge of the MOI.
2. No assumptions are made regarding the power–angle relationship in the series elements forming a cutset.
3. In the case of loss of synchronism of a single generator, this may be detected by applying the instability criterion to the series element made up of the leakage reactance of the generator transformer. In this case, the cutset consists of only a single element representing the generator transformer. However, this is not general even if only one generator goes out of step.

### 6.4.2 Modification of the Instability Criterion

Testing the instability criterion to detect the loss of synchronism in two test systems (10 machine system and IEEE 17 machine system) has shown the efficacy of this criterion in on-line detection of instability in most cases. However, in a couple of cases, there were errors of two types:

a. False alarm: a critically stable case was judged as unstable

b. False dismissal: a critically unstable case was judged to be stable

It is to be noted that errors occurred in borderline cases. The reason for the discrepancy in the results of simulation and detection algorithm is primarily due to the assumption that the individual areas are strictly coherent—the transient variations of bus frequencies in an area are identical. In fact, there are intermachine oscillations within an area that result in perturbations in the power flows in the lines forming the critical cutset.

There is a need to slightly modify the instability criterion as follows:

a. To avoid false alarms, a minimum threshold value $\delta_{min}$ is chosen for $\delta_k$, so that the criterion for instability is applied only when $\delta_k > \delta_{min}$.

b. To avoid false dismissals, a value $\delta_{max}$ is chosen for $\delta_k$ such that if $\delta_k > \delta_{max}$, the line $k$ is to be included in the critical cutset even if the instability criterion indicates otherwise. Note that we bypass the instability criterion if $\delta_k > \delta_{max}$.

Based on the studies carried out on the test systems, $\delta_{min}$ is chosen between 50° and 60°. $\delta_{max}$ is typically chosen as a value greater than 180° and less than 360°.

To summarize, the modifications in the instability criterion are as follows:

1. A candidate line $k$ is judged to be a member of the critical cutset if $\delta_k > \delta_{min}$ *and* the instability criterion is satisfied.

2. A candidate line $k$ becomes a member of the critical cutset if either the above condition is satisfied *or* $\delta_k > \delta_{max}$.

## Remarks

1. Condition (2) is always met as the angles across each series element of the critical cutset always increase without any bound when the system becomes unstable.

2. The modification in the instability criterion can delay the detection of instability.

## 6.5 Algorithm for Identification of Critical Cutset

The critical cutset depends on the operating condition and the disturbance, and is not known beforehand. Therefore, the condition for instability ($P_k < P_{ks}$ when $(d\delta_k/dt) > 0$) is checked in all the lines across which the angle exceeds the threshold value $\delta_{\min}$. As soon as the instability criterion is satisfied in a candidate line, it is added to the set of lines that satisfy the instability criterion. Also, it is checked whether the lines in which the condition for instability is met up to the instant form a cutset. When the condition for instability is detected in a series element, the information is transmitted to the central computing station where the identification of the critical cutset is carried out. Note that this information is in the form of a logical data (true or false).

A graph-theoretic algorithm based on fusion of adjacent buses is used to determine whether a given set of lines form a cutset. The algorithm presented in this section is a modified version of the algorithm given in Deo (1974) to check the connectedness of a graph. The network connectivity information is stored in the form of the adjacency matrix. The adjacency matrix of a network with $n$ buses is a $n \times n$ symmetric binary matrix whose $ij$th element is 1 if there is a line connecting the $i$th and $j$th buses and 0 if there is no line between the $i$th and $j$th buses. If there are two or more parallel lines connecting two buses, then these lines are treated as a single line since the variations of angle and power in these lines are similar and hence the detection of instability in these lines occurs at the same instant.

The postfault network is assumed to be connected initially. Let A be the adjacency matrix of the postfault network and S be an empty set. Whenever instability is detected in a line connecting buses $i$ and $j$, this line is included in the set S and the connectivity information about this line is removed from the adjacency matrix $A$ by setting $A_{ij} = A_{ji} = 0$; then, it is checked whether a path exists between buses $i$ and $j$. This is accomplished by fusing all buses adjacent to bus $i$ (of the adjacency matrix $A'$) repeatedly (the process of "fusion" of two buses is explained later). All buses with numbers equal to the column numbers (or row numbers) of the elements of the $i$th row (or $i$th column), which have a value of 1, are adjacent to bus $i$. If a path exists between buses $i$ and $j$, then at some stage, bus $j$ is adjacent to bus $i$ and hence the set S is still not a cutset. When instability is detected in the last line belonging to the critical cutset, there exists no path between buses $i$ and $j$, and then, the set S is a cutset.

Whenever instability is detected in a line, matrix $A$ is updated and a new matrix $A'$ is defined. This is done because the fusion of buses reduces the number of buses and the original connectivity information is lost; the original connectivity information is again required when instability is detected in a new line. Therefore, the original connectivity information is preserved in matrix $A$ and the fusions are performed on matrix $A'$.

The fusion of the $k$th bus to the $i$th bus is accomplished by OR-ing, that is, logically adding the $k$th row to the $i$th row as well as the $k$th column to the $i$th column of matrix $A'$. In logical adding, $1 + 0 = 0 + 1 = 1 + 1 = 1$ and $0 + 0 = 0$. Then all the elements of the $k$th row and the $k$th column of matrix $A'$ are set to zero.

Whenever instability is detected in a line, the maximum number of fusions that may have to be performed in this algorithm, in order to check whether

S is a cutset or not, is $n - 1$, where $n$ is the number of buses. The maximum number of logical additions that may have to be performed is $[n(n + 1) - 2]/2$. Therefore, the upper bound on the execution time is proportional to $n^2/2$. The flowchart of the algorithm for the identification of the critical cutset is given in Padiyar and Krishna (2006).

## 6.6 Prediction of Instability

For faster detection of instability, the variations of $P_k$ and $\delta_k$ are predicted by fitting a polynomial curve to the sampled measurements. The sampling period $\tau$ is chosen as one cycle. The measurements separated by two cycles are used for curve fitting. The algorithm for the prediction of instability in a line is as follows:

1. If $\delta_k$ measured at the current sampling instant is less than that measured at the previous sampling instant, stability is indicated in the line. If $\Delta P_k$ $(P_k - P_{ks}) < 0$ and $\delta_k > \delta_{min}$, or $\delta_k$ measured at the current sampling instant is greater than $\delta_{max}$, instability is indicated in the line.

2. If $\delta_k$ measured at the current sampling instant is greater than $\delta_{min}$, a quadratic curve is fitted to the three sampled measurements of $\Delta P_k$ and a cubic curve is fitted to the four sampled measurements of $\delta_k$. The samples of $\delta_k$ are measured at the instants $t_s - 6\tau, t_s - 4\tau, t_s - 2\tau$, and $t_s$ $(t_s \geq t_{cl} + 6\tau)$, where $t_s$ is current sampling instant and $t_{cl}$ is the fault clearing time. $\Delta P_k$ is measured at the instant $t_s - 4\tau, t_s - 2\tau$, and $t_s$:

$$\Delta P_k = a_1 t^2 + b_1 t + c_1 \tag{6.25}$$

$$\delta_k = a_2 t^3 + b_2 t^2 + c_2 t + d_2 \tag{6.26}$$

3. The following two equations are solved for real positive values to obtain the instant $t_1$ at which $\Delta P_k = 0$ and the instant $t_2$ at which $d\delta_k/dt = 0$:

$$a_1 t_1^2 + b_1 t_1 + c_1 = 0 \tag{6.27}$$

$$3a_2 t_2^2 + 2b_2 t_2 + c_2 = 0 \tag{6.28}$$

4. If $t_1 < t_2$ and $(t_1 - t_s) < h$, or if Equation 6.27 has a real positive solution with $(t_1 - t_s) < h$, where $h$ is the time step (sampling period) and Equation 6.28 does not have a real positive solution, instability is

indicated in the line; otherwise a new set of measurements are obtained at the next sampling instant and the procedure from step 1 is repeated.

The procedure is stopped as soon as stability or instability is indicated in the line. It is to be noted that curve fitting and solution of the quadratic equations are required only if $\delta_k$ exceeds $\delta_{\min}$. The condition $(t_1 - t_s) < 1$ is used to limit the error due to extrapolation. As soon as instability is predicted in a line, the information is sent to the central computing station. System instability is predicted when the lines for which instability is predicted form a cutset.

## 6.7 Case Studies

### 6.7.1 Ten-Generator New England Test System

The system diagram and the data are given in Appendix D. The network losses are neglected and the loads are assumed to be of constant impedance type. The algorithm for the detection of instability is tested by simulating three-phase faults at different locations and the fault is cleared at an instant when it is critically unstable. The time step is chosen as 10 ms. Thus, for a fault at bus 14, the critical clearing time lies between 0.27 and 0.28 s. The system is critically stable when the clearing time $(t_{cl})$ is 0.27 s and critically unstable when $t_{cl} = 0.28$ s. $\delta_{\min}$ is chosen as 50° and $\delta_{\max} = 200°$.

The application of the instability criterion works in all the cases. For a line $k$ belonging to the critical cutset, $\Delta P_k = P_k - P_{ks}$ and $d\delta_k/dt$ are monitored or predicted for (a) critically stable and (b) critically unstable cases. It was observed that $d\delta_k/dt < 0$ before $\Delta P_k$ becomes zero (during the swing) for the critically stable case. On the other hand, $d\delta_k/dt < 0$ when $\Delta P_k = 0$ for the critically unstable case. Similar behavior is observed in all the cases (of faults considered).

The critical cutsets and angles across them are given in Padiyar and Krishna (2006) for all the cases considered. Here, the angles between the COIs of the two areas at the instants of detection and prediction are given in Table 6.1 for all the cases (with classical models of generators) considered when the fault is cleared without tripping a line. Similarly, Table 6.2 gives the results for the cases when a line is tripped (to clear the fault). The effect of generator modeling is also studied by considering the detailed modeling with AVR and the results are given in Table 6.3 for some cases.

It was observed that the elements of the critical outset satisfy the instability criterion at different times (Padiyar and Krishna 2006). Thus, the instant of instability detection or prediction refers to the instant when the last line belonging to the cutset satisfies the instability criterion. The algorithm for

**TABLE 6.1**

Results from Case Study of the 10-Generator System (Classical Model, Fault Cleared without Line Tripping)

| Fault Bus | Fault Clearing Time (s) | Instants of Instability Detection and Prediction (s) | | Elements Belonging to Critical Cutset | Angles between COI of Two Areas at the Instant of Instability (°) | |
|---|---|---|---|---|---|---|
| | | Detection | Prediction | | Detection | Prediction |
| 37 | 0.220 | 1.220 | 0.637 | 11–12,18–19 | 194.7 | 131.8 |
| 27 | 0.200 | 0.950 | 0.950 | 29–9 | 175.5 | 175.5 |
| 38 | 0.260 | 1.743 | 0.693 | 11–12,18–19 | 202.1 | 131.7 |
| 36 | 0.190 | 1.157 | 0.640 | 11–12,18–19 | 192.7 | 132.9 |
| 24 | 0.230 | 1.197 | 0.680 | 11–12,18–19 | 192.7 | 133.2 |
| 21 | 0.240 | 1.240 | 0.707 | 11–12,18–19 | 194.4 | 131.6 |
| 39 | 0.210 | 1.193 | 0.760 | 11–12,18–19 | 190.3 | 132.7 |
| 35 | 0.250 | 1.667 | 1.283 | 11–12,18–19 | 201.7 | 158.0 |
| 34 | 0.280 | 1.313 | 0.813 | 11–12,18–19 | 201.7 | 145.6 |
| 33 | 0.270 | 1.653 | 1.020 | 11–12,18–19 | 197.4 | 147.6 |
| 28 | 0.150 | 0.700 | 0.700 | 29–9 | 160.6 | 160.6 |
| 29 | 0.130 | 0.697 | 0.697 | 29–9 | 162.8 | 162.8 |
| 26 | 0.140 | 1.023 | 1.023 | 29–9 | 165.8 | 165.8 |
| 23 | 0.230 | 1.230 | 0.763 | 11–12,18–19 | 192.5 | 132.7 |
| 22 | 0.220 | 1.620 | 0.787 | 11–12,18–19 | 204.5 | 130.1 |
| 20 | 0.250 | 1.417 | 1.017 | 11–12,18–19 | 203.7 | 150.9 |
| 31 | 0.270 | 1.153 | 0.737 | 11–12,18–19 | 199.5 | 130.6 |
| 18 | 0.300 | 1.400 | 0.933 | 11–12,18–19 | 206.7 | 147.6 |
| 17 | 0.310 | 1.243 | 0.860 | 11–12,18–19 | 202.4 | 144.0 |
| 15 | 0.260 | 1.460 | 0.943 | 11–12,18–19 | 203.8 | 148.4 |
| 14 | 0.280 | 1.247 | 0.797 | 11–12,18–19 | 202.1 | 146.0 |
| 13 | 0.270 | 1.320 | 0.670 | 11–12,18–19 | 191.8 | 134.1 |
| 12 | 0.230 | 1.847 | 1.147 | 11–12,18–19 | 414.6 | 148.2 |

detection/prediction of instability gave correct results in all the cases studied. The average time taken for the detection of instability (measured from the instant of fault clearing) is 1.004 s for detection and 0.569 s (using prediction) for the 10-generator system. The average value of the angles across the critical lines is 159.2° for detection and 54.7° for prediction. Thus, there is a significant improvement (in speeding up the process) by predicting the angles across the elements of the critical cutset.

For fault at bus 12 cleared without line tripping and fault at bus 37 cleared by tripping the line 37–27 (with classical models of the generators), there is false dismissal using only the criterion based on energy function. In both these cases, the instability is detected when $\delta_k > \delta_{max}$. However, for both cases, the prediction algorithm indicates instability at an earlier instant as there is no error of false dismissal with prediction (Padiyar and Krishna 2006).

**TABLE 6.2**

Results from Case Study of the 10-Generator System (Classical Model,
Fault Cleared by Line Tripping)

| Fault Bus | Line Cleared | Fault Clearing Time (s) | Instants of Instability Detection and Prediction (s) | | Elements Belonging to Critical Cutset | Angles between COI of Two Areas at the Instant of Instability (°) | |
|---|---|---|---|---|---|---|---|
| | | | Detection | Prediction | | Detection | Prediction |
| 37 | 37–27 | 0.210 | 2.293 | 0.693 | 11–12,18–19 | 366.1 | 130.6 |
| 27 | 37–27 | 0.170 | 1.137 | 1.120 | 26–29,26–28 | 164.0 | 160.2 |
| 38 | 37–38 | 0.250 | 1.350 | 0.833 | 11–12,18–19 | 191.9 | 141.7 |
| 36 | 36–24 | 0.190 | 1.107 | 0.657 | 11–12,18–19 | 192.9 | 135.8 |
| 24 | 36–24 | 0.220 | 1.603 | 0.737 | 11–12,18–19 | 202.2 | 132.2 |
| 21 | 36–21 | 0.200 | 1.417 | 0.983 | 11–12,18–19 | 199.7 | 140.2 |
| 34 | 34–35 | 0.270 | 1.337 | 0.670 | 11–12,18–19 | 196.7 | 126.3 |
| 33 | 33–34 | 0.240 | 0.957 | 0.957 | 20–3 | 165.7 | 165.7 |
| 28 | 28–29 | 0.030 | 1.330 | 1.047 | 26–29 | 123.8 | 106.2 |
| 29 | 28–29 | 0.030 | 1.163 | 0.913 | 26–29 | 124.3 | 105.9 |
| 26 | 26–29 | 0.070 | 1.037 | 1.003 | 26–28 | 171.4 | 142.3 |
| 23 | 23–24 | 0.200 | 1.400 | 0.900 | 11–12,18–19 | 189.8 | 135.0 |
| 22 | 22–23 | 0.220 | 1.520 | 0.787 | 11–12,18–19 | 200.4 | 130.9 |
| 20 | 20–33 | 0.230 | 0.713 | 0.713 | 20–3 | 154.9 | 154.9 |
| 31 | 20–31 | 0.260 | 1.243 | 0.810 | 11–12,18–19 | 203.8 | 135.1 |
| 18 | 18–19 | 0.250 | 1.967 | 1.283 | 11–12 | 190.3 | 152.0 |
| 17 | 17–18 | 0.300 | 1.517 | 0.967 | 11–12,18–19 | 203.2 | 146.3 |
| 15 | 15–18 | 0.260 | 1.193 | 0.777 | 11–12,18–19 | 205.9 | 140.3 |

**TABLE 6.3**

Results from Case Study of the 10-Generator System (with Detailed Generator Models)

| Fault Bus | Line Cleared | Clearing Time (s) | Instants of Instability Detection and Prediction (s) | | Elements Belonging to Critical Cutset | Angles between COI of Two Areas at the Instant of Instability (°) | |
|---|---|---|---|---|---|---|---|
| | | | Detection | Prediction | | Detection | Prediction |
| 27 | — | 0.300 | 0.817 | 0.583 | 26–29,26–28 | 212.8 | 181.6 |
| 36 | — | 0.250 | 0.900 | 0.517 | 11–12,18–19 | 199.9 | 150.3 |
| 24 | — | 0.310 | 0.977 | 0.560 | 11–12,18–19 | 198.3 | 150.6 |
| 21 | — | 0.310 | 1.110 | 0.610 | 11–12,18–19 | 195.1 | 149.6 |
| 28 | — | 0.200 | 0.567 | 0.467 | 26–29,26–28 | 185.7 | 162.0 |
| 14 | 14–34 | 0.350 | 1.117 | 0.533 | 11–12,18–19 | 221.3 | 143.2 |
| 37 | 37–27 | 0.300 | 1.300 | 0.517 | 11–12,18–19 | 215.0 | 148.6 |
| 26 | 26–29 | 0.150 | 0.767 | 0.467 | 26–28 | 178.4 | 141.6 |
| 27 | 37–27 | 0.280 | 0.913 | 0.713 | 26–29,26–28 | 202.0 | 184.1 |
| 22 | 22–23 | 0.270 | 1.053 | 0.670 | 11–12,18–19 | 198.4 | 152.7 |

When a fault is cleared by tripping a line, $P_{ks}$ (steady-state value in the post-fault configuration) is different from $P_{k0}$ (steady-state value in the prefault configuration). However, since variation in $P_k$ is large compared to $P_{k0}$ or $P_{ks}$, $P_{k0}$ may be used instead of $P_{ks}$ to simplify the computation.

### 6.7.2 Seventeen-Generator IEEE Test System

Studies are also carried out on the 17-generator IEEE test system. Only classical models of generators are considered. The results of the study are shown in Table 6.4 for four cases of three-phase faults at buses 112, 120, 72, and 129 followed by tripping of a line. For the first three cases, the critical cutset consists of only one element. However, in the last case (for the fault at bus 129), there are 19 elements in the critical cutset as mentioned earlier. The critical cutset and the angles for the last case are shown in Table 6.5.

**TABLE 6.4**

Results from Case Study of the IEEE 17-Generator System (Classical Model)

| Fault Bus | Line Cleared | Fault Clearing Time (s) | Instants of Instability Detection and Prediction (s) | | Elements Belonging to Critical Cutset | Angles between COI of Two Areas at the Instant of Instability (°) | |
|---|---|---|---|---|---|---|---|
| | | | Detection | Prediction | | Detection | Prediction |
| 112 | 4–112 | 0.240 | 0.423 | 0.340 | 112–121 | 182.1 | 155.4 |
| 120 | 5–120 | 0.260 | 0.443 | 0.360 | 112–121 | 179.3 | 153.4 |
| 72 | 14–72 | 0.450 | 0.650 | 0.583 | 110–114 | 193.0 | 177.6 |
| 129 | 5–129 | 0.310 | 1.077 | 0.993 | See Table 6.5 | 285.8 | 226.0 |

**TABLE 6.5**

Critical Cutset and Angles for Fault at Bus 129

| | Angles at the Instant of Instability (°) | |
|---|---|---|
| Cutset Elements | Detection | Prediction |
| 1–3, 111–115 | 107.7, 84.5 | 55.5, 55.4 |
| 2–13, 144–146 | 113.9, 64.3 | 64.6, 64.3 |
| 4–112, 8–10 | 71.6, 163.3 | 57.4, 57.0 |
| 8–13, 8–15 | 324.6, 218.6 | 60.9, 55.1 |
| 12–13, 25–26 | 327.9, 132.5 | 64.6, 55.4 |
| 26–74, 51–141 | 57.6, 57.5 | 57.6, 57.5 |
| 53–55, 55–57 | 84.0, 57.8 | 84.0, 57.8 |
| 71–85, 1–4 | 55.1, 55.8 | 55.1, 55.8 |
| 143–144, 3–14 | 58.9, 180.2 | 58.9, 55.1 |
| 20–53 | 351.0 | 125.2 |

The average time taken for the detection and prediction of instability is 0.333 s and 0.254 s, respectively. The average value of the angles across the series elements is 128.0° for detection and 63.2° with prediction. $\delta_{min}$ was selected as 55°.

### 6.7.3 Discussion

Unlike other techniques, the proposed instability criterion does not require the knowledge of the MOI and no assumptions are made regarding the power–angle relationship.

The only assumption made is that the system instability results in the initial separation into two areas. This assumption is used implicitly in all methods proposed earlier.

The main objective of the proposed instability detection criterion is system protection as distinct from equipment protection. If instability is detected, corrective actions need to be taken in order to maintain system integrity. A major feature of the detection criterion proposed here is that it is based on local measurements (of current and voltage) within a line. Only logical data need to be transmitted to the central location where it is processed to check for system instability. This is different from the techniques based on the use of phasor measurements and their telemetering (Centeno et al. 1997, Counan et al. 1993). The computational complexity in the detection of instability can be reduced from simplification of the system by identifying coherent groups and the use of dynamic equivalents from off-line studies (Counan et al. 1993).

For a given MOI, there are many possible cutsets connecting the two separating areas, but there is usually a unique cutset across which the angles become unbounded in case of instability. For the 10-generator system, there are few critical cutsets. For many contingencies, generator 2 separates from the rest of the system, since its inertia constant is high compared to other generators. For this MOI, the critical cutset consists of lines 18–19 and 11–12.

The detection criterion derived from the energy function analysis is based on the assumptions of coherent areas and constant power loads. The coherency assumption is made to neglect the fast oscillations within the areas and account only for interarea (slow) oscillations, which contribute to system separation. The detection criterion is effective even when the loads are modeled as constant impedances.

The main requirements of a method to detect instability are accuracy (no false alarms and false dismissals) and speed. The proposed method is capable of distinguishing between stable and unstable swings accurately. The stability criterion is accurate even for critically stable and critically unstable cases, which are considered in the case studies. Therefore, it is expected that the criterion is able to perform accurate stability classification for more stable or unstable cases. The detection is sped up by the extrapolation of system trajectories.

The MOI is the same for both classical and detailed models for all the cases studied. For a fault at buses 27 and 28 of the 10-generator system, the critical cutset is different for classical and detailed models of the generator.

The values of $\delta_{min}$ and $\delta_{max}$ are decided based on simulation studies. These values are the same for all the cases for a given system and the actual values used are not very critical. The detection/prediction procedure starts as soon as the angle across a line exceeds $\delta_{min}$. In most cases, instability is predicted as soon as the angle exceeds $\delta_{min}$. The method given in Counan et al. (1993) involves only checking whether the angle between the two areas exceeds $\delta_{max}$, whose value is not specified.

Transient instability may lead to uncontrolled tripping of generators and lines, and may finally result in the formation of islands. Controlled system separation is used as a protective measure in order to retain intact as much of the system as possible. The proposed instability detection method can be used to initiate this emergency control measure by tripping the elements of the critical cutset at the instant of instability detection/prediction. This is discussed in Section 6.9.

## 6.8 Study of a Practical System

Additional studies were carried out on the application of the novel algorithm for the detection of instability to a practical system adapted from a regional electricity board in India. The original system consists of 193 generators and 493 buses. For simplicity, the system is reduced by replacing the generators in a power station by an equivalent generator. Thus, the original system is represented by a reduced system, which has 55 generators, 355 buses, 587 transmission lines, 83 interconnecting transformers, and a total load of 22,676 MW.

All the generators are represented by classical models. The loads are modeled as constant impedances. The system is assumed to be lossless by neglecting mechanical damping and transmission line resistances. In total, 131 cases are studied for on-line detection of loss of synchronism. The results obtained are (a) the critical cutset, (b) the angle across each element of the critical cutset at the time of (i) detection and (ii) prediction of instability, (c) the time taken for both detection and prediction, and (d) the MOI. The elements of the critical cutset are conveniently labeled by their terminal buses. If there are more than one line (or series element) connecting two buses, all the parallel lines are clubbed together. It is to be noted that once the critical cutset is determined, MOI is easily obtained from a graph algorithm. MOI is displayed as the advanced generators that separate from the rest.

Table 6.6 shows results of four sample cases (out of 131). In two cases (cases 1 and 4), only one generator separates from the rest with a single member of the critical cutset, which happens to be the generator transformer. In the

**TABLE 6.6**

Sample Results for the Practical System

| Case Number | Faulted Bus | Line Cleared | Clearing Time (s) | Critical Cutset | Angle across Critical Cutset (°) | | Mode of Instability Group A | Instability Detection Instant (s) | |
|---|---|---|---|---|---|---|---|---|---|
| | | | | | Detect | Predict | | Detection | Prediction |
| 1 | 137 | 137–138 | 0.28 | 38–137 | 70.32 | 51.72 | 38 | 0.72 | 0.54 |
| 2 | 203 | 203–344 | 0.20 | 95–348 | 55.10 | 55.10 | 4,5,6,7,8,9, | 1.00 | 1.00 |
| | | | | 97–107 | 55.19 | 55.19 | 29,30,31, | | |
| | | | | 186–285 | 55.12 | 55.12 | 32,33,34, | | |
| | | | | 86–91 | 57.15 | 57.15 | 35,36,40, | | |
| | | | | 88–89 | 55.14 | 55.14 | 41,42,45, | | |
| | | | | 89–129 | 59.11 | 59.11 | 47,48,49, | | |
| | | | | 65–101 | 64.03 | 64.03 | 50,55 | | |
| 3 | 354 | 354–344 | 0.20 | 95–348 | 52.34 | 52.34 | 4,5,6,7,8,9, | 1.22 | 1.14 |
| | | | | 97–107 | 51.86 | 51.86 | 29,30,31, | | |
| | | | | 186–285 | 60.90 | 52.23 | 32,33,34, | | |
| | | | | 86–91 | 55.70 | 50.74 | 35,36,40, | | |
| | | | | 88–89 | 50.24 | 50.24 | 41,42,45, | | |
| | | | | 89–129 | 55.48 | 55.48 | 47,48,49, | | |
| | | | | 65–106 | 290.56 | 59.44 | 50,55 | | |
| 4 | 199 | 126–199 | 0.31 | 6–199 | 107.69 | 69.60 | 6 | 0.51 | 0.43 |

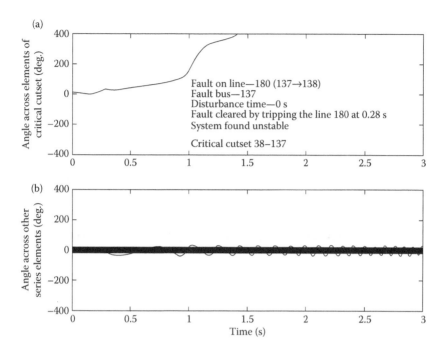

**FIGURE 6.5**
Angles across the series elements. (a) The critical cutset. (b) Other than the critical cutset (case 1).

second case, 23 generators separate from the rest with seven elements in the cutset. Figures 6.5 and 6.6 show the angles across (i) elements of the critical cutset and (ii) rest of the series elements for cases 1 and 2, respectively. For case 1, there are only two groups of generator that separate from each other (one group containing only one generator). For case 2, after the initial separation, there are two more separations that result in four groups of generators that separate from each other. This is displayed in the swing curves shown in Figure 6.7. The process of separation into four groups is shown in Figure 6.8. The system $S_0$ separates into two groups $S_1$ and $S_2'$. Subsequently, $S_2'$ splits into $S_2$ and $S_3'$. Finally, $S_3'$ separates into $S_3$ and $S_4$. It should be possible to extend the algorithm for on-line detection for the sequential separation into more than two groups.

## 6.8.1 Discussion

As with the previous case studies of the two test systems, the results are very encouraging. In all the 131 cases, the algorithm predicted correctly the initial MOI separation of the system into two groups of generators. Further refinements in the algorithm should attend to the problem of handling the inter-machine oscillations within a group of generators even when they remain in synchronism. The errors of false alarm and false dismissal introduced by

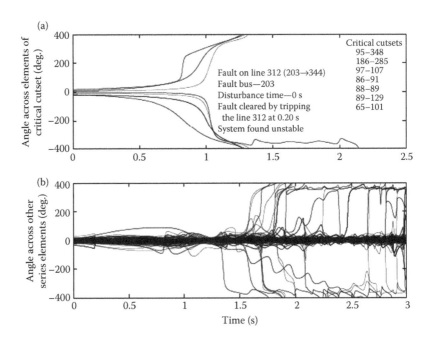

FIGURE 6.6

Angles across the series elements. (a) The critical cutset. (b) Other than the critical cutset (case 2).

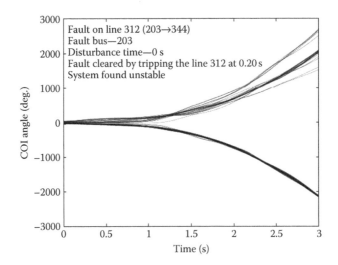

FIGURE 6.7

Swing curves for case 2.

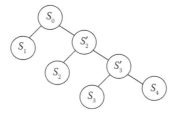

**FIGURE 6.8**
Process of system separation into four groups following instability.

these oscillations can perhaps also be tackled by applying signal process-
ing (filtering) techniques to the measured variables of $d\delta_k/dt$ (or $\delta_k$) and $\Delta P_k$.
However, this is not investigated here. The practical implementation includ-
ing refinement of the algorithm involves further developments.

The algorithm can be readily applied for off-line studies. In most of the
cases, it was observed that the prediction of instability was possible within
1 s. This is comparable to other approaches for detecting instability using
PEBS and controlling UEP method (that may require simulation of the system for about 1 s). Further, the algorithm gives accurate results on MOI and
critical cutset. The critical energy that leads to separation can be computed
based on the critical cutset. Ignoring the intermachine oscillations within
each group can facilitate fast computation of energy in the critical cutset. Out
of the 131 cases considered, there are 83 cases in which the initial instability
is due to loss of synchronism of a single generator. For such cases, the algorithm gives accurate results within a fraction of a second.

It is interesting to observe that some lines appear frequently in the critical
cutset. Table 6.7 shows such lines and the frequency of their occurrence. The
first seven lines appear to be very significant in the loss of system stability.
If FACTS controllers are to be deployed to prevent loss of synchronism, these
lines are the natural choice for the location of FACTS. Table 6.8 shows the frequency of occurrence when only a single generator goes out of step. Generator
15 appears to be very critical followed by generators 45, 55, 22, 5, and 36.

## 6.9 Adaptive System Protection

"The conventional relaying approach for detecting loss of synchronism is
by analyzing the variation in the apparent impedance as viewed at a line or
generator terminals. Following a disturbance, this impedance will vary as a
function of the system voltages and the angular separation between the systems. Out of step, pole slip or just loss of synchronism are equivalent terms
for the condition when the impedance locus travels through the generator.

**TABLE 6.7**

Frequency of Occurrence of Lines in the
Critical Cutset (Practical System)

| Sl. No. | Lines | | Frequency of Occurrence |
| | From Bus | To Bus | |
|---|---|---|---|
| 1 | 86 | 91 | 47 |
| 2 | 88 | 89 | 45 |
| 3 | 89 | 129 | 44 |
| 4 | 95 | 348 | 43 |
| 5 | 186 | 185 | 42 |
| 6 | 97 | 107 | 34 |
| 7 | 65 | 106 | 28 |
| 8 | 107 | 108 | 9 |
| 9 | 121 | 107 | 9 |
| 10 | 105 | 111 | 9 |
| 11 | 84 | 136 | 6 |
| 12 | 85 | 87 | 6 |
| 13 | 75 | 86 | 6 |
| 14 | 65 | 101 | 5 |

**TABLE 6.8**

Frequency of Occurrence of Loss of Synchronism
of a Single Generator (Practical System)

| Sl. No. | Generator Going Out of Step | Frequency of Occurrence |
|---|---|---|
| 1 | 15 | 20 |
| 2 | 45 | 7 |
| 3 | 55 | 7 |
| 4 | 22 | 6 |
| 5 | 5 | 5 |
| 6 | 36 | 5 |

When the impedance goes through the transmission line, the phenomenon is also known as power swing. However all of them refer to the same event: loss of synchronism" (Begovic et al. 2005).

Conventional out-of-step relays are located in some selected transmission lines and are based on impedance measurement. There are other quantities that are measured for out-of-step relaying described in the literature. In Taylor et al. (1983), the rate of change of resistance in addition to resistance is measured for out-of-step relaying. In Ohura et al. (1990), the phase angle difference between substations is measured in order to detect out-of-step condition.

In Marioka et al. (1993), the angular velocity of the generator is used for out-of-step relaying. EAC and its variations have been used to predict whether a swing is stable or not. This has been applied in Florida–Georgia interconnection to provide adaptive protection (Centeno et al. 1997). However, this approach has several drawbacks. Apart from modeling errors, there are problems of incorporating system dynamics. An improvement on EAC is EEAC (applied for direct stability evaluation) (Pavella and Murthy 1994). However, the initial development of EEAC considered only static OMIB whose parameters are not time varying. The development of dynamic OMIB improved the results, but the problem of ranking of critical machines (required as the first step in EEAC) is quite complicated and affects the accuracy.

Controlled system separation is used as a protective measure in order to retain intact as much of the system as possible. For a given MOI, there are many possible cutsets along which the system can be separated into two islands. The cutset should be selected such that the difference between generation and load within each island is minimum. Initially, the system can be separated into two islands by tripping the lines belonging to the critical cutset, when instability is detected.

For the fault at bus 20 (of the 10-generator system) cleared by tripping the line 20–33 at 0.23 s (critically unstable case), generator 3 initially separates from the rest of the system; instability is first detected in the transformer 20–3, which forms the critical cutset. Controlled system separation is simulated by tripping generator 3. The group containing nine generators when the system is separated at the instant of instability detection/prediction (0.713 s) remains stable (Padiyar and Krishna 2006).

For fault at bus 38 (of 10-generator system) cleared by tripping the line 37–38 at 0.25 s (critically unstable clearing time), generator 2 separates from the rest of the system (see Figure 6.9). There are two lines belonging to the critical cutset (11–12 and 18–19). Controlled system separation is simulated by tripping these lines at the instant of instability detection (1.350 s) and it was observed that the system (containing the nine generators) is stabilized (see Figure 6.10).

For the fault at bus 112 (of 17-generator system) cleared by tripping the line 4–112 at 0.24 s (critically unstable clearing time) generator at bus 121 initially separates from the rest of the system. The transformer 112–121 forms the critical cutset. By tripping the generator at bus 121, at the instant of instability prediction (0.340 s), the system is stabilized (see Figure 6.11). The system separation can also be done at the instant of instability detection (0.423 s). However, if prompt action is not taken in stabilizing the system by tripping the out-of-step generator, the system will be split into several groups (see Figure 6.12). In general, it was observed that if a single generator going out of step is disconnected at the time of detection/prediction, the system remains stable; otherwise, the system may split into several parts (uncontrolled islands).

Figure 6.2 shows the swing curves for fault at bus 129 (of 17-generator system) cleared by tripping the line 5–129 at 0.31 s (critically unstable clearing time).

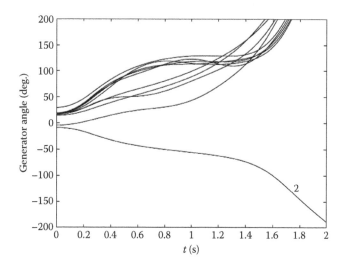

**FIGURE 6.9**
Swing curves for fault at bus 38 (10-generator system).

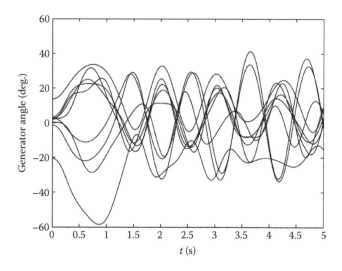

**FIGURE 6.10**
Swing curves for the nine generators following system separation.

Seven generators separate from the rest of the system. There are 19 lines belonging to the critical cutset. Controlled system separation is simulated by tripping these lines. Figures 6.13 and 6.14 show the plot of swing curves of the two groups (A and B) when the system is separated at the instant of instability prediction (0.933 s); the generator angles are with respect to COI of the respective

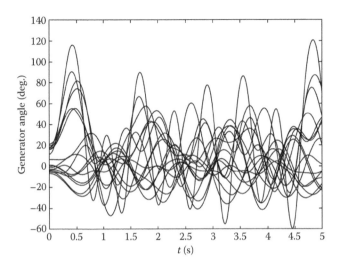

**FIGURE 6.11**
Swing curves for the 16 generators following tripping the generator at bus 121 (17-generator system).

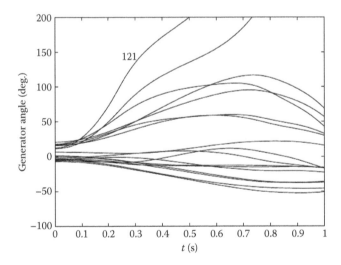

**FIGURE 6.12**
Swing curves for fault at bus 112 cleared by tripping line 4–112 (17-generator system).

groups. The two groups become unstable with generator at bus 114 stepping out in group A and generator at bus 118 stepping out in group B. Further system separation (by tripping the generators at buses 114 and 118 at the instants of instability prediction) stabilizes the two groups as shown in Figures 6.15 and 6.16. The instant of instability prediction is 1.593 s for group A and 1.810 s for group B.

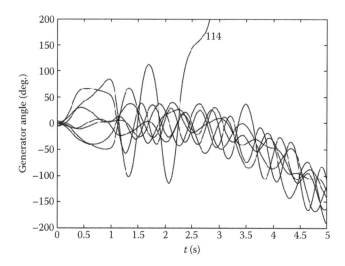

**FIGURE 6.13**
Swing curves for the seven advanced generators (group A) following controlled system separation (17-generator system).

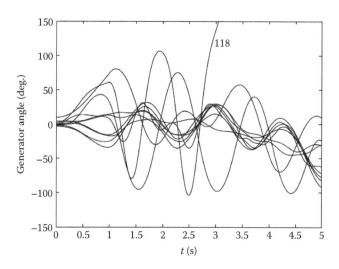

**FIGURE 6.14**
Swing curves for the remaining 10 generators (group B) following controlled system separation (17-generator system).

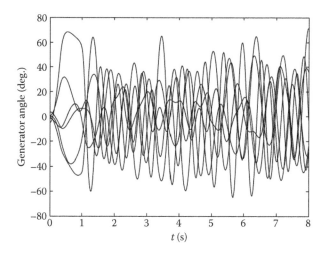

**FIGURE 6.15**
Swing curves of group A generators following tripping of generator at bus 114.

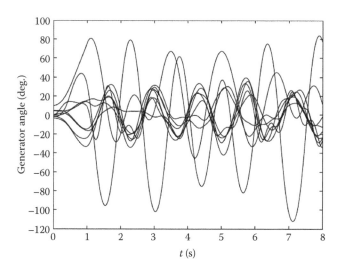

**FIGURE 6.16**
Swing curves of group B generators after tripping generator at bus 118.

### 6.9.1 Discussion

The main objective of the novel instability detection criterion is system protection as distinct from equipment protection. If instability is detected, corrective actions need to be taken in order to maintain system integrity. The instability detection criterion can also be used to initiate control, which is discussed in Chapters 8 and 9.

Most of the stability detection techniques reported in the literature involve global phasor measurements. The adaptive out-of-step relaying requires phasor measurements for all the buses at the interconnection (Centeno et al. 1997). The relaying scheme also requires the knowledge of the equivalent system parameters that have to be reasonably accurate, particularly with changing system conditions.

The proposed method of instability detection is in contrast to the protection system developed for Electricite de France (Counan et al. 1993), where the system is divided into many areas and the generators in any area form a coherent group. The boundaries separating the areas are predetermined. Instability is determined by comparison of the phase angle measurements from different areas.

# 7

## Sensitivity Analysis for Dynamic Security and Preventive Control Using Damping Controllers Based on FACTS

### 7.1 Introduction

In this chapter, we will review the sensitivity analysis techniques applied for DSA and preventive control. The power system security is defined as "the ability of the bulk power system to withstand sudden disturbances such as electric short circuits or unanticipated loss of system components" (Fouad 1988). This implies that following the occurrence of a sudden disturbance, (i) the power system will survive the ensuing transient and transit to an acceptable steady-state condition, and (ii) in this new steady-state condition, all power system components operate within established limits.

The second requirement is checked by static security assessment, based on the analysis of algebraic equations and constraints that apply in steady state. The first requirement is tested from transient analysis, which is the basis of DSA. On-line DSA is not performed; off-line studies are carried out for different operating conditions and system configurations, for a specified set (or sequence) of contingencies (or events that could threaten dynamic security). From these studies, secure operating conditions are determined, for example, loading of a generator and power flow at a critical transmission interface.

Transient energy function methods have been applied for DSA. By defining energy margin (EM) for specific disturbances, it is possible to determine the limits on interface flows or generator outputs, based on sensitivity analysis. It is possible to devise preventive control strategies to ensure that the system continues to remain in "normal" and "secure" state. This will prevent the transition to emergency state with a credible list of contingencies.

The concept of preventive control (based on stability constraints) discussed in the existing literature is premised on the first swing transient stability. However, the stability of the equilibrium point during normal operating conditions is more important. The loss of small signal stability can lead to growing oscillations that will finally result in loss of synchronism or voltage collapse. For a prescribed range of normal operating conditions, the system must remain stable even if major disturbances do not occur. By applying FACTS controllers, we will derive the general control strategies to provide adequate damping of small oscillations that are caused by small, random perturbations in the load. The control strategies are based on linearized system models as the damping controllers are designed essentially to damp small oscillations by modulating the network parameters.

## 7.2 Basic Concepts in Sensitivity Analysis

Consider the DAE that describe a power system for stability analysis given by

$$\dot{x} = f(x, y, p) \tag{7.1}$$

$$0 = g(x, y, p) \tag{7.2}$$

where $x$ is the $n$-dimensional state vector, $y$ is the $n_a$-dimensional vector of algebraic variables (typically voltage phasors expressed in polar or rectangular coordinates), and $p$ is the $r$-dimensional vector of parameters that specify operating variables (e.g., loads or generator outputs). $f$ and $g$ are the $n$- and $n_a$-dimensional vectors. If the Jacobian $[\partial g/\partial y]$ is nonsingular, then, according to implicit function theorem, there exists a locally unique and smooth function $F$ such that

$$\dot{x} = F(x, p) \tag{7.3}$$

Note that Equation 7.3 is obtained by eliminating the algebraic variable $y$ in Equation 7.1 by expressing $y$ (by solving Equation 7.2) as

$$y = h(x, p) \tag{7.4}$$

Pavella et al. (2000) define three types of sensitivity analysis

i. Analysis of the linearized system
ii. Analysis of supplementary motion for study of parameter variations on the system trajectory
iii. Analysis based on performance indices rather than system variables

The second type of sensitivity analysis is also termed as trajectory sensitivity (Laufenberg and Pai 1998, Tomovic 1963). The EM is a performance index used for DSA and its sensitivity analysis can be used for preventive control (to prevent loss of first swing transient stability).

We will first review the application of EM and trajectory sensitivities for DSA and preventive control. The application of FACTS controllers for damping oscillations will be taken up next. The control strategies for the series and shunt FACTS controllers are developed using energy concepts based on linearized models.

## 7.3 Dynamic Security Assessment Based on Energy Margin

### 7.3.1 Transient Energy Margin

The transient EM, $\Delta W$, is defined as the difference between the critical energy and the energy at the fault clearing time. Thus

$$\Delta W = W_{cr} - W_{cl} \tag{7.5}$$

where $W_{cr}$ and $W_{cl}$ are the critical energy and the energy at the fault clearing time, respectively. If $\Delta W > 0$, then the system is stable; $\Delta W < 0$ implies that the system is unstable. $\Delta W = 0$ implies a critically stable system.

An alternative measure of transient stability is the ratio of PE margin to the KE at the fault clearing time. Thus

$$\Delta W_n = \frac{W_{cr} - W_{2cl}}{W_{1cl}} \tag{7.6}$$

where $\Delta W_n$ is the normalized EM (Fouad and Vittal 1983, Padiyar and Ghosh 1989). $W_1$ and $W_2$ refer to the kinetic and potential energy, respectively. If $\Delta W_n > 1$, the system is stable; if $\Delta W_n < 1$, the system is unstable.

It is to be noted that $W_{cr}$ is dependent on the fault or disturbance in addition to the system characteristics. When we consider different contingencies, those with the lowest values of $\Delta W_n$ constitute the most severe disturbances.

### 7.3.2 Computation of Energy Margin

The computation of EM involves the estimation of the critical energy. In the literature, there are two approaches to the computation of $W_{cr}$. These are (a) CUEP and (b) the PEBS method. Unfortunately, both approaches have limitations (Maria et al. 1990, Vittal et al. 1988).

In the first approach, the problems are (i) determination of MOI and (ii) convergence problems in numerical computation of UEP, particularly under the stressed system conditions. In the second approach (also called as the Kyoto approach), the computational complexity is reduced as the critical energy based on the sustained fault trajectory can be computed easily. The method is reasonably successful in computing critical clearing times; however, when the MOI for the sustained fault trajectory is different from the MOI for the critically unstable case, the method gives erroneous results (see case 6 for the 10-generator system presented in Chapter 4). Maria et al. (1990) present a hybrid approach that combines the desirable features of both time domain simulation and transient energy function method. The criterion given by Athay et al. (1979) is used to detect PEBS crossing. The system is simulated for the postfault condition after the fault is cleared. At each step of numerical integration of the system dynamic equations, it is checked whether the PE reaches a maximum or the projection of the system trajectory in the angle space crosses PEBS. If PEBS crossing is detected, the system is judged to be unstable and the PE at the PEBS crossing is taken as the critical energy. If the PE reaches a maximum value, the system is stable. The maximum PE along the line joining the point (in the angle space) and the SEP is taken as the critical energy. Liu et al. (1996) extend the hybrid transient stability analysis for the SPM.

Chiang et al. (1988, 1994) define a gradient system based on the following equation:

$$M_i \frac{d\omega_i}{dt} = P_{mi} - P_{ei} = f_i(\delta_1, \delta_2, ..., \delta_m), \quad i = 1, 2, ..., m \tag{7.7}$$

It is assumed that there is an infinite bus in the system so that there is no need to express the rotor angle with reference to COA ($\delta_{COI}$). In the absence of an infinite bus, Equation 7.6 needs to be replaced by the following equation:

$$M_i \dot{\bar{\omega}}_i = P_{mi} - P_{ei} - \frac{M_i}{M_T} P_{COI} = f(\theta_1, \theta_2, ..., \theta_m) \tag{7.8}$$

where $\theta_i = \delta_i - \delta_{COI}$, $\bar{\omega}_i = \omega_i - \omega_0$, and $\omega_0 = d\delta_{COI}/dt$.

Note that $\theta_m$ can be expressed in terms of $(m-1)$ rotor angles referred to COI. If $W_{PE}$ is referred with respect to the postfault SEP, PEBS is defined to be the surface formed by a continuous set of all first maxima along the rays emanating from $\theta_S$ (postfault SEP). The PEBS is characterized by the equation obtained from equating the directional derivative of $W_{PE}$ to zero (Pai 1981, Sauer et al. 1983)

$$(\nabla W_{PE})^t \theta = 0 \tag{7.9}$$

where $\nabla$ denotes the gradient with respect to the $\theta$ vector.

Note that the *i*th component of the gradient vector is $-f_i$. Also note that $f_i$ refers to the postfault system (however, the PE is evaluated along the sustained fault trajectory). The computations are simplified if the fault-on trajectory defined by the angles during the fault is approximated by

$$\theta_i = \alpha_i + \beta_i t^2, \quad \alpha_i = \theta_{i0}, \quad \beta_i = \frac{f_i^F}{2M_i}(t_0^+)$$

where $f_i^F(t_0^+)$ refers to the value of $f_i$ (for the faulted system) at the instant soon after the fault has occurred at $t = t_0$.

### 7.3.2.1 Evaluation of Path-Dependent Integrals

Conceptually, the region of stability surrounds the postfault SEP ($\theta_s$), which is taken as a reference point. There is no loss of generality if $\theta_s$ is assumed to be zero. The computation of path-dependent integrals in $W_{PE}$ is normally carried out by assuming linear paths (from $\theta_s$). However, this can result in significant error, particularly in evaluating SPEF and hence, it is desirable to compute the path-dependent integrals by using trapezoidal rule of integration, starting from $\theta_o$ (the machine angles at $t = t_0$) (Pai et al. 1993).

### Remarks

1. The computation of $W_{cr}$ described here assumes classical models of generators and constant-impedance-type loads. With these assumptions, it is possible to compute the bus voltages $V_i \angle \phi_i$ ($i = 1,2,...,n$) in terms of $\theta_i$ ($i = 1,2,...,m$). Hence, the PEBS method (as described above) also applies for SPEF.

   As a matter of fact, it is shown in Chapter 4 that

$$W_{PE} = \sum_{i=1}^{m} \int_{\delta_{i0}}^{\delta} (P_{ei} - P_{mi})d\theta_i$$

   is applicable even for detailed generator and load models.

2. The KE at the clearing time has to be corrected to account accurately for the component that leads to the system separation. This requires the knowledge of MOI (which is also important for computing $W_{cr}$).

### 7.3.2.2 Computation of Energy Margin Based on Critical Cutsets

It was shown in Chapter 6 that there is a unique cutset for a specified disturbance in a power system along which the system separates into two groups. Thus, it should be possible to compute $W_{cr}$ by the following equation:

$$W_{cr} = \sum_{k=1}^{nc} \int_{\phi_{ks}}^{\phi_{ku}} (P_k - P_{ks})d\phi_k \tag{7.10}$$

where $\phi_{ks}$ is the equilibrium (postfault) angle across the line or series element $k$. $\phi_{ku} > \phi_{ks}$ is the angle at which $P_k(\phi_{ku}) = P_{ks}$. If the terminal voltages are constant, then, for a lossless line (without controllers), $\phi_{ku} = \pi - \phi_{ks}$.

## Remarks

1. Chandrashekhar and Hill (1986) proposed a cutset stability criterion using a SPM. However, their method requires checking all possible cutsets to determine the cutset with the lowest energy. This is not practical in a large system. Also, they assume constant voltage magnitudes (assumed to be 1.0 p.u.) at all buses.

2. Even if the MOI is known, there are many possible cutsets. However, there is only one unique cutset the members of which experience unbounded angles across them, if the system is unstable.

3. The network controllers are not considered in the formulation. In Chapter 5, we have seen how the FACTS and HVDC converter controllers can be accommodated in computing the energy function (even if controller dynamics are neglected).

## 7.4 Energy Margin Sensitivity

Sauer et al. (1983) apply EM computations to determine the load supply capability (LSC) and simultaneous interchange capability (SIC) subject to stability constraints. This is an extension of the analysis presented in Garver et al. (1979) and Landgren and Anderson (1973) to compute the maximum LSC or interchange subject to the operating and economic constraints. Typically, DC load flow equations are used, which are linear and simplify the analysis.

El-Kady et al. (1986) compute sensitivities of the energy function using repetitive analysis and applying them for determining the interface power flow limits using distribution factors. The first-order sensitivity of the PE is used in Debs and Dominguez (1990) to predict directly a margin for voltage dip stability. Vittal et al. (1989) derive stability limits using analytic sensitivities of the EM. The stability limits are derived for (i) generation, (ii) load, and (iii) network parameters. Tong et al. (1993) present a sensitivity-based BCU method for derivation of stability limits. They claim that their method relaxes the assumption that the MOI or controlling UEP does not change with change in the parameters.

### 7.4.1 Application to Structure Preserving Model

The literature on DSA does not consider SPEF for sensitivity analysis. Krishna and Padiyar (2000) apply SPEF for the computation of the EM sensitivities for the network parameters to determine the best locations for installation of TCSC.

Assuming classical models for generators and constant-impedance-type loads, the PE can be simplified to the following expression:

$$W_{PE}(\theta, V, \phi) = -\sum_{i=1}^{m} P_{mi}(\theta_i - \theta_{is}) + \sum_{i=1}^{n} G_{li} \int_{\phi_{is}}^{\phi_i} V_i^2 d\phi_i + \frac{1}{2}\sum_{i=1}^{m}(Q_{gi} - Q_{gis}) \quad (7.11)$$

where $G_{li}$ is the conductance of the load at bus $i$. $Q_{gi}$ is the reactive power supplied by the generator $i$ (at the internal bus) and can be expressed as

$$Q_{gi} = \frac{E_i}{x_i'}\left[E_i - V_i\cos(\theta_i - \phi_i)\right]$$

The path-dependent term in the PE function (involving the load power) can be approximated as

$$G_{li}\int_{\phi_{is}}^{\phi_i} V_i^2 d\phi_i = \frac{1}{2}G_{li}(V_i^2 + V_{is}^2)(\phi_i - \phi_{is}) \quad (7.12)$$

It is possible to compute the sensitivities of the EM with respect to the capacitive reactance of TCSC inserted in a line immediately after the fault clearing. The assumption is that the CUEP is not changed by the insertion of the TCSC capacitance. On the basis of sensitivity information, it is possible to decide the best locations for the TCSC. From the case study of the 10-generator New England test system, it is observed that the critical cutset 11–12 and 18–19 are among the best locations for faults where generator 2 separates from the rest. Other lines that belong to the cutset (other than the critical cutset) also show high sensitivity values. For a fault at bus 29 that leads to the separation of generator 9 from the rest, the best locations for TCSC are in lines 28–29 and 26–29. Shubhanga and Kulkarni (2002) derive the EM sensitivity for a given degree of series compensation in line $k$ as

$$\frac{\Delta W_m}{\Delta \bar{x}_{ck}} = \frac{1}{2}\left[I_k^2(\delta_{uep}) - I_k^2(\delta_{cl})\right]x_k \quad (7.13)$$

The series compensation $\bar{x}_{ck}$ is defined as

$$\bar{x}_{ck} = \frac{x_{ck}}{x_k}$$

where $x_c$ is the capacitive reactance inserted in line $k$.

Note that $\delta_{cl}$ and $\delta_{uep}$ are vectors. The following assumptions are made in the derivation:

1. The generators are represented by classical models. The real power loads are constant power type and the reactive power loads are constant reactance type.

2. A fixed amount of compensation is introduced soon after the fault is cleared and removed after adequate time. It is assumed that the postfault uncompensated system is stable.

It is to be noted that the objective of the compensation is to improve first swing transient stability. The line current $I_k$ corresponds to the postfault uncompensated system.

**Remarks**

1. For a specified disturbance, there is a unique MOI and critical cutset for a system. The installation of series FACTS controllers should be in the lines that form a cutset.

2. The effectiveness of the insertion of series compensation in a line depends on the EM sensitivity.

3. The provision of more than one controller in different lines belonging to a cutset tends to improve the stability in an additive fashion. In other words, total increase in the critical energy tends to be the sum of the increases in critical energy by controllers in individual lines (assuming MOI is unaltered).

4. The EM sensitivity can also be expressed as

$$\frac{\Delta W_m}{\Delta \bar{x}_{ck}} = \frac{(V_{iu}^2 + V_{ju}^2 - 2V_{iu}V_{ju}\cos\phi_{ku}) - (V_{icl}^2 + V_{jcl}^2 - 2V_{icl}V_{jcl}\cos\phi_{kcl})}{2x_k}$$

where $V_i$ and $V_j$ are the terminal voltages of line $k$. The subscripts $u$ and $cl$ refer to the values at UEP and clearing time, respectively. If $V_i$ and $V_j$ can be held constant, then

$$\frac{\Delta W_m}{\Delta \bar{x}_{ck}} = \frac{V_i V_j (\cos\phi_{ks} - \cos\phi_{kcl})}{x_k}$$

It is assumed that $\cos\phi_{ku} = -\cos\phi_{ks}$.

## 7.5 Trajectory Sensitivity

Expanding the vector Equation 7.3, we get

$$\dot{x}_i = F_i(x_1,...,x_n; p_1,...,p_r) \tag{7.14}$$

If the parameters are changed by $\Delta p_k$ ($k = 1,2,...r$), Equation 7.14 is changed to

$$\dot{\tilde{x}} = F_i(\tilde{x}_1,...,\tilde{x}_n; p_1 + \Delta p_1,...,p_r + \Delta p_r) \tag{7.15}$$

By defining

$$\Delta x_i(t) = \tilde{x}_i(t) - x_i(t)$$

we can obtain

$$\Delta x_i(t, \Delta p) = \sum_{j=1}^{r} \frac{\partial x_i(t)}{\partial p_j} \Delta p_j \tag{7.16}$$

By defining the sensitivity function $S_{pj}^{xi}(t)$ as

$$S_{pj}^{xi}(t) = \frac{\partial x_i(t)}{\partial p_j}$$

we can obtain the differential equation for trajectory sensitivity as

$$\dot{S}_{pj}^{xi} = \sum_{k=1}^{n} \frac{\partial F_i}{\partial x_k} S_{pj}^{xk} + \frac{\partial F_i}{\partial p_j} \tag{7.17}$$

There are *nr* sensitivities that are defined by as many differential equations of the type given above.

### 7.5.1 Sensitivity to Initial Condition Variations

Trajectory sensitivity with respect to the initial conditions is referred to as $\beta$ sensitivity (the sensitivity to a parameter is called $\alpha$ sensitivity). Consider the following system:

$$\dot{x} = F(x), \quad x(t_0) = \beta \tag{7.18}$$

Defining sensitivity with respect to variations in $\beta$ as $x_\beta(t) = \partial x/\partial \beta$, we obtain

$$\dot{x}_\beta = \left[\frac{\partial f}{\partial x}\right] x_\beta, \quad x_\beta(t_0) = I \tag{7.19}$$

where $I$ is the identity matrix of dimension $n$.

The trajectory sensitivities have applications in power system stability analysis (Hiskens and Pai 2000, Laufenberg and Pai 1998). Shubhanga and Kulkarni (2004) apply sensitivity calculations with respect to clearing time of fault, to infer the effectiveness of change of parameters on transient stability and its applications for generation rescheduling.

### 7.5.2 Discussion

The computations of trajectory sensitivities increase in complexity as the number of parameters increase. In contrast, the EM sensitivity calculations involve less effort. However, the accuracy of EM is limited on account of difficulties in determining MOI and critical energy. In this context, the procedure given in the previous chapter to detect loss of synchronism is quite accurate and the prediction can be performed within a second or so. The determination of critical energy using PEBS method could also require similar amount of simulation time with lesser accuracy.

Another important aspect is the increasing complexity in system operation, not only due to its size and vulnerabilities but also due to the variability in the network power flows caused by the introduction of deregulation or restructuring. There is increasing uncertainty about system parameters that diminishes the effectiveness of operational planning to maintain security. However, technological developments such as WAMS enable the deployment of control strategies based on real-time measurements.

The design of damping controllers based on energy functions, using linear control (similar to PSS) will be taken up next as a means of preventive control to maintain small signal stability of the equilibrium state as load or network parameters vary in the normal system state.

## 7.6 Energy Function-Based Design of Damping Controllers

Gronquist et al. (1995) presented an approach to derive control strategies for damping of electromechanical power oscillations using an energy function method. Gandhari et al. (2001) applied control Lyapunov functions to derive control laws for damping oscillations. However, these papers consider nonlinear control strategies, which may not be required during normal operation and may not be fully effective during large disturbances that threaten system security. In the latter case, switching-type control strategies are more effective as they make full use of the controller rating (which will be discussed in Chapters 8 and 9).

### 7.6.1 Series FACTS Controllers

The objectives of the analysis presented here is to present a unified framework for the design of damping controllers on series FACTS controllers such as TCSC, SSSC, and series converters in UPFC. The analysis is based on simplified models that enable the derivation of control laws and formulation of indices to evaluate the effectiveness of locations of the controllers. The analysis is validated using case studies with detailed system models (Kulkarni and Padiyar 1999, Padiyar and Sai Kumar 2005).

The following assumptions are made in the analysis:

1. A synchronous generator is represented by the classical model of a constant voltage source behind the transient reactance.
2. The real power drawn by a load is independent of the bus voltage magnitude but varies linearly with bus frequency.
3. The electrical network is assumed to be lossless.
4. The mechanical torques acting on the generator rotors are constant and the mechanical damping is neglected.
5. The bus voltage magnitudes are assumed to be constants.
6. There is only one line connected between a pair of specified nodes (say $i$ and $j$). Note that the parallel lines can be clubbed together and represented by an equivalent line.

We have seen (in Chapter 3) that a lossless system can be represented by an analogous network consisting of nonlinear inductors (corresponding to series elements in the electrical network) and linear capacitors representing machine inertias. Linearizing the equation for a series branch $k$, we obtain

$$\Delta P_k = \frac{V_i V_j}{x_k} \cos \phi_{k0} \Delta \phi_k \tag{7.20}$$

where $\Delta P_k$ is analogous to the current and $\Delta \phi_k$ is analogous to the flux linkage. Since the linear inductance is defined as the ratio of flux linkage and current, we obtain

$$L_k = \frac{x_k}{V_i V_j \cos \phi_{k0}} \simeq x_k \tag{7.21}$$

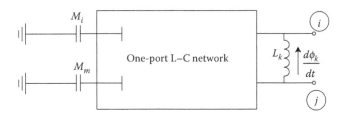

**FIGURE 7.1**
A one-port network.

Note that the terminal voltages of line $k$ are assumed to be constant. If in addition, we assume $V_i = V_j = 1.0$ and $\cos \phi_{k0} \simeq 1.0$, we get $L_k \simeq x_k$.

The network of linear inductors and $m$ linear capacitors (corresponding to $m$ generator inertias) can be treated as one-port network across nodes $i$ and $j$ as shown in Figure 7.1.

It is assumed that line $k$ is connected across the port $i-j$. The capacitors are connected at the internal buses of the generators. Owing to small perturbations, there will be voltages (deviations in the frequencies) across the capacitors, which can be found from the solution of the linearized system equations.

### 7.6.1.1 Linearized System Equations

It is possible to express the linearized system equations as

$$\begin{bmatrix} M \dfrac{d^2 \Delta \delta}{dt^2} \\ 0 \end{bmatrix} = -\begin{bmatrix} K_{gg} & K_{gl} \\ K_{lg} & K_{ll} \end{bmatrix} \begin{bmatrix} \Delta \delta \\ \Delta \phi \end{bmatrix} \tag{7.22}$$

where

$$[M] = \text{Diag}[M_1, M_2, \quad M_m], \quad [K_{gg}] = \text{Diag}[K_{g1}, K_{g2}, \quad K_{gm}]$$

$$[K_{gl}] = [-K_{gg} \quad 0] = [K_{lg}]^t, \quad K_{gi} = \frac{E_{gi} V_i \cos(\delta_i - \phi_i)}{x'_{gi}}$$

$[K_{gl}]$ has dimensions $m \times n$. The matrix $[K_{ll}]$ is an $n \times n$ symmetric matrix with elements given by

$$K_{ll}(i, j) = -K_{ij}, \quad K_{ll}(i, i) = \sum_{j=1}^{n} K_{ij}, \quad K_{ij} = \frac{V_i V_j}{x_k} \cos \phi_{k0}$$

It is possible to eliminate $\Delta\delta$ from Equation 7.22 and derive

$$[M]\frac{d^2\Delta\delta}{dt^2} = -[K]\Delta\delta \tag{7.23}$$

where

$$[K] = [K_{gg}] - [K_{gl}][K_{ll}]^{-1}[K_{lg}]$$

It can be shown that $[K]$ is a singular matrix. The system eigenvalues are the square roots of the eigenvalues of matrix $[A_2]$ given by

$$[A_2] = -[M]^{-1}[K]$$

The eigenvalues of $[A_2]$ are real negative, except for one eigenvalue, which is zero. Note that Equation 7.23 can be expressed in the state space form given by

$$\begin{bmatrix} \Delta\dot\delta \\ \Delta\dot\omega \end{bmatrix} = [A_1]\begin{bmatrix} \Delta\delta \\ \Delta\omega \end{bmatrix} = \begin{bmatrix} 0 & I \\ A_2 & 0 \end{bmatrix}\begin{bmatrix} \Delta\dot\delta \\ \Delta\dot\omega \end{bmatrix} \tag{7.24}$$

where $I$ is identity matrix of order $m$. The eigenvalues of $[A_1]$ are $\pm j\omega_i^m$, $i = 0,1,2,\ldots(m-1)$. The superscript $m$ indicates modal frequency in radians per second. For $i = 0$, $\omega_i^m = 0$, which corresponds to the frequency of COI defined by

$$\delta_{COI} = \frac{1}{M_T}\sum M_i\delta_i, \quad \omega_{COI} = \frac{1}{M_T}\sum M_i\omega_i \tag{7.25}$$

where $M_T = \sum M_i$.

The eigenvector $V_i$ for the matrix $[A_1]$ corresponding to the eigenvalue $\lambda_i = j\omega_i^m$ can be expressed as

$$V_i = [V_{i\delta}^t \quad V_{i\omega}^t]^t$$

It can be shown that

$$V_{i\omega} = j\omega_i^m V_{i\delta} \tag{7.26}$$

The voltage across the capacitor $(M_j)$ for a specified swing mode $i$ is given by

$$V_{Cj} = \sum_{i=1}^{m-1} V_{i\omega}(j)c_j \cos(\omega_i^m t + \beta_i) \qquad (7.27)$$

where $V_{i\omega}$ is the component of the eigenvector $V_i$ defined in Equation 7.26. $c_j$ and $\beta_i$ are determined from the initial conditions on $\Delta\delta$ and $\Delta\omega$. Since the system model is linear for small signal stability analysis, we can apply the superposition theorem by considering one mode at a time.

By applying the compensation theorem, we can replace the capacitor $M_j$ by a voltage source $V_{i\omega}(j)$ corresponding to mode $i$ ($j = 1,2,\ldots,n_g$). At the port in Figure 7.1, we can replace the network by the Thevenin equivalent circuit shown in Figure 7.2. The Thevenin voltage $V_{Ck}$ is sinusoidal of frequency $\omega_i^m$, obtained by removing $L_k$. The Thevenin inductance $L_k^{th}$ is the equivalent inductance of the network, viewed from the port when the voltage sources are short circuited.

Assuming the presence of a series FACTS controller in line $k$ (TCSC or SSSC), we can express the variation of power $P_K$ in the line as

$$\Delta P_k = \frac{\partial P_k}{\partial \phi_k}\Delta\phi_k + \frac{\partial P_k}{\partial u_k}\Delta u_k = \Delta P_{k0} + \Delta P_{kc} \qquad (7.28)$$

$$= \frac{1}{L_k}\Delta\phi + B_k\Delta u_k$$

where $B_k = P_{k0}/x_k$ for a TCSC. $B_k \simeq 1/x_k$ for an SSSC.

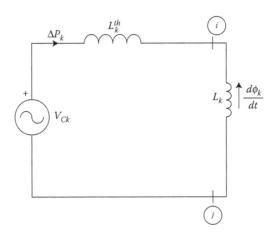

**FIGURE 7.2**
Thevenin equivalent circuit.

It is possible to replace the voltage source $V_{Ck}$ by an equivalent capacitor $C_k$ by using the relationship given by

$$V_{Ck} = \frac{1}{\omega_i^m C_k} \Delta P_k \tag{7.29}$$

The equivalent circuit shown in Figure 7.2 is redrawn (see Figure 7.3a) with the contribution of the controller represented as a current source $\Delta P_{kc} = B_k \Delta u_k$. If we propose a control law given by

$$\Delta u_k = G_k \left( \frac{d \Delta \phi_k^{th}}{dt} \right) \tag{7.30}$$

where $G_k$ (the proportional gain) is equivalent to a conductance ($G_k^{eq} = G_k B_k$) across the capacitance as shown in Figure 7.3b. For simplicity, the dependence of $C_k$ on a specified mode ($m_i$) is not shown in the figure. It is to be noted that both $L_k$ and $L_k^{th}$ are independent of the swing mode.

Note that for a particular mode ($m_i$), the decrement factor ($\sigma_i$) is given by $G_k^{eq}/2C_k$. Hence, the lower the value of $C_k$, the lower the gain of the controller required for a specified damping. Alternatively, for a specified gain, the damping achieved is maximum for the lowest-value $C_k$.

Since it is convenient to compute Thevenin voltage for a line $k$, we can observe that the best location for a series damping controller is one for which $|V_{Ck}|$ is maximum. It is convenient to define a normalized location factor (NLF) as follows:

$$NLF(m_i) = \frac{|V_{Ck}|}{|V_{i\omega}|} \tag{7.31}$$

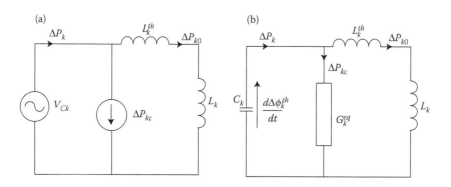

**FIGURE 7.3**
Equivalent circuit with controller. (a) With voltage and current sources. (b) Representation by R–L–C circuit.

where $|V_{i\omega}|$ is a norm of the eigenvector component $V_{i\omega}$ corresponding to mode $i$. We can use the Euclidean norm or the sum norm. The NLF defined here is different from what is given in Kulkarni and Padiyar (1999).

The procedure for computing NLF ($m_i$) is simpler and is given below:

1. For a specified critical mode ($m_i$), compute $V_{i\omega}$.
2. Treating $V_{i\omega}$ vector as the vector of sinusoidal voltage sources at the generator internal buses, compute Thevenin voltage across different lines.

It is convenient to consider the original network (of series elements) in which the reactances are calculated (in p.u.) at the base system frequency. Since $L_k \simeq x_k$ and although the equivalent network (made up only of series elements) impedances have to evaluated at the modal frequency $\omega_i^m$, the calculation of NLF for any mode is unaffected even if we use $x_k$ instead of $\omega_i^m L$.

Another approximation is to compute Thevenin voltage across the line (under consideration) from the original network including all the lines.

### 7.6.1.2 Synthesis of the Control Signal

From the network analogy, we have identified the appropriate control signal as $d\Delta\phi_k^{th}/dt$, which can be synthesized from the locally measured quantities. If the voltage magnitudes $V_i$ and $V_j$ are assumed to be equal to unity and $\sin\phi_k \simeq \phi_k$, we have

$$\Delta\phi_k = \Delta P_k x_k, \quad \Delta\phi_k^{th} = \Delta P_k(x_k + x_k^{th}) \tag{7.32}$$

A practical scheme for the damping controller is shown is Figure 7.4. This shows a washout circuit (which will result in zero offset in steady state) and a practical derivative controller. The washout circuit time constant ($T_w$) can be chosen as 10 s and $T_m$ can be set at 0.01 s.

### Remarks

1. The nominal value of $x_k^{th}$ can be obtained from the electrical networks of series elements. However, since there are approximations made in computing $x_k^{th}$, it is convenient to treat this as a tunable parameter.

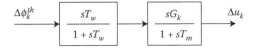

**FIGURE 7.4**
Damping controller for a series FACTS device.

2. The gain $G_k$ of the controller is usually obtained from the root locus. This is similar to the tuning of a PSS.

3. There is no need for phase compensation in the controller although the control scheme is obtained based on a simplified system model using classical models of generators. This will be evident when we present a case study on the design of damping controllers for a TCSC or UPFC.

4. It can be shown that the use of the control signal $d\Delta\phi_k/dt$ instead of $d\Delta\phi_k^{th}/dt$ limits the effectiveness of the controller as gain $G_k$ is increased beyond a limit. If $\Delta u_k = G_k s\Delta\phi_k$, then since $\Delta\phi_k = \Delta\phi_k^{th} - \Delta P_k x_k^{th}$, we have

$$\Delta u_k = \frac{G_k s\Delta\phi_k^{th}}{1 + sG_k B_k x_k^{th}}$$

by substituting for $\Delta P_k = B_k\Delta u_k$. $s$ is the Laplace operator. The damping will increase as $G_k$ is increased from zero and reaches a maximum when $G_k = G_{k\,max}$ given by

$$G_{k\,max} = \frac{1}{B_k x_k^{th}\omega_i^m}$$

If $G_k > G_{k\,max}$, the damping will reduce. There is no limit on the gain if $\Delta\phi_k^{th}$ is used as the control signal (unless limited by the unmodeled dynamics).

5. In Equation 7.32, one could also use the line current as measured signal instead of the power flow.

6. The capacitance $C_k$ can be viewed as the modal inertia (corresponding to a specified mode, $m_i$) viewed from the series element $k$ (Padiyar 2007). The concept of modal inertia (originally used in the context of multi-mass torsional system of rotor masses connected by flexible shafts) was extended to the analysis of low-frequency swing modes by Padiyar and Saikumar (2003). The modal inertia of a specified swing mode varies with the location (of the generator) where PSS is to be designed to damp that mode. The effectiveness of the location of a PSS can be evaluated from the modal inertia as the decrement factor ($\sigma$) is inversely proportional to the modal inertia (for a given damping torque). The concept of modal inertia is extended here to evaluate the effectiveness of the locations of series network controllers (for damping swing modes).

## Example 1

Consider the three-machine system (Anderson and Fouad 1977). The data are essentially the same except for the following modifications. All generators are equipped with static exciters with $K_A = 200$, $TA = 0.05$. A shunt susceptance of 0.5 p.u. is provided at bus 5 for voltage support. The loads are assumed to be constant impedance type and mechanical damping is assumed to be zero.

The eigenvalues for the two swing modes at the operating point are calculated as

$$\lambda_{1,2} = -0.4510 \pm j13.1570 \text{ (Mode 1)}$$

$$\lambda_{3,4} = -0.0392 \pm j8.5809 \text{ (Mode 2)}$$

The computation of location factors gave line 5–4 as the best line for damping the swing mode 2. The series FACTS controller considered an SSSC.

The tuning of the two parameters associated with a damping controller ($G$ and $x^{th}$) was done using the sequential linear programming (SLP) optimization technique to maximize the damping of mode 2 subject to the following constraints: (a) the damping ratio of all the eigenvalues is greater than 0.02 and (b) the real part of all eigenvalues is less than −0.65. The optimal values of the controller parameters are $x^{th} = 0.3391$ and $G = 0.2698$ for the damping controller in line 5–4. The eigenvalues for the two swing modes with the controller are

$$\lambda_{1,2} = -0.6631 \pm j13.1510 \text{ (Mode 1)}$$

$$\lambda_{3,4} = -4.6609 \pm j7.4558 \text{ (Mode 2)}$$

This clearly shows the role of the controller in damping mode 2.

With multiple controllers, there is a need for coordinated control, that is, simultaneous tuning of all the control parameters.

### 7.6.1.3 Case Study of a 10-Machine System

The generators are represented by one-axis machine model with static exciters (Kulkarni and Padiyar 1999). The eigenvalues with constant-power-type loads are shown in Table 7.1 only for swing modes. It is observed that modes 6 and 9 are unstable. From the computations of normalized location factors, it was observed that the installation of a series damping controller (TCSC) in line 11–12 is most effective in damping mode 9 of about 4 rad/s. However, it is ineffective in damping the higher frequency mode 6. The line 26–29 is the best location for damping mode 6. Providing TCSC damping controllers in both lines (11–12 and 26–29) stabilized the system. With TCSCs in lines 11–12 (with gain of 0.3, $x_k^{th} = 0.06$) and 26–29 (with gain of 0.15, $x_k^{th} = 0.04$), the swing modes are also shown in Table 7.1. It is interesting to observe that in addition to the modes 6 and 9, the two damping controllers in the two lines

**TABLE 7.1**

Swing Modes (10-Generator System)

| Number | Eigenvalues without Controller | Eigenvalues with Controller |
|--------|-------------------------------|------------------------------|
| 1 | −0.2740 ± j8.6879 | −0.2746 ± j8.6878 |
| 2 | −0.2039 ± j8.2640 | −0.5516 ± j8.2797 |
| 3 | −0.2056 ± j8.3455 | −0.2051 ± j8.3397 |
| 4 | −0.1580 ± j7.1570 | −0.1584 ± j7.1566 |
| 5 | −0.1627 ± j6.9797 | −0.1637 ± j6.9807 |
| 6 | 0.1999 ± j5.8935 | −0.6350 ± j5.8410 |
| 7 | −0.0844 ± j6.2703 | −0.0944 ± j6.2770 |
| 8 | −0.1900 ± j6.1831 | −0.2480 ± j6.1338 |
| 9 | 0.0619 ± j3.9243 | −0.9535 ± j3.9121 |

also provide damping of modes 2 and 8. The remaining five modes are unaffected. It is also observed that the damping controllers essentially provide a damping torque without affecting the synchronizing torque.

### 7.6.2 Shunt FACTS Controllers

In the previous section, we assumed that the bus voltage magnitudes are constants. This assumption is essential to relate the linearized model of the network to a linear $L$–$C$ network with bus frequency deviation analogous to voltage and power (torque) analogous to the current. In this section, we will relax the assumption as a shunt FACTS controller injects reactive current to control the voltage. In the previous section, we derived the control law from the consideration of introducing a linear damping term, which is proportional to the Thevenin frequency, so that we insert conductances in a linear $L$–$C$ network. This is also equivalent to introducing dissipation in the total energy such that the energy decreases along a trajectory from the initial state to the stable equilibrium state. We take this approach in deriving the control law for the damping controller associated with the shunt FACTS device. The other assumptions mentioned earlier also apply here. In addition, we will also neglect the presence of a series FACTS controller unless the series controller is used only to regulate constant impedance (inserted) by the controller.

The network analogy described is also applicable here. However, we will first consider the nonlinear system before the linearization. The lossless network with classical models of the generator can be viewed as a network with nonlinear inductors and linear capacitors (corresponding to the generator rotor masses). See Chapters 3 and 4. The equations of the nonlinear inductor representing the series impedance of line $k$ is given by

$$P_k = \frac{V_i V_j}{x_k} \sin \phi_k \Rightarrow i_k = f(\psi_k) \tag{7.33}$$

where $i_k$ is the current in the inductor (corresponding to $P_k$ in the line) and $\Psi_k$ is the flux linkage of inductor $k$ (corresponding to the angle across the line). The energy stored in the inductor described by Equation 7.33 is given by

$$W_k = \int i_k \, d\psi_k = \int f(\psi_k) \, d\psi_k \tag{7.34}$$

Substituting Equation 7.33 in Equation 7.34 and noting that $\Psi_k = \phi_k$, we get the following result after integration by parts:

$$\int P_k \, d\phi_k = \frac{-V_i V_j \cos\phi_k}{x_k} + \int \frac{\cos\phi_k}{x_k}(V_i dV_j + V_j dV_i) \tag{7.35}$$

Adding and subtracting $V_i^2 + V_j^2/2x_k$ to the RHS, we can write Equation 7.35 as

$$\int P_k \, d\phi_k = \frac{V_k^2}{2x_k} - \int I_{ri}^k dV_i - \int I_{rj}^k dV_j \tag{7.36}$$

where $V_k^2 = V_i^2 + V_j^2 - 2V_i V_j \cos\phi_k$ is the square of the voltage drop in line $k$. $I_{ri}^k$ and $I_{rj}^k$ are the reactive currents injected in line $k$ at the two ends of the line. These are defined as

$$I_{ri}^k = \frac{V_i - V_j \cos\phi_k}{x_k}, \quad I_{rj}^k = \frac{V_j - V_i \cos\phi_k}{x_k} \tag{7.37}$$

When we add the energy in all the inductors (lines and transient reactances of the generators), we get

$$W_L = \sum_{k=1}^{n_b} W_k = \sum_{k=1}^{n_b}\left(\frac{V_k^2}{2x_k}\right) - \sum_{j=1}^{N}\int I_{Rj} dV_j \tag{7.38}$$

where $I_{Rj}$ is the reactive current injected from bus $j$ in the network, which has $N(n+m)$ buses. $n_b$ is the total number of series branches in the network.

The total energy stored in the capacitors (representing generator masses) is obtained as

$$W_C = \frac{1}{2}\sum_{i=1}^{m} M_i \left(\frac{d\delta_i}{dt}\right)^2 \tag{7.39}$$

We can express the reactive current ($I_{Rj}$) at bus $j$ as the sum of three components

$$I_{Rj} = I_{Rj0} + \frac{\partial I_{Rj}}{\partial V_j} \Delta V_j + \Delta I_{Rj} \tag{7.40}$$

where the second term represents the voltage-dependent component of $I_{Rj}$ (for small perturbations) and the third term represents the contribution of the SMC. For a conservative system of $L$–$C$ network, we can show that (see Chapters 3 and 4)

$$\frac{d}{dt}(W_L + W_C) = 0 \tag{7.41}$$

However, when $\Delta I_{Rj} \neq 0$, and is controlled as

$$\Delta I_{Rj} = -K_{rj}\left(\frac{dV_j}{dt}\right) \tag{7.42}$$

The total energy does not remain constant and we get

$$\frac{d}{dt}(W_L + W_C) = -\sum K_{rj}\left(\frac{dV_j}{dt}\right)^2 \tag{7.43}$$

For positive $K_{rj}$, the average value of the RHS in the above equation is negative. Hence, we introduce dissipation in the system if the reactive current injection at bus $j$ is controlled as given by Equation 7.42. However, this control law is not optimum. For deriving the optimum control law, we need to express $\Delta V_j = (V_j - V_{j0})$ in terms of the perturbations in state variables $\Delta \delta$ and $\Delta \dot{\delta}$. For this, we need to model the system to relate $\Delta V_j$ with $\Delta \delta_j$.

### 7.6.2.1 Linear Network Model for Reactive Current

Since we are essentially dealing with small perturbations, a linear model is adequate. At the generator internal buses, we have

$$\Delta \hat{E}_{gi} = j\hat{E}_{gi0}\Delta \delta_i \tag{7.44}$$

where $\hat{E}_{gi0} = |E_{gi}|e^{j\delta_{i0}}$. (Note that $|E_{gi}|$ is a constant as the generator is represented by a classical model.)

Since we are considering only reactive currents injected at various buses and their effect on voltage magnitudes, we can introduce the approximation of $\Delta\phi_k \simeq 0$ (the bus angles are all equal for the lossless network). With this assumption, we get $\Delta I_{Ri}^k = -\Delta I_{Rj}^k$ and described by

$$\Delta I_{Ri}^k = b_k(\Delta V_i - \Delta V_j) \tag{7.45}$$

where $b_k = 1/x_k$. Equation 7.45 represents the current flow through an equivalent conductance ($b_k$), which is connected between nodes $i$ and $j$. At any node $j$, there is one equivalent conductance ($b_j^s$) connected between the node $j$ and the reference (ground) node given by

$$b_j^s = -\frac{\partial I_{Rj}}{\partial V_j}, \quad j = 1,...,N \tag{7.46}$$

where $N = n + m$ is the total number of buses, including the generator internal buses. The analogous network relating reactive currents to the bus voltage magnitudes is termed as RCFN (Padiyar 1984). Similarly, the network relating variations in the reactive currents and voltage magnitudes is termed as incremental RCFN (IRCFN) (Padiyar and Suresh Ras 1996). This is a linear resistive network excited by sinusoidal voltage sources at the generator internal buses. The shunt susceptances are linear and are unchanged after linearization. Note that the shunt reactor is represented by a positive resistance while the capacitor is represented as a negative resistor in RCFN.

The linear IRCFN relating the variations in the voltage magnitudes and the reactive currents is shown in Figure 7.5a. The Thevenin equivalent at node $j$ is shown in Figure 7.5b.

For simplicity, we will consider only one modulation controller at bus $j$. Thevenin voltage $\Delta V_j^{th}$ at bus $j$ is a function of the state variables $\Delta\delta$. If we

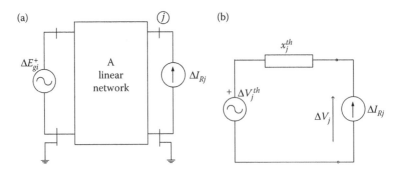

**FIGURE 7.5**
An equivalent network relating voltage magnitudes to reactive current injections: (a) IRCFN and (b) Thevenin equivalent at node $j$.

substitute $\Delta\delta = V_{i\delta}$ where $V_{i\delta}$ is the component of the eigenvector corresponding to mode ($m_i$), we get the equivalent circuit for mode $m_i$ described by the equation

$$\Delta V_j^{th} + x_j^{th}\Delta I_{Rj} = \Delta V_j \qquad (7.47)$$

From Equation 7.47, we can obtain the optimal control law for the modulation of the reactive current at bus $j$ as

$$\Delta I_{Rj} = -K_{rj}\frac{d}{dt}(\Delta V_j^{th}) \qquad (7.48)$$

Note that $\Delta V_j^{th}$ is a function of mode $m_i$. In computing Thevenin voltage, we assume $\Delta E_{gi} = |E_{gi}V_{i\delta}(i)|$, where $v_{i\delta}(i)$ is the component of the eigenvector $V_{i\delta}$ corresponding to the generator $i$. If we assume $|E_{gi}| = 1.0$, then $\Delta E_{gi} = |V_{\delta}(i)|$.

Just as in the case of the series FACTS controllers, we define an index (NLF) to evaluate the effectiveness of the location of the damping controller, based on Thevenin voltage as follows:

$$\text{Normalized location factor (NLF)} = \frac{|\Delta V_j^{th}|}{V_{\delta}} \qquad (7.49)$$

where $|V_{i\delta}|$ is the norm of eigenvector $V_{i\delta}$ computed for a specified (critical) mode $m_i$. The best location for the modulation controller for mode $m_i$ is the bus $j$ for which $|\Delta V_j^{th}|$ is maximum.

## Remarks

1. Although $x_j^{th}$ can be obtained from the linear network, it is convenient to treat it as a tunable parameter in addition to the proportional gain $K_{rj}$. The tuning and performance evaluation of damping controllers for STATCOMs are presented in Padiyar and Swayam Prakash (2003).

2. The linear network (IRCFN) used to compute $\Delta V^{th}$ and $x^{th}$ has both series and shunt elements. The series elements correspond to the (constant) reactances of the lines and generators. The shunt elements correspond to the shunt susceptances at various buses. At load buses, the shunt elements are given by Equation 7.46.

3. The use of the control signal ($dV_j/dt$) instead of ($dV_j^{th}/dt$) limits the effectiveness of the modulation control of shunt FACTS controllers as the gain $K_{rj}$ is increased.

## Example 2

Here, we consider the same three-machine system considered in Example 1. The best location for damping mode 2 is bus 4. By selecting $K_r = 5.789$ and $X^{th} = 0.0450$, we get the closed-loop eigenvalues for the swing modes, after installing the SMC at a STATCOM connected to bus 4, as

$$\lambda_{1,2} = -0.3641 \pm j13.1610$$

and

$$\lambda_{3,4} = -0.5006 \pm j8.1286$$

It was observed that installing an SMC leads to the reduction of damping of an exciter mode. This is a generic phenomenon of strong resonance (Padiyar and Sai Kumar 2006) in which change in parameters lead to interaction or coupling between two oscillatory modes. The modes that are far away initially move close to each other and collide in such a way that one of the modes may subsequently become unstable. If the system matrix is not diagonalizable at the point of collision of the eigenvalues, the phenomenon is termed "strong resonance," and it is termed "weak resonance" if it is diagonalizable. Using multimodal decomposition and a reduced model, it is possible to predict the behavior of the interacting modes near a strong resonance. Depending on the value of $x^{th}$, the critical mode can be the swing mode or the exciter mode. Even a small change in $x^{th}$ can affect the critical mode. This indicates the importance of strong resonance in the design of damping (modulation) controllers.

## Remarks

1. The variations introduced by the linear controllers for damping small oscillations can be viewed as $\lambda$-variations (Kokotovic and Rutman 1965), which increase the system order.

2. The auxiliary damping controllers with FACTS can be viewed as a supplement to PSS with power and speed inputs (PSS2B) (IEEE PES Digital Excitation Task Force 1996). However, there are problems associated with the application of PSS. Kamwa et al. (2005) state "Although it (PSS) was introduced and extensively used a long time ago, and despite its inherent simplicity, it may still be one of the most misunderstood and misused pieces of generator control equipment." They also claim that "PSS technology still has great days ahead."

3. As a matter of fact, PSS can be viewed as a shunt FACTS controller that can modulate the reactive current injected by the generator at its

internal bus. Since a component of the PE can be expressed as (see Chapter 4)

$$W_{25} = \int_{E'_{q0}}^{E'_q} i_d \, dE'_q$$

and the reactive current ($i_R$) injected by the generator depends on $-i_d$, it is possible to arrange such that $\Delta i_d \propto (dE'_q/dt)$ to provide damping. However, it is to be noted that a PSS can only modulate $E_{fd}$ directly.

## 7.7 Damping Controllers for UPFC

A four-machine, two-area system (see Figure 7.6) is selected for the case study. This system is the same as that described in Klein et al. (1991) except that the damping of all the generators is assumed to be 1.0 p.u. (instead of zero) and the armature resistances are neglected. The generators are represented by two-axis models and loads are modeled as constant impedances. The UPFC is installed in one of the three tie lines connecting the two areas (between buses 3 and 13) with the shunt branch connected at bus 3.

The UPFC has three independent control variables—real (series) voltage ($V_p$), reactive (series) voltage ($V_r$), and reactive (shunt) current ($I_R$). We can apply modulation controllers for all the three variables. The block diagram of these controllers is shown in Figure 7.7. The two-series voltage modulation controllers use Thevenin angle as the control signal while the shunt reactive current modulation controller uses Thevenin voltage ($\Delta V^{th}$) as a control signal, which is synthesized from Equation 7.47 as

$$\Delta V^{th} = \Delta V - \Delta x_{sh}^{th} \Delta I_R$$

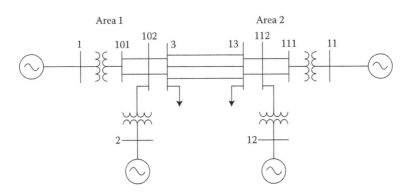

**FIGURE 7.6**
Single-line diagram of two-area, four-machine system.

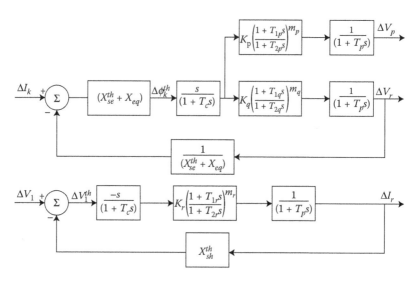

**FIGURE 7.7**
Block diagram of UPFC controllers.

It is to be noted that $\Delta I_R$ is the reactive current injected at the specified bus and $x_{sh}^{th}$ is different from what is defined in deriving a control law for series FACTS controllers ($x_{se}^{th}$). The control signal for the series modulation controllers is synthesized as

$$\Delta \phi^{th} = \Delta I(x_{eq} + x_{se}^{th}) - \Delta V_r \qquad (7.50)$$

where $x_{eq}$ is the equivalent line reactance, taking into account the fixed series compensation (if any) and the operating value of the reactive voltage injected.

There are five control parameters (two Thevenin reactances and three gains) associated with the three modulation controllers. This is in addition to the three parameters ($T_1$, $T_2$, and $m$) associated with each phase compensation circuit. The parameters of the phase compensation circuits are computed from the knowledge of the phase angle of the residue corresponding to the critical mode. The selection of the controller parameters is based on a constrained optimization technique whose objective is to maximize the damping of the low-frequency interarea mode while satisfying constraints on the location of other eigenvalues (Padiyar and Sai Kumar 2005).

The eigenvalues for the three swing modes for the system without modulation controllers are given below. The operating values of $V_p$, $V_r$, and $I_R$ are assumed to be zero.

$$\lambda_{1,2} = -0.0088 \pm j4.4450 \text{ (Interarea mode)}$$

$$\lambda_{3,4} = -0.7644 \pm j7.2940 \text{ (Local mode 1)}$$

**TABLE 7.2**

Details of Phase Compensation (UPFC Damping Controllers)

| Controllers | Phase of Residue | Required Phase Compensation | $T_1$ (s) | $T_1$ (s) | Number of Stages $m$ |
|---|---|---|---|---|---|
| $V_p$ | 128.5° | 51.5° | 0.6449 | 0.0785 | 1 |
| $V_r$ | 179.6° | 0.0° | — | — | 0 |
| $I_R$ | −107.4° | −72.6° | 0.1139 | 0.4445 | 2 |

**TABLE 7.3**

Eigenvalues of Interarea Mode (Four-Generator, Two-Area System)

| Modulated Variables | Eigenvalues without Phase Compensation | Eigenvalues with Phase Compensation |
|---|---|---|
| $I_R$ | −0.5392 ± j4.4632 | −0.5909 ± j4.4583 |
| $V_r$ | −1.9455 ± j4.8991 | −1.9455 ± j4.8991 |
| $V_r + I_R$ | −2.1033 ± j4.8613 | −3.0892 ± j4.2701 |
| $V_r + V_p$ | −2.8736 ± j5.4176 | −3.4847 ± j5.2239 |
| $V_r + V_p + I_R$ | −3.6951 ± j5.8982 | −3.9056 ± j4.2940 |

$$\lambda_{5,6} = -0.7410 \pm j6.6904 \quad \text{(Local mode 2)}$$

It is obvious that the interarea mode is very lightly damped (having a damping ratio of 0.002). The details of the phase compensation circuits required are given in Table 7.2.

It is to be noted that no phase compensation is required for the series reactive voltage modulation controller. The requirements of phase compensation for the other two controllers indicate the influence of the unmodeled dynamics on the control law derived in the previous section. The eigenvalues corresponding to the interarea mode for different combinations of the controllers with and without phase compensation are shown in Table 7.3. These indicate that phase compensation improves the damping (as expected). It was also observed that the optimal gains of the controllers were less for the case with phase compensation compared to that without phase compensation.

Table 7.3 shows that the modulation controller for the series reactive voltage is most effective. The influence of the active voltage ($V_p$) modulation controllers is to improve both damping and synchronizing torques. It is more effective than the shunt reactive current modulation.

## Remarks

1. The three damping controllers for a UPFC are decoupled and each contribute to the damping of the interarea mode.

2. No phase compensation is required for the design of modulation controllers for the series reactive voltage (even though the control law is based on simplifying assumptions).

3. The damping controllers for modulation $V_r$ and $V_p$ are more effective than that for $I_R$ (shunt-connected FACTS controller).

### 7.7.1  Discussion

It is interesting to observe that damping controllers associated with the series FACTS controllers do not require lead–lag networks for phase compensation, even though the control law was derived based on the simplified system model. Also, the series damping controllers are much more effective than the damping controllers connected with the shunt-connected FACTS controllers whose performance can be improved by phase compensation.

Gama et al. (2000) describe the operating experience of damping power oscillations using TCSC in Brazilian north–south interconnection. Their controller is quite complex with lead–lag filters equipped with non-windup limiters to overcome the effects of drifts caused by DC offsets. There is also the problem of not maintaining the desired phase shift during large disturbances. As mentioned earlier, it is necessary to prevent the damping controller action during large disturbances by limiting the output to ensure that the controller essentially damps small oscillations. For damping large oscillations during major disturbances, nonlinear damping based on switching control (described in the next chapter) is more effective.

# 8

## Application of FACTS Controllers for Emergency Control—I

### 8.1 Introduction

In the previous chapter, we considered the application of FACTS controllers for damping small oscillations caused by small disturbances. The system equations were linearized and control strategies for series and shunt FACTS controllers were formulated to design damping controllers. These controllers can also help in increasing the region of small signal stability in the parameter space. From dynamic security considerations, a system will be operating in the alert state if a credible contingency can result in the loss of small signal stability by growing oscillations in power flows in lines that would finally lead to loss of synchronism and system breakup. Thus, the provision of damping controllers and their operation can be viewed as a preventive control to maintain dynamic security.

In this chapter, we consider the application of FACTS controllers for improving dynamic security under large disturbances (such as faults followed by their clearing). In such cases, we have to retain the nonlinear system models in developing control strategies. The objectives of the controllers are (1) to ensure first swing stability by switching in spare capacity of the FACTS device and (2) to steer the system trajectory toward postfault stable equilibrium point by a suitable control strategy. The second objective is important to ensure that multiswing instability is avoided.

The size and complexity of power systems have increased due to the increase in load demand and system interconnections. The economic and environmental constraints on transmission expansion have resulted in power systems operating under stressed conditions with low security margins. This will ultimately lead to increased dependence on control for optimal and secure operation of power systems.

Maintaining system security is an important factor in the operation of a power system. Security is the ability to maintain system integrity by preventing cascading outages when the system is subjected to disturbances. The on-line static security assessment implemented in the energy control centers assumes that the system reaches a new steady-state condition after being subjected to a disturbance, and checks whether in this steady-state condition, any equipment or transmission line is overloaded and also whether any bus voltage magnitude is outside the permissible range. The DSA deals with the determination of whether a large disturbance will lead to a transition from the present operating condition to a new satisfactory operating condition. The transition depends on the system dynamics and the disturbance. The purpose of DSA is to take preventive action or decide remedial (corrective) action if the contingency actually occurs.

The corrective actions to maintain transient stability are initiated only after the detection of a large disturbance. Most of these corrective actions known as discrete supplementary controls (IEEE Committee Report 1978) are designed to contribute to first swing stability by essentially increasing the electrical power output or decreasing the mechanical power input of the advanced generators.

In this chapter, we describe the basic concepts of applying FACTS controllers for emergency control and present algorithms that lead to bang-bang-type controllers. The control strategies are illustrated with the help of examples of a single- or two-machine system. The extension to multimachine systems is described in the next chapter.

## 8.2 Basic Concepts

The "normal" operating state of a power system implies that (a) the load demands are met by available generators that are committed and operate with adequate spinning reserve and (b) the operating equipment in power stations and the transmission network are not overloaded and the bus voltages lie in the acceptable band (say 0.90–1.05 p.u.). The tripping of a large generator due to the protective action should not lead to cascade tripping of other generators by protective action. Similarly, tripping of a heavily loaded transmission line should not result in cascade tripping of other lines.

Unlike steady-state or small signal stability, which has to be continuously maintained at all times, the transient stability is a function of the disturbance. To prevent loss of synchronism, the control action need not be continuous. It is initiated following a disturbance and can be stopped when the system reaches close to the postfault stable equilibrium. The assumption is that the postfault equilibrium is stable. According to the IEEE Committee Report (1978), the following discrete controls are listed:

1. Dynamic braking
2. High-speed circuit breaker reclosing
3. Independent pole tripping
4. Discrete control of excitation systems

5. Controlled system separation and load shedding
6. Series capacitor insertion
7. Power modulation of HVDC lines
8. Turbine bypass valving
9. Momentary and sustained fast valving
10. Generator tripping

With the introduction of SVCs and FACTS devices such as TCSC, thyristor-controlled phase-angle regulators (TCPAR), or static phase-shifting transformer (SPST), fast control to maintain system security is feasible. Along similar lines such as HVDC converter controls, FACTS controllers using high-power semiconductor devices such as thyristor, GTO, and IGBT can be programmed to provide a discrete control action in the event of a major disturbance that may threaten transient stability of the system. For example, dynamic braking using braking resistors has been employed to enhance transient stability limit (Ellis et al. 1966, Shelton et al. 1975). Thyristor-controlled braking resistor (TCBR) is a FACTS controller that improves system response (Donnelly et al. 1993, Hingorani and Gyugyi 2000).

Thus, the control action is temporary in nature. Transient stability controllers are termed as discrete supplementary controllers as opposed to primary controllers such as speed governor and excitation systems in addition to protective relaying.

## 8.3 Switched Series Compensation

The use of switched series capacitors or TCSC for improving transient stability and time-optimal control of power systems has been considered in Kimbark (1966), Smith (1969), Rama Rao and Reitan (1970), Kosterev and Kolodziej (1995), Padiyar and Uma Rao (1997), and Chang and Chow (1998). The nature of the control action is of bang-bang type, where a single control variable (series compensation) is switched between two limits. The control strategy is not obvious in the case of a UPFC, which has three independent control variables. In this section, we will introduce a general control algorithm that is valid for all FACTS controllers, including UPFC. Before presenting this algorithm, we will briefly review some concepts on time-optimal control involving switchings.

### 8.3.1 Time-Optimal Control

Time-optimal control is based on Pontryagin's minimum principle (Athans and Falb 1966) and involves bang-bang control. Its application to a single-machine system, considered the series capacitor control to transfer the system from a disturbed state to a stable state by a single switching of the capacitor. Although the concept has been proposed as early as in 1970 (Rama Rao and Reitan 1970), the introduction of TCSC has enabled such control strategies to be applied in practice.

To illustrate time-optimal control, consider an SMIB system in which series capacitors are used. The swing equations, neglecting damping, are given by

$$\begin{rcases} \dot{\delta} = \omega = f_1(\omega) \\[2ex] \dot{\omega} = \dfrac{1}{M}\left[ P_m - \dfrac{E_g E_b \sin\delta}{(X - X_C)} \right] = f_2(\delta, \omega, X_C) \end{rcases} \qquad (8.1)$$

where $E_g$ and $E_b$ are generator internal voltage and infinite bus voltage, respectively. $X$ is the fixed reactance (between the generator internal bus to the infinite bus) and $X_C$ is the reactance of the switchable capacitor. It has a nominal value of $X_{C0}$ and is constrained by $X_{Cmin} \leq X_C \leq X_{Cmax}$.

Let Equation 8.1 represent the postfault system and $t_{cl}$ be the initial time of the postfault system (the fault clearing time). Let $T$ be the final time at which the system trajectory reaches an SEP ($\delta_{sep}$, 0) of the postfault system.

The objective of the time-optimal control is to minimize the performance index

$$J = T - t_{cl} = \int_{t_{cl}}^{T} f_0 \, dt, \quad f_0 = 1 \qquad (8.2)$$

To develop the conditions for optimality, we define the Hamiltonian for the system of Equation 8.1 as

$$H = f_0 + p_1 f_1 + p_2 f_2 = 1 + p_1\omega + \frac{p_2}{M}\left( P_m - \frac{E_g E_b \sin\delta}{(X - X_C)} \right) \qquad (8.3)$$

where the costates $p_1$ and $p_2$ satisfy

$$\begin{rcases} \dot{p}_1 = -\dfrac{\partial H}{\partial \delta} = \dfrac{p_2}{M}\dfrac{E_g E_b \cos\delta}{(X - X_C)} \\[3ex] \dot{p}_2 = -\dfrac{\partial H}{\partial \omega} = -p_1 \end{rcases} \qquad (8.4)$$

From the minimum principle, the optimal control satisfies the conditions

$$X_C = \begin{cases} X_{Cmax}, & \text{if } p_2 \sin \delta > 0 \\ X_{Cmin}, & \text{if } p_2 \sin \delta < 0 \end{cases} \quad (8.5)$$

If $p_2 \sin \delta = 0$, the control cannot be determined. Equation 8.5 suggests bang-bang control.

The stability region of the postfault system with $X_C = X_{Cmax}$ in the $\delta - \omega$ plane is shown in Figure 8.1. This figure also shows the system trajectory starting from $\delta_0$ the SEP of the prefault system. At point $P_1$, the fault is cleared and $X_C$ in switched to $X_{Cmax}$. The trajectory with $X_C = X_{Cmax}$ intersects a trajectory ($S$) that passes through $\delta_{sep}$ (the SEP of the postfault system) and has $X_C = X_{Cmin}$. At the point of intersection $P_2$, the capacitor is switched to $X_{Cmin}$ and the system reaches $\delta_{sep}$ in minimum time. The capacitor is switched to the nominal value $X_{C0}$ when the EP is reached.

The trajectory $S$ is defined by the equation

$$W_T = W_{KE} + W_{PE} = \frac{1}{2} M \omega^2 - P_m(\delta - \delta_{sep}) - \frac{E_g E_b (\cos \delta - \cos \delta_{sep})}{(X - X_{Cmin})} = 0 \quad (8.6)$$

Note that the equation is based on the energy function defined for the postfault system with $X_C = X_{Cmin}$ (Kosterev and Kolodziez 1995).

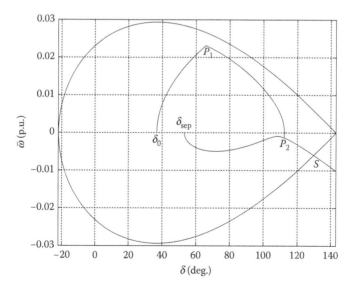

**FIGURE 8.1**
System trajectory with single switching.

Complications arise if the assumption that $S$ intersects the trajectory $P_1P_2$ is not true (particularly when the difference between $X_{Cmax}$ and $X_{Cmin}$ is small). In such situations, multiple switchings of a series capacitor are required (Chang and Chow 1998). Chang and Chow also mention several issues that may affect the robustness of the time-optimal control in realistic power systems. These relate to the complexity of models, dependence of capacitor switching on the postfault EP, control signals that are required for the implementation of the control scheme, and effects of noise present in these signals. They also state, "Suboptimal schemes applicable to a range of power transfer conditions have to be developed. Because time-optimal is open-loop control, its integration with a feedback control to achieve small-signal stability should be investigated."

The control strategy developed in the next subsection addresses some of these issues. In the first place, the strategy is a suboptimal in that the SEP is not necessarily reached in minimum time. The strategy is seamlessly integrated with the functioning of the damping controller described in Chapter 7. The discrete control scheme that is described in the next subsection ceases to function when the energy in the system falls below a threshold and then allows the linear damping controller to act.

## 8.4 Control Strategy for a Two-Machine System

Consider a two-machine system shown in Figure 8.2. This can be reduced to an equivalent single machine described by

$$M_{eq}\frac{d^2\delta_{12}}{dt^2} = P_m^{eq} - P_e^{eq} \tag{8.7}$$

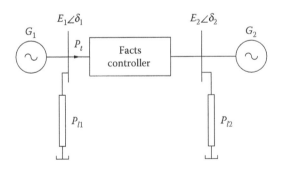

**FIGURE 8.2**
A two-machine system.

where

$$M_{eq} = \frac{M_1 M_2}{M_1 + M_2}, \quad P_m^{eq} = \frac{M_2 P_{a1} - M_1 P_{a2}}{M_1 + M_2},$$

$$P_{a1} = P_{m1} - P_{l1}, \quad P_{a2} = P_{m2} - P_{l2}, \quad P_e^{eq} = P_t, \quad \delta_{12} = \delta_1 - \delta_2 = \delta$$

In steady state, $P_m^{eq} = P_t$.

In the presence of a FACTS controller, $P_e^{eq} = P_t(\delta_{12}, u_c)$, where $u_c$ is the control vector associated with the FACTS controller connected in the line. $u_c$ can be a single variable such as $X_{TCSC}$ or $V_r$ if a TCSC or SSSC is connected in series with the line. It can be $B_{SVC}$ or $I_r$ if an SVC or a STATCOM is connected in shunt at an appropriate location. It can be a vector of three variables ($V_C$, $\beta$, and $I_r$) if a UPFC is connected (where $V_C \angle \beta$ is the injected series voltage and $I_r$ is the reactive current drawn by the shunt converter).

The objective of the control strategy is to steer the system to a state in the vicinity of postfault SEP. The control scheme is independent of the size of the FACTS controller and is designed to work reliably.

If the reactances of the two machines are neglected in comparison with the reactance of the tie line, we can express the PE of the system ($W_{PE}$) as energy associated with the tie line. This follows from

$$W_{PE} = \int_{\delta_0}^{\delta} (P_e - P_m) d\delta = \int_{\delta_0}^{\delta} (P_t - P_{t0}) d\delta \tag{8.8}$$

For convenience, we can assume that $\delta_0$ is also the postfault SEP (this is not a restriction and can be relaxed). The first swing stability can be optimized by maximizing the power in the line ($P_t$) as soon as the large disturbance is detected. This is achieved by selecting $u_c = u_{cm}$ that corresponds to maximum power. If we assume that this will stabilize the system, then $\delta_{max}$ will be less than or equal to the UEP ($\delta_{um}$) that results from $u_c = u_{cm}$.

If $\delta_{max} \leq \delta_u$ (where $\delta_u$ is the UEP with $u_c = u_{c0}$ where $u_{c0}$ is the (nominal) control vector for the postfault system), then setting $u_c = u_{c0}$ when $\delta = \delta_{max}$ will ensure that the system will reach the SEP due to the inherent damping in the system. However, if we wish to damp the oscillations by the discrete control where $u_c$ can take three values—$u_{c0}$, $u_{cm}$, and $u_{cl}$ (where $u_{cl}$ corresponds to the minimum power flow in the line)—the following control strategy is proposed (Krishna 2003, Krishna and Padiyar 2005):

1. As soon as a large disturbance is detected (such as a fault followed by its clearing), the control variables are selected such that the power flow $P_t$ is maximized.

2. The control variables are switched to their nominal operating values ($u_{c0}$) when $W_{PE}(u_{c0})$ is maximum and ($d\delta/dt) \leq 0$.

3. The control variables are selected ($u_c = u_{cl}$) such that the power flow is minimized when $d\delta/dt$ is minimum and $(d\delta/dt) < -\varepsilon$.

4. The control variables are switched to their nominal operating values when $(d\delta/dt) = 0$.

5. The control variables are selected such that the power flow is maximized when $d\delta/dt$ is at a maximum value and $(d\delta/dt) > \varepsilon$.

6. Go to step 2.

The control action is disabled when $|d\delta/dt| < \varepsilon$ (where $\varepsilon$ is appropriately specified).

Step 2 needs clarification. Since $W_{PE} = \int_{\delta_0}^{\delta} (P_t - P_{t0}) d\delta$, $(dW_{PE}/dt) = 0$ wherever $\delta = \delta_{max}$ or $P_t = P_{t0}$. If $\delta_{max} > \delta_u$, then although $(dW_{PE}/dt) = 0$ at $\delta = \delta_u$, $(d\delta/dt) > 0$ at that point. The switching of the control variables has to be delayed until $\delta = \delta_u$ in the reverse swing when $(d\delta/dt) < 0$. Thus, the condition for switching covers both cases (i) $\delta_{max} < \delta_u$ and (ii) $\delta_u < \delta_{max} \leq \delta_{um}$.

Note that $W_{PE} (u_c = u_{c0})$ is maximum when $\delta = \delta_u$ or $\delta = \delta_{max}$ ($\delta_{max} \leq \delta_u$). $W_{PE}$ ($u_{c0}$) starts decreasing as $\delta$ increases beyond $\delta_u$.

The control strategy can be explained with the help of power angle curves shown in Figure 8.3. If we consider a major disturbance such as a three-phase fault at the generator terminals, the angle $\delta$ increases during the fault from $\delta_0$ to $\delta_{cl}$ (when the fault is cleared). The accelerating area is "abcd." Owing to the KE stored in the rotor, it swings beyond $\delta_{cl}$ and reaches a peak of $\delta_{max}$ when the decelerating area "defg" is equal to the area "abcd." To minimize $\delta_{max}$, it is necessary to maximize the power flow. Assuming $\delta_{max} \leq \delta_u$, switching the control variables to the nominal operating values at the instant when $\delta = \delta_{max}$ results in steering the rotor angle $\delta$ to the stable equilibrium value of $\delta_0$. When $\delta = \delta_0$, $d\delta/dt$ is negative and is at the minimum. Selecting the control variables such that power flow is minimum results in reducing the undershoot in the angle. The rotor angle reaches a minimum value of $\delta_{min}$ such that the accelerating area "aijk" is equal to the decelerating area "gha." At $\delta = \delta_{min}$, $(d\delta/dt) = 0$ and switching the control variables to the nominal operating values at this instant results in steering the rotor angle to $\delta_0$. When $\delta = \delta_0$, $d\delta/dt$ is maximum and the control variables are selected such that the power flow in the line is maximum. To summarize our discussion, both overshoot and undershoot of the rotor angle can be minimized if (a) the power flow is maximized when $\delta > \delta_0$, $(d\delta/dt) > 0$ and (b) the power flow is minimized when $\delta < \delta_0$, $(d\delta/dt) < 0$.

The application of the control strategy requires the control signal $\dot{\delta}_{12} = \dot{\delta}$ and the nominal power flow in the line, which can be computed (with the knowledge of the nominal control variables $u_{c0}$). The angle $\delta_{12}$ is the difference in the angles of terminal buses of the line, which can be computed from the local measurements of voltage and current in the line. The knowledge of $P_{t0}$ and $u_{c0}$ enables the computation of $W_{PE} (u = u_{c0})$. Since the switching (from nominal power to maximum (or minimum) power and back) is

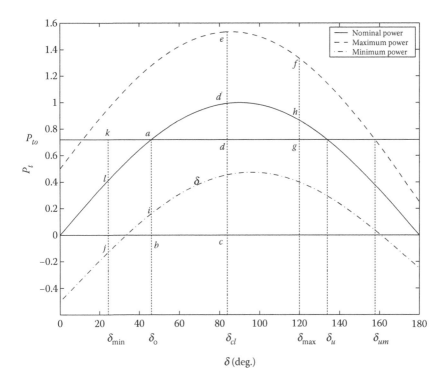

**FIGURE 8.3**
Power angle curves.

performed at the instants when $d\delta/dt$ or $W_{PE}$ is at their peak values (or when $d\delta/dt$ is minimum), the exact computation of these control signals is not essential.

## Remarks

1. It is convenient to apply the control strategy to the discrete control of a TCSC. However, a TCSC can operate only in the capacitive region. In this case, the control variable is a scalar and $X_{TCSC}$ can be switched only between $X_{TCSC}^{min}$ and $X_{TCSC}^{max}$. If we consider $X_{TCSC}^{min}$ as a fixed capacitance and merge it with the line reactance, the control variable becomes $\Delta X_{TCSC} = X_{TCSC} - X_{TCSC}^{min}$. $\Delta X_{TCSC}$ can assume two values—zero and $\Delta X_{TCSC}^{max}$. The lower value corresponds to the nominal operating power and $\Delta X_{TCSC}^{max}$ corresponds to the maximum power flow in the line.

2. In contrast with TCSC, an SSSC injects a reactive voltage that can vary from inductive to capacitive. If we consider $V_{r0}$ (the operating value of the SSSC voltage) as zero, $V_{rmin} < 0$ corresponds to minimum

power flow in the line while $V_{rmax} > 0$ corresponds to maximum power. We assume here, that $V_r$ is positive with capacitive compensation. It is anticipated that SSSC controller performance is better than that of TCSC as switching between three levels of controller output results in faster decay of the oscillations following a major disturbance.

3. When only a scalar control variable is to be switched, there is direct correlation between the variable and the power flow in the line. For example, capacitive reactive voltage injected by an SSSC or capacitive current drawn by an SVC or STATCOM will result in increased power flow in the line, compared to inductive voltage injection by an SSSC or inductive current drawn by a STATCOM. However, if a UPFC is used to improve stability, there are three independent variables—shunt reactive current ($I_r$), series active voltage ($V_p$), and reactive voltage ($V_r$) injection. It is not obvious how these variables need to be controlled to program maximum or minimum power flow in the line.

4. From Equation 8.8, it is obvious that the PE in the line is increased or decreased as the power flow ($P_t$) is increased or decreased. However, the PE also depends on the nature of the power angle curve. We will see later that when we analyze TCPAR or SPST, $W_{PE}$ can be maximized even when the maximum power is not altered.

## 8.5 Comparative Study of TCSC and SSSC

The system considered is shown in Figure 8.4.
   The generator and network data on a 1000 MVA base are

$$f_B = 60\,\text{Hz}, H = 5, X_d = 1.6, X_d' = 0.32, X_q = X_q' = X_d',$$

$$X_{tr} = X_b = 0.1, X_{L1} = X_{L2} = 0.2$$

The initial operating data are

$$V_g = 1.0, \ P_g = 1.0, \ E_b = 1.0$$

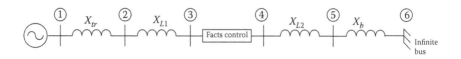

**FIGURE 8.4**
An SMIB system with a series FACTS controller (TCSC or SSSC).

When the generator is represented by a classical model (neglecting field flux decay and damper winding), the generator is represented by the following equations (neglecting damping):

$$\left.\begin{array}{l} \dot{\delta} = \omega_B S_m \\[2mm] \dot{S}_m = \dfrac{1}{2H}(P_m - P_e) \end{array}\right\} \tag{8.9}$$

$X_0$ is assumed to be zero and $X_{TCSC}$ is switched between 0 and $X_{TCSC}^{max}$. For the sake of comparison with a TCSC, SSSC is modeled as a variable capacitive reactance with symmetrical limits (negative (inductive) and positive (capacitive)).

### 8.5.1 Control Strategy

For the system to remain stable following a fault, it is essential that the generator speed (and hence the KE) become zero at some time after the fault clearing. This is possible only if the KE gained during the fault can be completely converted to PE.

The PE of the system will be greater with a capacitor. This fact is utilized in devising the switching strategy for the TCSC, implemented as follows:

1. The capacitor is inserted for the first time as soon as the disturbance is detected.
2. The capacitor is switched off when $(dW_{PE}/dt) = 0$ and $(d\delta/dt) \le 0$.
3. The capacitor is reinserted when $d\delta/dt$ is at a maximum, provided $(d\delta/dt)_{max} \ge \varepsilon$.

In comparing the above switching strategy with that given in Section 8.4, we note that steps 3 and 4 given in the latter are omitted here. In contrast, SSSC is switched between nominal (zero), maximum, and minimum values of reactances assumed to be capacitive (the minimum is negative of the maximum). The maximum (capacitive) reactance for both TCSC and SSSC is assumed to be 0.3 p.u. Note that we are emulating SSSC as a variable reactance for comparison with TCSC. Normally, an SSSC is modeled as a variable reactive voltage source.

The TCSC used in the SMIB system is switched between 0.3 and 0 p.u. A three-phase fault at the generator terminals is assumed. Figure 8.5 shows the plots of the (a) system trajectories, (b) generator rotor angles for the TCSC and SSSC, and (c) variation of the energies (total, potential, and kinetic) for SSSC with a clearing time of 0.170 s. (The variation of the energies for the TCSC is given in Padiyar and Uma Rao (1997)). The classical model of the generator is used. It is seen that the PE reaches a maximum at

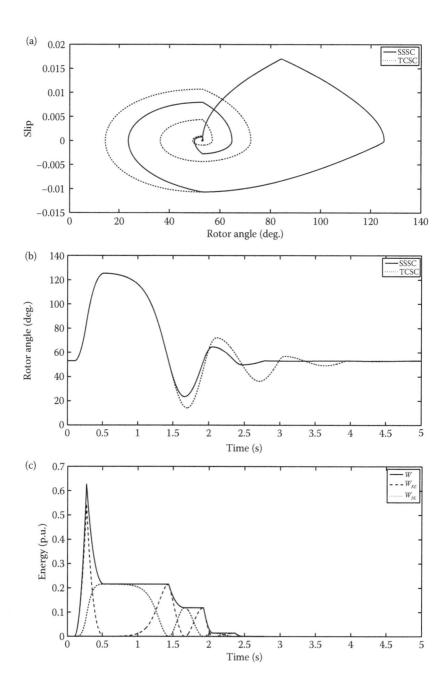

**FIGURE 8.5**
Comparison of TCSC and SSSC (for clearing time = 0.170 s). (a) System trajectories. (b) Rotor angles. (c) Variation of energies for SSSC.

the same time as the KE becomes zero. Hence, the capacitor is switched off at that instant.

Figure 8.6 shows the plots of the (a) system trajectories, (b) generator rotor angles for TCSC and SSSC, and (c) the energies for SSSC, for the same fault, for a clearing time of 0.178 s. It can be observed that the PE reaches a maximum before the KE becomes zero. The TCSC and SSSC are switched to the nominal operating values (of zero) only when the PE again reaches a maximum.

For the classical model of the generator, the CCT is increased from a value of 0.1 s without a FACTS controller to 0.178 s with TCSC/SSSC and the proposed switching strategy. The UEP for the system without the TCSC/SSSC is 127°. With the switching, the system is able to maintain stability up to an angle of 143° (UEP with the capacitor of 0.3 p.u.). The CCT with the line compensated by a fixed capacitor of 0.3 p.u. is 0.191 s. This is because, with the capacitor in the line, the initial rotor angle is less (35°) than the case without the capacitor (53°). Therefore, it can withstand the fault for a longer duration before it reaches the UEP and becomes unstable.

As seen from these studies, the steady state is reached within 4 s for TCSC and 3 s for SSSC. This clearly shows the advantage of switching control (with three levels—nominal power, maximum power, and minimum power). The energy is measured relative to the postfault equilibrium, which corresponds to the nominal value of the controller parameter (and hence, power).

The transient stability limit is increased from 1060 MW (without a controller) to 1285 MW with a discrete control of TCSC/SSSC. (It is assumed that the three-phase fault is cleared in 60 ms).

**Remarks**

1. We can estimate the required control signal $d\delta/dt$ by measuring the bus frequencies at the terminals of the line and computing the frequencies of Thevenin voltages at the two ends of the line. This also requires the computation of Thevenin impedance viewed from the terminal of the lines. Perhaps, the better approach may be to extract the desired control signal by applying the signal processing techniques to the measurements.

2. The transient energy is given by

$$W = W_1(\dot{\delta}) + W_2(\delta, u_{c0}) = \frac{1}{2} M(\dot{\delta})^2 + \int_{\delta_0}^{\delta} [P_e(u_{c0}) - P_m] d\delta \qquad (8.10)$$

where $u_{c0}$ is the operating value of the FACTS controller. In this example, the TCSC/SSSC quiescent reactance is assumed to be zero.

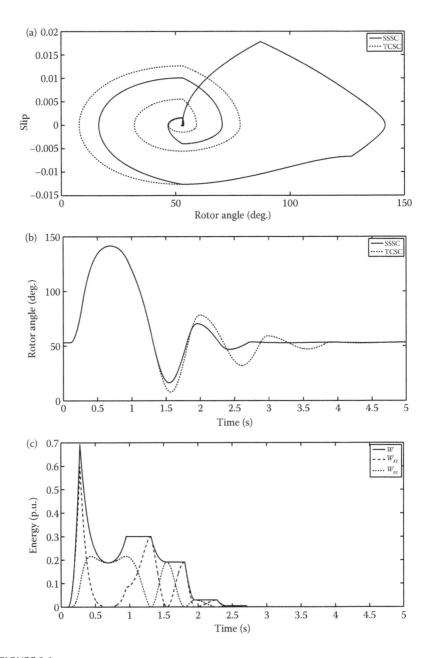

**FIGURE 8.6**
Comparison of TCSC and SSSC (for clearing time = 0.178 s). (a) System trajectories. (b) Rotor angles. (c) Variation of energies for SSSC.

The derivative of the total energy $W$ is given by

$$\frac{dW}{dt} = M\dot{\delta}\ddot{\delta} + [P_e(u_{c0} - P_m)]\dot{\delta} \tag{8.11}$$

Since $M\ddot{\delta}$ is equal to $[P_m - P_e(u_c)]$, we get

$$\frac{dW}{dt} = \dot{\delta}[P_e(u_{c0}) - P_e(u_c)] \tag{8.12}$$

If $P_e(u_c) > P_e(u_{c0})$ when $\dot{\delta}$ is positive, $(dW/dt) < 0$. Similarly, to ensure $(dW/dt) < 0$, we have to arrange $P_e(u_c) < P_e(u_{c0})$ when $\dot{\delta} < 0$.

Assuming that the postfault equilibrium is at the origin in the phase plane (there is no loss of generality in this assumption), we note that the faulted trajectory is in the first quadrant, after the fault is cleared and when $\dot{\delta} > 0$, we switch the TCSC/SSSC (capacitive) reactance to maximum to ensure the postfault trajectory crosses the $\delta$ or $x$-axis of the phase plane and moves to the fourth quadrant. At the crossing of the $x$-axis, $d\delta/dt = 0$. At this point, we arrange $u_c = u_{c0}$ such that $d\delta/dt$ continues to decrease. When $d\delta/dt$ is minimum (and negative), switching $u_c = u_{cl}$ ensures that the trajectory moves to the third quadrant and $d\delta/dt$ continues to increase until it becomes zero. Arranging $u = u_{c0} > u_{cl}$ ensures that the trajectory crosses the $x$-axis and moves to the second quadrant. When $d\delta/dt$ is positive maximum, switching $u_c = u_{cm}$ reduces the total energy.

We note that the total energy reduces in the first and third quadrants, while it remains constant in the second and fourth quadrants. As the angle and speed continue to oscillate, the energy associated with the oscillations continue to decrease until the trajectory approaches the origin and the control action can be stopped. The linear damping controller described in Chapter 7 can take over the job of ensuring stability of the postfault equilibrium.

## 8.6 Discrete Control of STATCOM

The control strategy given in Section 8.4 is applied for a STATCOM connected at the midpoint of the external impedance connected between the generator and the infinite bus as shown in Figure 8.7. The system data are given below:

*Generator data:* $f_B = 60\,\text{Hz}$, $x_d = 1.79$, $x'_d = 0.169$, $T'_{do} = 4.3$, $R_a = 0.0$, $x_q = 1.71$, $x'_q = 0.169$, $T'_{q0} = 0.85$, $H = 5.0$, $D = 0.0$

*AVR data:* $K_A = 200$, $T_A = 0.05$, $E_{fdmax} = 6$, $E_{fdmin} = -6.0$, $R_E = 0.02$, $X_E = 1.0$, $E_b = 1.0$

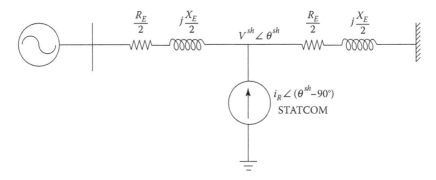

**FIGURE 8.7**
An SMIB system with a STATCOM.

The STATCOM is modeled in two ways:

a. A simplified model of the constant-magnitude reactive current (which can be switched from zero to (capacitive) maximum or (inductive) maximum source of $i_R = 0.336$)

b. A detailed model in $D$–$Q$ variables as described in Chapter 5

The reactive current injected in steady state is assumed to be zero and during a transient, $i_R$ can be switched to $\pm i_{R\max}$.

*(a) Simplified model*

The discrete control strategy is applied to this system by creating a large disturbance. The generator is initially supplying a power of 0.03, with its terminal voltage being 1.0. The STATCOM does not inject any current in steady state. A step change of 0.69 in the mechanical torque at 0.1 s is considered a large disturbance. The system is unstable without STATCOM. The application of the discrete control strategy stabilizes the system as shown in Figure 8.8a. The control action is initiated at the instant of disturbance. The system trajectory projected on the phase plane $(\delta,\omega)$ is plotted in Figure 8.8b.

*(b) Detailed model of STATCOM*

To validate the results obtained using the simplified model, we consider a detailed model in $D$–$Q$ variables. The converter is assumed to be a type 2 converter where the magnitude of the voltage injected is not directly controlled using PWM. Only the phase angles $(\alpha)$ of the injected voltage is controlled (Schauder and Mehta 1993). The controller is shown in Figure 8.9. The expression for the reactive current in steady state is given by the following equation:

$$i_{R0} = \frac{-V(\omega L + (k^2/2G)\sin 2\alpha_0)}{R^2 + k^2(R/G) + \omega^2 L^2} \tag{8.13}$$

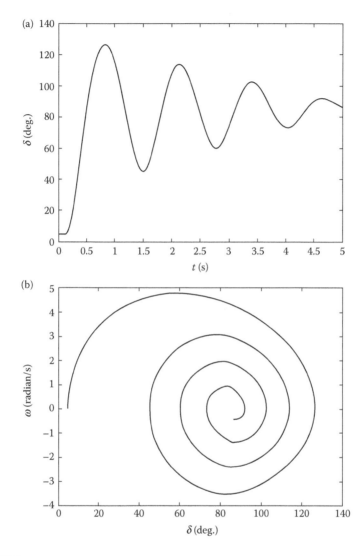

**FIGURE 8.8**
Results with STATCOM (simplified model). (a) Swing curve. (b) System trajectory in a phase plane.

where $V = |V^{sh}|$ and $\alpha$ is the angle by which the injected voltage by STATCOM leads $\hat{V}^{sh}$.

Since $\omega L \ll (k^2/2G)$, the limits on $\alpha$ are selected as $\pm (\pi/4)$ radian. The controller parameters are $K_p = -0.25$, $K_i = -2.5$.

With a discrete control of STATCOM, the system is stabilized for the disturbance involving a step change in $T_m$. It was observed that the swing curve is the same as that shown for the simplified model of the STATCOM. This

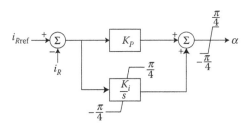

**FIGURE 8.9**
A type 2 controller for STATCOM.

shows that for stability analysis, a simplified model of STATCOM is adequate (just as in the case of SVC).

## 8.7 Discrete Control of UPFC

Consider an SMIB system with a UPFC connected to the line as shown in Figure 8.10. The generator is represented by the voltage source $E_1 \angle \zeta$ in series with the direct axis transient reactance, which is absorbed in $X_1$. For the classical generator model, $\zeta = \delta$, where $\delta$ is the rotor angle. When the system, including UPFC, is assumed to be lossless ($R_1 = 0.0$, $R_2 = 0.0$), the power flow in the line can be expressed as

$$P_e = a_0 + a_1 V_C \cos \beta + a_2 V_c \sin \beta + a_3 I_C \cos \psi + a_4 I_C \sin \psi \qquad (8.14)$$

The equality constraint equation is obtained from the fact that the power loss in UPFC is zero. Thus

$$\text{Re}[\hat{V}_1 I_C \angle -\psi + \hat{I}_2^* V_C \angle \beta] = 0 \qquad (8.15)$$

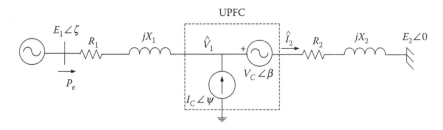

**FIGURE 8.10**
An SMIB system with UPFC.

This equation can be expressed as

$$c_1 V_C \cos \beta + c_2 V_C \sin \beta + c_3 I_C \cos \psi + c_4 I_C \sin \psi = 0 \tag{8.16}$$

Note that Equations 8.14 and 8.16 can be derived by applying the superposition theorem (to determine the currents and voltages in the network due to the three voltage sources ($E_1 \angle \delta$, $E_2 \angle 0$, and $V_C \angle \beta$) and one current source ($I_C \angle \psi$)).

There are inequality constraints imposed by the ratings of the series and shunt converters. These are

$$\left. \begin{array}{l} 0 \leq V_C \leq V_{C\max} \\ 0 \leq I_C \leq I_{C\max} \end{array} \right\} \tag{8.17}$$

The optimization problem is to minimize or maximize $P_e$ (given by Equation 8.14) subject to the constraints Equations 8.16 and 8.17. Kuhn–Tucker conditions are the necessary conditions for the optimality (Luenberger 1984) and when applied here result in the following equations:

$$\frac{\partial P_e}{\partial V_C}(V_C - V_{C\max}) = 0 \tag{8.18}$$

$$\frac{\partial P_e}{\partial I_C}(I_C - I_{C\max}) = 0 \tag{8.19}$$

$$\frac{\partial P_e}{\partial \beta} = 0 \tag{8.20}$$

## Example

For the special case of $X_1 = X_2 = X/2$, $E_1 = E_2 = E$, we can derive the following conditions (Krishna 2003):

For maximum power ($P_e$)

$$V_C = V_{C\max}, \quad I_C = I_{C\max}$$

$$\beta = \frac{\zeta}{2} - \frac{\pi}{2}, \quad \psi = \begin{cases} \dfrac{\zeta}{2} + \dfrac{\pi}{2}, & -\pi < \zeta < 0 \\[2mm] \dfrac{\zeta}{2} - \dfrac{\pi}{2}, & 0 < \zeta < \pi \end{cases}$$

For minimum power ($P_e$)

$$V_C = V_{C\max}, \quad I_C = I_{C\max}$$

$$\beta = \frac{\zeta}{2} + \frac{\pi}{2}, \quad \psi = \begin{cases} \dfrac{\zeta}{2} - \dfrac{\pi}{2}, & -\pi < \zeta < 0 \\[2mm] \dfrac{\zeta}{2} + \dfrac{\pi}{2}, & 0 < \zeta < \pi \end{cases}$$

Substituting these conditions, we can derive

$$P_e^{max} = \frac{EV_{C\,max}}{X} \cos\frac{\zeta}{2} - \frac{EI_{C\,max}}{2} \sin\frac{\zeta}{2} + P_0, \quad -\pi < \zeta < 0$$

$$= \frac{EV_{C\,max}}{X} \cos\frac{\zeta}{2} + \frac{EI_{C\,max}}{2} \sin\frac{\zeta}{2} + P_0, \quad 0 < \zeta < \pi$$

$$P_e^{min} = \frac{-EV_{C\,max}}{X} \cos\frac{\zeta}{2} + EI_{C\,max} \sin\frac{\zeta}{2} + P_0, \quad -\pi < \zeta < 0$$

$$= \frac{-EV_{C\,max}}{X} \cos\frac{\zeta}{2} - \frac{EI_{C\,max}}{2} \sin\frac{\zeta}{2} + P_0, \quad 0 < \zeta < \pi$$

where $P_0 = (E^2/X) \sin\delta$ is the power flow in the absence of UPFC.

The effective series impedance $(R_{se} + jX_{se})$ and shunt admittance $(G_{sh} + jB_{sh})$ of UPFC for maximum and minimum power are shown in Figures 8.11 through 8.14 (for $X_1 = X_2 = 0.5$, $E = 1.0$, $V_{Cmax} = 0.5$, and $I_{Cmax} = 0.5$).

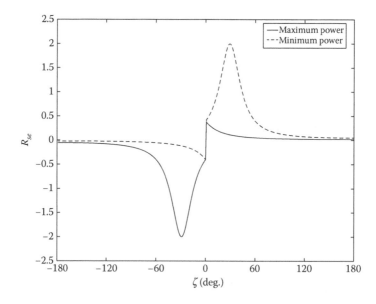

**FIGURE 8.11**

Variation of the effective series resistance of the UPFC.

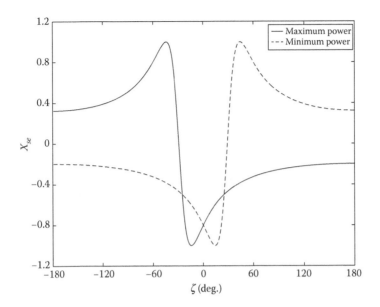

**FIGURE 8.12**
Variation of the effective series reactance of the UPFC.

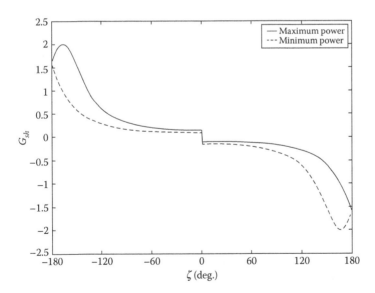

**FIGURE 8.13**
Variation of the effective shunt conductance of the UPFC.

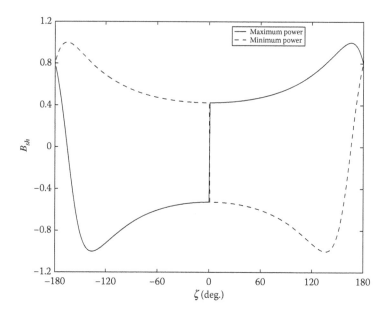

**FIGURE 8.14**
Variation of the effective shunt susceptance of the UPFC.

It can be seen from these figures that the plots of various quantities for the minimum power flow are inverted mirror images (about $y$-axis) of the respective parameters applicable to maximum power flow. Also, the shunt susceptance of the UPFC is not at the maximum value, but varies with the angle $\zeta$. In Mihalic et al. (1996), the control variables are chosen as series voltage magnitude, series voltage angle, and effective shunt susceptance. The maximum value of the effective shunt susceptance is assumed as the optimum value to maximize power. But the effective shunt susceptance need not be constant (at the limit) at all values of $\zeta$ as shown in Figure 8.14.

The power angle curves for (i) without UPFC, (ii) UPFC parameters set at maximum power, and (iii) UPFC parameters at minimum power are shown in Figure 8.15. For improving transient stability, the power flow should be maximized as soon as disturbance is detected. The effect of the location of UPFC on transient stability can be quantified by computing the area under the power angle curve $\left( \int_0^\pi P_e d\zeta \right)$ for various locations. It was observed (Krishna 2003) that there is no significant variation in the area below the power angle curve with respect to the location of UPFC in a line. For the values of the system parameters chosen, the optimal location of the UPFC is at the midpoint of the line. The area below the power angle curve without UPFC is 2.0 and about 3.5 with UPFC at the midpoint (note that the power flow expressions given earlier do not apply unless $X_1 = X_2$).

The power through the DC link (power transferred from the series branch to the shunt branch) for maximum and minimum power can be derived for

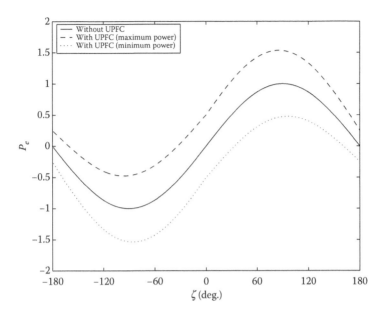

**FIGURE 8.15**
Power angle curves with and without UPFC.

the special case ($E_1 = E_2 = E$ and $X_1 = X_2 = X/2$) as follows. The voltage $\hat{V}_1$ at the node at which the shunt branch of UPFC is connected is obtained by superposition.

$$\hat{V}_1 = \frac{E\angle\zeta}{2} + \frac{E}{2} + \frac{V_{C\max}\angle\beta}{2} + j\frac{I_{C\max}\angle\psi}{4}X \tag{8.21}$$

The power through the DC link $P_{dc}$ is given by

$$P_{dc} = \text{Re}[\hat{V}_1 I_{C\max}\angle{-\psi}] = \frac{EI_{C\max}}{2}\cos(\zeta - \psi) + \frac{EI_{C\max}}{2}\cos\psi$$

$$+ \frac{V_{C\max}I_{C\max}}{2}\cos(\beta - \psi) \tag{8.22}$$

By substituting the expressions for $\beta$ and $\psi$ given earlier, the following expression for $P_{dc}$ is obtained. For both maximum and minimum power

$$P_{dc} = \begin{cases} -\dfrac{V_{C\max}I_{C\max}}{2}, & -\pi < \zeta < 0 \\[4mm] \dfrac{V_{C\max}I_{C\max}}{2}, & 0 < \zeta < \pi \end{cases} \tag{8.23}$$

It can be seen that for this special case, the power through the DC link for maximum and minimum power depends only on the ratings of the shunt and series converters and does not depend on the system parameters. Note that the power in the DC link flows from the series converter to the shunt-connected converter.

If the power through the DC link is constrained to be zero, the series and the shunt branches inject reactive voltage and reactive current, respectively. It can be shown that the optimal values of $V_C$ and $I_C$ are not at their maximum values for all values of $\zeta$ ($0 < \zeta < \pi$). It can be shown that the optimal value of $I_C$ increases from zero to $I_{Cmax}$ as $\zeta$ increases from 0 to $\zeta_1$. On the other hand, $V_C$ decreases from $V_{Cmax}$ to 0 as $\zeta$ increases from $\zeta_2$ to $\pi$. Not constraining $V_C$ and $I_C$ to be at their limits will be helpful at lower values of steady-state power.

### 8.7.1 Application of Control Strategy to SMIB System with UPFC

The discrete control is applied to the SMIB system with UPFC, the data for which are given earlier (Section 8.6). The generator is represented by the two-axis model, with a static exciter. The network and stator dynamics are neglected. The UPFC is located in the midpoint of the transmission line, such that $X_1 = X_2 = 0.5$. The reactance $X_1$ includes the generator direct axis transient reactance $x'_d$. The UPFC is represented by series voltage and shunt current sources; the series voltage rating ($V_{Cmax}$) is 0.2 and the shunt current ration ($I_{Cmax}$) is 0.1.

The control strategy is applied to this system by creating a large disturbance. The generator is initially supplying power of 0.03, with its terminal voltage being 1.0. The series voltage and shunt current of UPFC are zero in steady state. A step change of 0.69 in the mechanical torque at 0.1 s is considered as a large disturbance. The system is unstable without any control. The application of a control strategy stabilizes the system as shown in Figure 8.16a; the control is initiated at 0.1 s. MATLAB® optimization toolbox (Coleman et al. 1999) is used to obtain optimum values of the UPFC control variables. The series reactive voltage and the shunt reactive and active currents of the UPFC are plotted in Figure 8.16b–d. The variation of the shunt active current is similar to that of the power through the DC link flowing from the series to the shunt converter.

### 8.7.2 Discussion

The UPFC has three independent control variables. The set of independent control variables is not unique. The control variables are chosen as magnitude and angle of the injected series voltage and magnitude of the shunt current. The constraints on the ratings of the series and the shunt converters can be easily expressed due to the selection of these control variables.

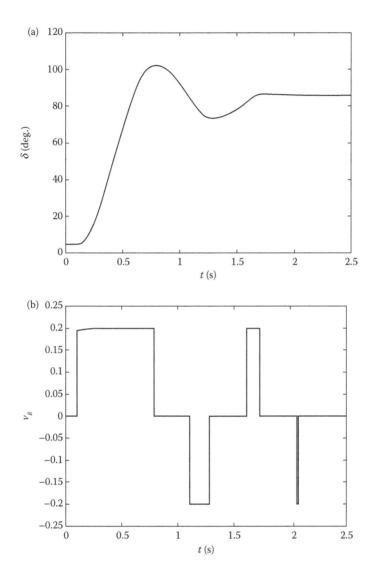

**FIGURE 8.16**
Effect of UPFC. (a) Swing curve. (b) Series reactive voltage. (c) Shunt reactive current (injected). (d) Shunt active current (injected).

When the power through the DC link is constrained to be zero, the optimal magnitudes of the injected voltage and current are not at the limits for large values of $\zeta$ ($\zeta$ near $\pi$) and small values of $\zeta$ ($\zeta$ near 0), respectively. This illustrates that for small values of $\zeta$, an SSSC is helpful in increasing the power, whereas for large values of $\zeta$, a STATCOM is more effective.

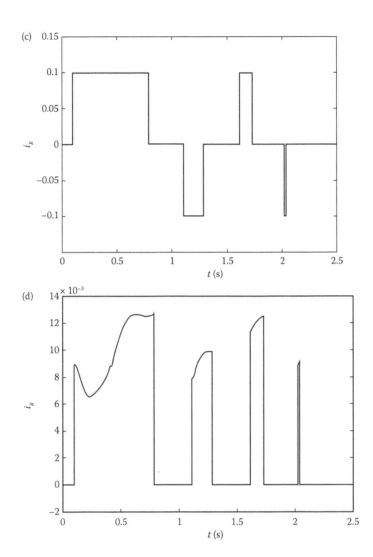

**FIGURE 8.16**
Continued.

## 8.8 Improvement of Transient Stability by Static Phase-Shifting Transformer

It was mentioned earlier that a TCPAR or SPST can help improve the transient stability limit even though it does not increase the power flow in a line. In steady state, a PST helps control the power flow in a line to ensure that it is not overloaded and the power outputs from power stations are redistributed to maintain static security. However, with power electronic control, it is feasible to improve dynamic security (Edris 1991).

Consider the SMIB system where the generator transformer is also configured to work as an SPST. Assume a three-phase fault occurs at the sending end of one of the transmission lines, which is cleared by tripping the faulted line section. The power angle curves for (i) prefault and (ii) postfault cases are shown in Figure 8.17 for the case without SPST. If area $A_2$ < area $A_1$, then the system will be unstable. Since the electrical power output $P_e$ changes to

$$P_e = \frac{E_g E_b}{X} \sin(\delta \pm \phi) \tag{8.24}$$

in the presence of the SPST, the power angle curves for this case are shown in Figure 8.18. Here, it is assumed that

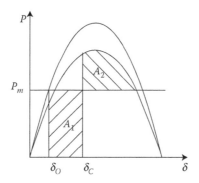

**FIGURE 8.17**
Power angle curves without SPST.

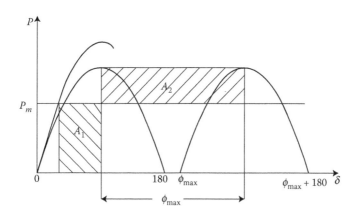

**FIGURE 8.18**
Power angle curves with controlled SPST.

a. The SPST is activated when the electrical power output is maximum and $dP_e/dt$ changes from positive to negative

b. The control algorithm is given by

$$\frac{d\phi}{dt} = \frac{d\delta}{dt} \tag{8.25}$$

The decelerating area ($A_2$) available now is (shown in Figure 8.18) much larger than $A_1$ shown in Figure 8.17.

Figure 8.18 shows only the condition for the first (forward swing). The control algorithm of Equation 8.25 can also be applied during the back swing to improve system stability.

Of course, this simple explanation indicates mainly the capability of the fast control of the phase angle shift of the SPST in improving the first swing stability of a power system. In practice, different control algorithms can be used to realize the benefits while taking into account the various constraints.

It is interesting to note that a generator connected to a transmission line through an SPST with a range of 360° phase shift can operate in asynchronous mode just as if it was connected to an HVDC line. However, it is cheaper to install a BTB HVDC link to provide asynchronous interconnection. Interestingly, a variable frequency transformer (VFT) is a new device that uses a rotary transformer with three-phase windings on both stator and rotor (McNabb et al. 2005). A drive system adjusts the VFT rotor to control the phase shift between the networks connected by the VFT. The first installation of this new technology is located at Langlois substation, interconnecting the New York (USA) and Hydro Quebec (Canada) system. It controls power transfer up to 100 MW in both directions. The development of VFT shows that a 360° phase shift is feasible with a rotary device, which is essentially based on the induction regulator—an old technology. However, the control and drive system for VFT are based on modern technology.

Another simple approach of providing phase shift when there is imminent loss of synchronism is to disconnect two groups of generators and then reclose with 120° phase rotation (Cresap et al. 1981). Transient stability is also improved because fast valving or excitation control is allowed more time to become effective.

## 8.9 Emergency Control Measures

There are occasions when the discrete controllers discussed in the previous sections are unable to prevent instability of the system. If no further action is taken, then the system may break apart due to uncontrolled tripping of generators (by out-of-step protection) and formation of islands, which are unstable.

There are two emergency control measures (to be used as last resort) to prevent a catastrophic failure of power systems leading to blackouts. These are

1. Controlled system separation and load shedding
2. Generator tripping

### 8.9.1 Controlled System Separation and Load Shedding

The objective here is to achieve a near balance between load and generation within each island formed by tripping the transmission lines connecting the areas. Controlled system separation is applied in the following situations:

a. To prevent transmission line overloading following disturbances and loss of other lines.
b. When oscillatory instability occurs between load/generation areas during certain system conditions. Interties between areas can be programmed to open if the rate and magnitude of power oscillations over the tie exceed permissible values.

However, interties should not be opened until the benefits of maintaining power interchange among areas are exhausted and the need of the hour is to retain as much of the system intact as possible. Thus, controlled system separation goes together with load shedding when there is deficit of generation, and generator tripping when there is excess.

The controlled system separation may often be done manually since the response time for the operator intervention is adequate to prevent collapse in many systems. When automatic separation is applied, voltage, current, power, or frequency transducers are used to detect limit violations. Controlled separation is not widely used to improve system stability. This is because the boundaries for system separation are not well defined for all possible system conditions. It has been applied in the western system in the United States to reduce the effects of severe disturbances in one area on the rest of the areas that are connected as a ring (or doughnut).

In Chapter 6, it was shown that controlled system separation by tripping the lines belonging to the cutset is a viable option. However, more studies are required before practical implementation.

Load shedding programs have been used by many utilities, in distribution systems or major industrial loads. The objective is to prevent frequency decay and maintain equilibrium between generation and load when there is loss of generation. Load shedding can help in preventing interties from opening due to transmission overloads. Load shedding is initiated by underfrequency relays based on discrete underfrequency values or rates of frequency decay. Load shedding is generally done in 3–6 steps to prevent excessive load dropping after frequency levels off at an acceptable value. The underfrequency relay settings are based on the limitations on the underfrequency operation of turbine generators and power plant auxiliaries. When voltage instability is a problem, undervoltage-actuated relays are also used for load shedding (Taylor 1994).

### 8.9.2 Generator Tripping

The selective tripping of generators for transmission line outages has been used extensively to improve stability. Generator tripping is one form of power control and can be compared to fast valving and dynamic braking. If $N$ number of identical generating units in parallel are connected to an infinite bus through an external reactance of $x_e$, the tripping of one unit is equivalent to decreasing $x_e$ by the ratio $(N-1)/N$. This improves both steady-state and transient stability.

Generator tripping has been mainly employed to improve stability of remote generation. It can also be used to improve interconnected system operation where tripping of an intertie can lead to instability. In the western United States, the tripping of Pacific HVDC intertie can lead to instability under certain conditions. The automatic tripping of certain hydrogenerators in the northwest helps in controlling the power flow on the parallel 500 kV AC ties, thereby averting transient instability.

Generator tripping is initiated from a transfer trip scheme or by arranging the protection scheme at the power plant such that when a transmission line is tripped following a line fault, one or more generators are also automatically tripped.

The impact of tripping and the consequent full load rejection on a thermal unit needs to be studied because of the response of the prime mover and the action of the overspeed controller can vary. When a thermal unit is tripped, the unit will normally go through its standard shutdown and startup cycle and full power may not be available for some hours. To overcome this problem, one method is to connect the station load to the unit tripped. The unit can then be rapidly reloaded after the disturbance is cleared. This requires that the unit and its controls be specifically designed for this mode of operation in which case, the unit can be resynchronized to the system and full load restored within about 15–30 min. However, it is essential that frequent tripping of thermal generators be avoided.

It is possible to apply energy function-based methods to determine the optimum strategy for generator dropping as well as load shedding in emergency situations. Some early attempts are reported by Zaborsky et al. (1981) and Fink (1985). Takahashi et al. (1988) describe a technique for fast generation shedding based on the observations of swings of generators.

# 9

## Application of FACTS Controllers for Emergency Control—II

### 9.1 Introduction

In the previous chapter, we considered the application of series and shunt FACTS controllers for stabilizing an SMIB system when subjected to a major disturbance (three-phase fault followed by clearing). The objectives of the controllers are to (a) ensure that the system is transiently stable in the first swing and (b) damp the nonlinear oscillations (of power flow, generator rotor angle) and steer the system trajectory close to the postfault equilibrium state, which is assumed to be stable. Once the system state is close to the stable equilibrium point, the emergency control action can be discontinued. In the literature, such control action has been termed as discrete in contrast to the continuous control action by power swing damping controllers, which are auxiliary or supplementary controllers in conjunction with power scheduling or voltage regulating controls. It is to be noted that damping controllers are designed to stabilize the system under normal operating conditions when generation is able to meet the varying load demand, without violating any inequality or transient stability constraints. The control action by these damping controllers is limited to prevent adverse interaction with voltage regulation or power scheduling control functions.

In this chapter, we extend the emergency control function to a multimachine system. In particular, we take up case studies on the New England 10-generator test system. Two FACTS controllers are considered—TCSC and UPFC. It will be shown that the control algorithm is general enough to consider more than one FACTS controller in appropriate locations in the system.

### 9.2 Discrete Control Strategy

From Proposition 1 (given in Chapter 6), a multimachine system initially breaks up into two groups of generators with each group remaining in synchronism.

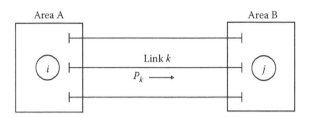

**FIGURE 9.1**
A two-area system separated by the critical cutset.

From Proposition 2, the system separates along a unique cutset (called the critical cutset). The objective of the discrete control action is to stabilize the system by utilizing one or more FACTS controllers located in one or more lines belonging to the cutset. Figure 9.1 shows a two-area system (areas A and B) separated by a critical cutset. Assuming that a fault occurs within area A, the generators within area A accelerate with respect to the generators in area B. The COI in the two areas are described by the following equations:

$$M_A \frac{d^2\delta_A}{dt^2} = \sum_{i\in G_A}(P_{mi} - P_{ei}) - \sum_{k=1}^{nc} P_{tk} \tag{9.1}$$

$$M_B \frac{d^2\delta_B}{dt^2} = \sum_{j\in G_B}(P_{mj} - P_{ej}) + \sum_{k=1}^{nc} P_{tk} \tag{9.2}$$

where $M_A = \sum_{i\in G_A} M_i$, $M_B = \sum_{j\in G_B} M_j$, and $\delta_A$ and $\delta_B$ are COA of areas A and B, respectively.

Multiplying Equation 9.1 by $M_B/M_T$ and Equation 9.2 by $M_A/M_T$, and subtracting Equation 9.2 from Equation 9.1, we get

$$M_{eq}\frac{d^2\delta_{eq}}{dt^2} = \frac{M_B}{M_T}P_{aA} - \frac{M_A}{M_T}P_{aB} - \sum_{k=1}^{nc} P_{tk} \tag{9.3}$$

where

$$P_{aA} = \sum_{i\in G_A}(P_{mi} - P_{ei}), \quad P_{aB} = \sum_{j\in G_B}(P_{mj} - P_{ej})$$

and

$$M_T = M_A + M_B, \quad M_{eq} = \frac{M_A M_B}{M_T}, \quad \delta_{eq} = \delta_A - \delta_B$$

Note that $G_A$ and $G_B$ are sets of buses where generators in areas A and B are connected. $P_{tk}$ is the power flow in line $k$ belonging to the cutset and $nc$ is the number of series elements in the cutset.

Since, the generators in area A (as well as B) are assumed to be strictly coherent, it can be assumed that the bus frequencies (during the transient following a disturbance) in area A are identical and equal to $d\delta_A/dt$. Similarly, the bus frequencies in area B are also identical and equal to $d\delta_B/dt$. The PE of the total system (including areas A, B, and the tie lines) is given by

$$W_{PE} = \sum_{i \in G_A} \int_{\theta_{i0}}^{\theta_i} (P_{ei} - P_{mi})d\theta_i + \sum_{j \in G_B} \int_{\theta_{j0}}^{\theta_j} (P_{ej} - P_{mj})d\theta_j$$

$$= \sum_{k \in L_A} \int_{\phi_{k0}}^{\phi_k} (P_k - P_{ks})d\phi_k + \sum_{r \in L_B} \int_{\phi_{r0}}^{\phi_r} (P_r - P_{rs})d\phi_r + \sum_{k=1}^{nc} \int_{\phi_{k0}}^{\phi_k} (P_{tk} - P_{tks})d\phi_k \qquad (9.4)$$

where $L_A$ and $L_B$ are the sets of lines in areas A and B, respectively.

We can express the total PE as

$$W_{PE} = W_{PEA} + W_{PEB} + W_{PET} \qquad (9.5)$$

where $W_{PEA}$ and $W_{PEB}$ are the potential energies evaluated in the lines of areas A and B, respectively. $W_{PET}$ is the PE evaluated for the lines in the cutset. If we assume frequencies of all the buses in area A as identical (neglecting intermachine oscillations), $W_{PEA} = 0$. Similarly, $W_{PEB} = 0$ (assuming identical bus frequencies in area B). Thus, we can write

$$W_{PE} = W_{PET} = \sum_{k=1}^{nc} \int_{\phi_{k0}}^{\phi_k} (P_{tk} - P_{tks})d\phi_k \qquad (9.6)$$

Note that $P_{tks}$ is the power flow in line $k$ (of the cutset) in postfault steady state.

Following a fault in area A, the generator rotors accelerate and gain KE. The system can be stabilized (prevent loss of synchronism) in the first swing if the power flow in one or more lines in the cutset can be maximized using FACTS controllers connected in these lines. If there is a choice available, the FACTS controller can be located in the lines that have high energy margin sensitivities (see Chapter 7).

The control strategy described in the previous chapter can be extended to the multimachine power systems. The control algorithm is applied to individual lines in the cutset and requires measurements of

$$\frac{d\phi_k}{dt} = \frac{d\phi_i}{dt} - \frac{d\phi_j}{dt}$$

where $\phi_k$ is the angle across line $k$ connected between node $i$ (in area A) and node $j$ (in area B).

Since the improvement of transient stability requires increase in the PE associated with the cutset (which is the sum of energies of the lines belonging to the cutset), the control action is decoupled and no adverse interactions are anticipated if multiple FACTS controllers are used.

To summarize, the control algorithm applicable to individual FACTS controller in a line is given below:

1. As soon as a large disturbance is detected, the control variables (1–3) associated with a FACTS controller are set at values that result in maximum power flow in the line.
2. The control variables are switched to the postfault operating values when

$$\frac{d\phi_k}{dt} = 0 \quad \text{or} \quad P_k = P_{ks} \quad \text{and} \quad \frac{d\phi_k}{dt} < 0$$

3. The control variables are set at values such that power flow is minimized when $d\phi_k/dt$ is minimum and $d\phi_k/dt < -\varepsilon$.
4. The control variables are switched to their operating values when $d\phi_k/dt = 0$.
5. The control variables are reset to maximize the power flow when $d\phi_k/dt$ is maximum and $d\phi_k/dt > \varepsilon$.
6. Go to step 2 and continue unless $|d\phi_k/dt| < \varepsilon$.

The detection of a major disturbance can be done by monitoring the bus frequencies. If there is a fault in area A, there will be a rapid increase in frequency of bus $i$.

## Remarks

1. It is not necessary to locate the FACTS controllers in the critical cutset. Note that the identification of the critical cutset (described in Chapter 6) helps to accurately determine the MOI. With the knowledge of MOI, we can choose the cutset lines where FACTS controllers can be located based on sensitivity information.
2. The discrete control action can be initiated even when the system is first swing stable. This is to ensure that the system does not lose synchronism during subsequent swings and reaches stable equilibrium at the earliest. This strategy improves system security as the contingencies are often unanticitipated in a large system.

3. In an SMIB system, the rotor velocity deviation ($d\delta/dt$) signal can be synthesized (based on local measurements) and applied for the implementation of the discrete control strategy. However, in a multimachine system, the desired control signal $d\delta_{eq}/dt$ is not directly available. Note that $d\phi_k/dt$ includes higher-frequency intermachine oscillations in the two areas. However, it is feasible to apply filtering techniques to extract the desired signal.

4. On the basis of case studies carried out, it is observed that in step 2 of the control algorithm, the condition of $P_k = P_{ks}$, $(d\phi_k/dt) \leq 0$, is not encountered. This simplifies the control strategy as the knowledge of $P_{ks}$ is not required.

## 9.3 Case Study I: Application of TCSC

The system considered is the New England 10-generator test system whose data are given in Appendix D. The generators are represented by classical models and the loads are modeled as constant impedances. A three-phase fault is applied at bus 26 and two cases are considered: (a) the fault is cleared without any line tripping and (b) the fault is cleared followed by tripping of the line connected between buses 26 and 28.

The mode of instability for this fault is the separation of generator 9 from the rest. The critical cutset for case (a) is the generator transformer connected between buses 9 and 29. For case (b), with line tripping, the critical cutset is the line connecting buses 26 and 29. However, based on the MOI of separation of generator 9, the following lines can be considered as part of one or more cutsets:

(i) 26–29; (ii) 26–28; (iii) 28–29; (iv) 26–27; and (v) 26–25

Providing a switchable TCSC in one or more lines is considered. The operating value of TCSC (capacitive) reactance is assumed to be zero and the maximum (switched) reactance is assumed to be 50% of the individual line reactance. As discussed in the previous chapter, steps 3 and 4 in the control algorithm are omitted as the TCSC reactance is switched between two values—operating and the maximum.

### 9.3.1 Fault at Bus 26: Without Line Tripping

From studies carried out (Uma Rao 1996), it was observed that a TCSC in line 26–29 is the most effective compared to other locations. The output of the generator 9 is 830 MW in the base case. If the fault at bus 26 is cleared in three cycles without any line outage, the system remains stable if the output of the generator is less than 1120 MW. Inserting a TCSC in line 26–29 with

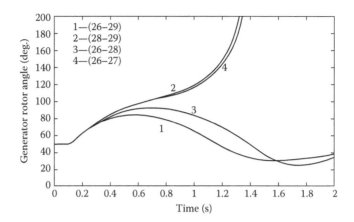

**FIGURE 9.2**
Swing curves of generator 9 with one TCSC connected in one of the four lines.

maximum (capacitive reactance) of 50% soon after the fault clearing results in increasing the transient stability limit by 100 MW (note that TCSC reactance is assumed to be zero in steady state).

Comparisons of the performance of TCSCs located in other three lines (26–28, 28–29, and 26–27) are depicted in Figure 9.2, which shows the swing curves of generator 9 with one TCSC located in any one of the four lines. It is observed that the system is also stabilized with a TCSC in line 26–28, but the TCSCs in other two lines, 28–29 and 26–27, are not effective. Applying two TCSCs in lines 26–29 and 26–28 increases the stability limit by about 150 MW to over 1260 MW. Installing TCSC in four lines (26–29, 26–28, 26–27, and 26–25) can increase the transient stability limit up to 1320 MW.

The total energy $(W_T)$, PE $(W_{PE})$, and KE $(W_{KE})$ in the system is computed and plotted in Figure 9.3 as a function of time with one TCSC in line 26–29. It is observed that the total energy decreases from 0.5 to 0.3 within 5 s of the fault clearing. With four TCSCs, the energy decreases to about 0.25 (see Figure 9.4). The switching instants of TCSCs in line 26–29 and 26–28 were found to be nearly coincident, whereas the switching instants of TCSCs in lines 26–27 and 26–25 were not coincident (not shown here). This is due to the fact that frequencies of buses 25 and 27 are not strictly coherent (identical) even though the nine generators in the system (excluding generator #9) are in synchronism. This is due to the fact that the bus frequencies are affected by the intermachine oscillations. In contrast, frequencies of buses 29 and 28 are nearly identical.

## Remarks

1. Unlike in the SMIB case, the damping of oscillations is slow, particularly when one TCSC is utilized for discrete control. However,

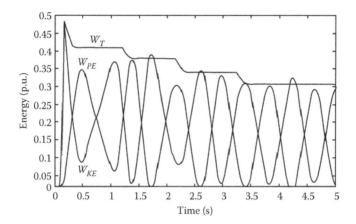

**FIGURE 9.3**
Variation of total, potential, and kinetic energies with TCSC in line 26–29.

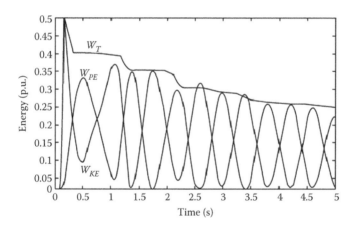

**FIGURE 9.4**
Variation of total, potential, and kinetic energies with TCSCs in four lines.

application of multiple TCSC improves the situation. It is to be noted that the generator or load damping were neglected in the simulation study.

2. The damping can be improved by application of SSSC instead of TCSC, as SSSC enables minimization of power flow in the line by injecting inductive reactive voltage, when required. However, SSSC is a costly FACTS controller as the VSC is interfaced with the line using a series-connected transformer. The problems of harmonic cancellation and protection of GTO/IGBT devices during line faults can impact costs and reliability. In contrast, TCSC has been applied extensively in many systems worldwide.

3. The CA control used for power scheduling (see Chapter 5) regulates voltage drops across all lines connected in parallel (in addition to the line where the TCSC is applied). It was observed (Uma Rao 1996) that the CA control applied for a single TCSC connected in line 26–29 resulted in increasing the stability limit of generator 9 to the same level as the TCSC with discrete control in that line. However, CA (or strictly speaking, constant voltage drop) control does not provide damping of the oscillations.

### 9.3.2 Fault at Bus 26: Cleared by Line Tripping

When the three-phase fault at bus 26 for three cycles is cleared by tripping line 26–28, the TCSC in line 26–29 is more effective than the previous case without line tripping. The stability limit of generator 9 in this case is only 890 MW without any TCSC. Providing a TCSC in line 26–29 (with 50% compensation), which is switched on at the time of fault clearing, results in increasing the stability limit to more than 1050 MW (an increase of over 160 MW). Providing an additional TCSC in line 26–25 results in increasing the stability limit to over 1130 MW.

In contrast, the application of an SVC at bus 29 (for the same disturbance) results in increasing the stability limit to over 1080 MW (see Chapter 5). It is to be noted that the SVC is assumed to be permanently connected (not switched) providing voltage regulation. In contrast, the TCSC is assumed to provide discrete control in response to a major disturbance and is switched off in steady state. However, the voltage regulation in an SVC does not provide extra damping following a major disturbance unless additional damping controller is provided. The transient stability limit is essentially based on the first swing stability.

## 9.4 Case Study II: Application of UPFC

The application of a UPFC in SMIB systems is described in the previous chapter. Unlike a TCSC with one control variable, a UPFC has three control variables: real and reactive voltages injected by the series-connected VSC and reactive current drawn by the shunt-connected VSC. The three variables have to be determined at each switching instant when the power flow in the line has to be maximized or minimized or set at the operating value. This requires a constrained optimization (with both equality and inequality constraints) procedure, unless explicit expressions for the power flow are obtained. It was shown that when the UPFC is located at the midpoint of a lossless line with terminal voltage phasors specified (independent of the power flow in the line and with equal voltage magnitudes), it is possible to express the power flow and other quantities explicitly in terms of the control variables. The expressions for maximum and minimum power flow are given in Chapter 8.

Here, we will consider the application of the control strategy to multimachine systems taking the case study of a 10-generator system. For a fault at bus 34 cleared by tripping the line 34–35 at 0.267 s, the system is unstable; generator 2 separates from the rest of the system. The lines 11–12 and 18–19 form the critical cutset.

### 9.4.1 Single UPFC

The control strategy is evaluated by a simulation study (Krishna 2003, Krishna and Padiyar 2005). The generators are represented by classical models. Loads are assumed to be of constant impedance type. Network losses are ignored.

First, we consider a single UPFC connected in line 11–12 at the bus 11. A rating of 0.2 is used for both series voltage and shunt current of UPFC. It is assumed that the UPFC is represented by a two-port network shown in Figure 9.5. The external network can be represented by its Thevenin equivalent at the two ports of UPFC. By considering Thevenin voltage at port 2 as reference, the two-port network equation can be written as

$$\begin{bmatrix} E_1 \angle \delta \\ E_1 \angle 0 \end{bmatrix} = - \begin{bmatrix} Z_{11} & Z_{12} \\ Z_{12} & Z_{22} \end{bmatrix} \begin{bmatrix} \hat{I}_1 \\ \hat{I}_2 \end{bmatrix} + \begin{bmatrix} \hat{V}_1 \\ \hat{V}_2 \end{bmatrix} \tag{9.7}$$

For optimization of power flow with UPFC, the open circuit voltages at the UPFC ports are computed at every step. The optimal values of $V_C \angle \beta$ and $I_C \angle \psi$ are obtained and from these values, the currents $\hat{I}_1$ and $\hat{I}_2$ are computed. These currents are used in the following network equation to obtain the network voltages:

$$[Y]\hat{V} = \hat{I} \tag{9.8}$$

where $[Y]$ is the bus admittance matrix, $\hat{V}$ is the vector of network voltages, and $\hat{I}$ is a vector of currents that includes generator currents (generator being

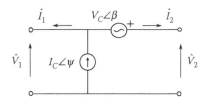

**FIGURE 9.5**
A UPFC as a two-port network.

represented by its Norton's equivalent circuit) and the currents at the UPFC ports.

MATLAB® optimization toolbox (Coleman et al. 1999) is used for the optimization of power. At each step of simulation, the solution obtained for the control variables at the previous step is taken as the starting value for optimization. The analytical solution obtained in the previous chapter for the lossless circuit is taken as the starting value for optimization at the first step of simulation.

UPFC does not inject any voltage or current in steady state. With control initiated at the instant of fault clearing, the system is stable. The system is also stable when the control is initiated at the instant of instability prediction (using the algorithm described in Chapter 6. Figure 9.6 shows the plots of total system energy for the two cases: (a) the control initiated at the instant of fault clearing and (b) the control initiated at the instant of fault prediction (0.7 s). Without any control, the system energy in the postfault period would have been constant due to the absence of damping whereas every switching of the UPFC control variables to maximize or minimize the power flow in the line contributes to damping, which can be seen by the decrease in energy during these periods. The decay of the total energy (with respect to the postfault SEP) is delayed when the control is initiated 0.7 s after the fault clearing. However, the system remains stable and the trajectories are expected to reach the postfault stable equilibrium state.

Figure 9.7 shows the plot of power flow in the line in which UPFC is situated (when the control is initiated at the instant of fault clearing). The

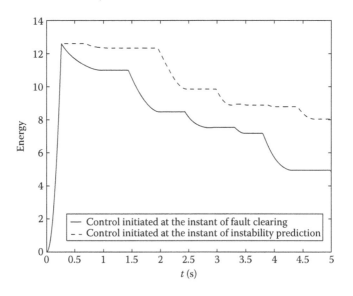

**FIGURE 9.6**
Plots of variations of total energies.

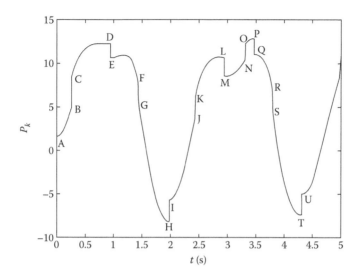

**FIGURE 9.7**
Variation of power flow in line 11–12 with UPFC.

discontinuities in the curves are due to the switchings between operating values of power, maximum power, and minimum power. Curve AB is the power during fault. The fault is cleared at point B. The power is maximized as soon as the fault is cleared. Maximization of power at the instant of fault clearing results in the discontinuity BC. Curve CD corresponds to the setting of UPFC control variables to obtain maximum power. The control variables are switched to their operating values at point D when the rate of change of phase angle $d\delta_k/dt$ across line 11–12 becomes zero. $d\delta_k/dt$ for the postfault network (with base values of UPFC control variables) is plotted in Figure 9.8. The discontinuities in the plot of power correspond to points in the plot of $d\delta_k/dt$. Curve EF corresponds to the operating values (with $V_C = 0$ and $I_C = 0$). The power is minimized at point F when $d\delta_k/dt$ reaches a minimum. Curve GH corresponds to the minimum power flow. The control variables are switched to their operating values at point H when $d\delta_k/dt$ becomes zero. This switching sequence is continued till the oscillations are damped. With reference to Figure 9.7, the curves IJ, MN, and QR are for base values ($V_C = 0$, $I_C = 0$) of control variables, the curves KL and OP are for maximum power, and the curve ST is for minimum power flow. It is to be noted that at point L, $d\delta_k/dt$ becomes zero and control variables are switched to their operating values. There is a discontinuity LM in power due to this switching. After this instant, $d\delta_k/dt$ increases and reaches a maximum at point N. Therefore, the power flow is again maximized at point N.

The swing curves are plotted in Figure 9.9 for case (b) when the control is initiated at the instant of instability detection (0.7 s after the fault clearing). The control variables $V_P$, $V_R$, and $I_R$ of the UPFC are plotted in Figure 9.10a–c.

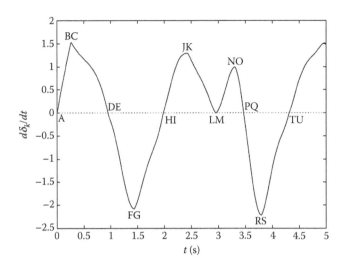

**FIGURE 9.8**
Variation of rate of change of phase angle across line 11–12.

Here, $V_P$ is the voltage drop in the direction of the line current and $I_R$ is the reactive current injected by the shunt-connected VSC. The power through the DC link (power transferred from the series branch to the shunt branch) of UPFC is shown in Figure 9.10d. Note that these variables are obtained when the control action is initiated at the instant of instability prediction, whereas the control variables shown in Krishna and Padiyar (2005) are obtained for the

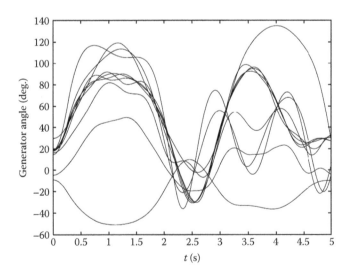

**FIGURE 9.9**
Swing curves with UPFC control action initiated at the instant of instability prediction.

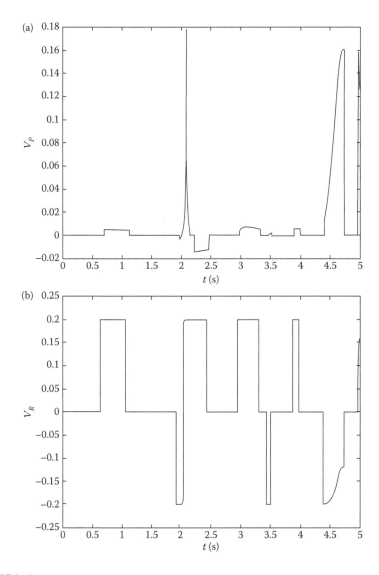

**FIGURE 9.10**
UPFC control variables. (a) Series active voltage ($V_P$), (b) series reactive voltage ($V_R$), (c) shunt reactive current injected ($I_R$), and (d) DC link power.

case when the control action is initiated at the instant of fault clearing. There are no major qualitative differences between these two cases.

It is to be noted that since phase angles of the voltage and current of the UPFC are with respect to a common reference and are obtained by the optimization procedure, there are no nonlinear algebraic equations to be solved during simulation.

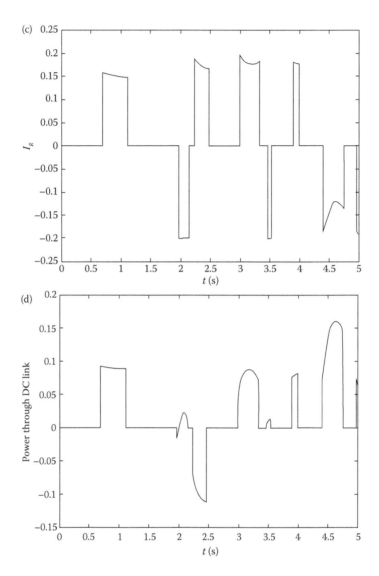

**FIGURE 9.10**
Continued.

## 9.4.2 Multiple UPFC

Here, we consider two UPFCs installed in two or more lines belonging to a cutset. For the disturbance considered in the 10-generator system, there are two lines in the cutset 11–12 and 18–19. If UPFCs are to be installed in both lines, we can extend the approach outlined for a single UPFC. Each UPFC is represented by a two-port network. The external network can be represented by a four-port Thevenin equivalent network. If the transfer impedances

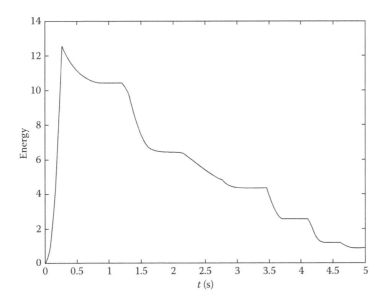

**FIGURE 9.11**
Variation of total energy with two UPFCs (lines 11–12 and 18–19).

between the ports of one UPFC and the ports of other UPFC are neglected, the network can be modeled by two decoupled two-port Thevenin equivalent networks. This enables the application of the decoupled optimization of power flows in the cutset lines to determine the control variables in individual UPFCs. The instants of switchings are also determined in a similar manner. However, if transfer impedances between ports of different UPFCs are not negligible, simplification of the computations by decoupling may not be feasible (Krishna 2003).

The results from the case study of applying two UPFCs with identical ratings of $V_C = 0.2$ and $I_C = 0.2$ are obtained when the controls in both UPFCs are initiated immediately (at the instant) after the fault clearing. The total system energy is shown in Figure 9.11. The application of two UPFCs results in improvement of both first swing stability and damping of oscillations.

### 9.4.3 Practical Implementation

The application of one or more UPFC in a multimachine power system is demonstrated from the simulation studies on the New England 10-generator system. The results from the study indicate that UPFCs, as versatile FACTS controllers, can provide significant improvement in transient stability and damping of oscillations following a major disturbance. The performance with UPFC is superior to that with TCSC in stabilizing the system during emergencies that threaten system integrity.

There are two major issues in the practical implementation:

1. On-line computation of the control variables associated with a UPFC to maximize and minimize the power flow in the line (belonging to a cutset, not necessarily a critical cutset)
2. Determination of the instants at which the switching of the control variables (from operating to maximum power and vice versa, from operating to minimum power and vice versa)

It is not practical to adopt constrained optimization techniques for on-line application. What is presented in the previous subsection is a proof of the concept based on simulation studies. For practical implementation, simpler and decentralized control strategies are required.

Consider a UPFC connected at the sending end of a lossless transmission line. If we assume the voltage magnitudes at the two ends of the line are maintained constant at $V$ (note that the reactive current injected by the shunt-connected VSC can help in regulating the sending end voltage). Let

$$V_S = V\angle\frac{\delta}{2}, \quad V_R = V\angle-\frac{\delta}{2}, \quad \hat{I} = I\angle-\phi$$

The injected voltage $\hat{V}_C$ can be expressed as

$$\hat{V}_C = V_C\angle\beta = (V_p + jV_r)e^{-j\phi} \tag{9.9}$$

We can derive (Padiyar 2007) the expressions for the power and reactive power at the receiving end as

$$P_R = P_0 + \frac{VV_p}{X_L}\sin\left(\frac{\delta}{2} - \phi\right) + \frac{VV_r}{X_L}\cos\left(\frac{\delta}{2} - \phi\right) \tag{9.10}$$

$$-Q_R = Q_0 - \frac{VV_p}{X_L}\cos\left(\frac{\delta}{2} - \phi\right) + \frac{VV_r}{X_L}\sin\left(\frac{\delta}{2} - \phi\right) \tag{9.11}$$

where $P_0$ and $Q_0$ are the active and reactive power (at the receiving end) in the absence of the series connected converter, defined as

$$P_0 = \frac{V^2}{X_L}\sin\delta, \quad Q_0 = \frac{V^2}{X_L}(1 - \cos\delta)$$

It can also be shown that

$$\sin\phi = \frac{V_p}{2V\sin(\delta/2)} \tag{9.12}$$

We can express $V_p$ and $V_r$ (from Equation 9.9) as

$$V_p = V_C \cos(\beta + \phi), \quad V_r = V_C \sin(\beta + \phi)$$

and finally derive

$$P_R = P_0 + \frac{VV_C}{X_L} \sin\left(\frac{\delta}{2} + \beta\right) \tag{9.13}$$

$$-Q_R = Q_0 - \frac{VV_C}{X_L} \cos\left(\frac{\delta}{2} + \beta\right) \tag{9.14}$$

$P_R$ is maximum when $\beta = (\pi/2) - (\delta/2)$ and

$$P_{R\max} = P_0 + \frac{VV_{C\max}}{X_L} \tag{9.15}$$

Similarly, $P_R$ is minimum when $\beta = -(\pi/2 + \delta/2)$ and

$$P_{R\min} = P_0 - \frac{VV_{C\max}}{X_L} \tag{9.16}$$

Thus, the series converter can be controlled to set $V_C = V_{C\max}$ and $\beta$ at an appropriate value to give maximum or minimum power flow in the line. If we want the operating value of power $P_0$, then $V_C = 0$.

The instants at which the power flow is maximized or minimized are to be determined strictly from monitoring the value of $(d\delta_A/dt - d\delta_B/dt)$, which is the difference in the frequencies of COIs in areas A and B. This is also equal to $d\delta_k/dt$, where $\delta_k$ is the difference in the angles of terminal buses of line $k$, if there is strict coherency in areas A and B. The intermachine oscillations will affect $d\delta_k/dt$. A practical way of overcoming this problem is to apply signal processing techniques to filter out intermachine oscillations.

## Remarks

1. In simulation studies, $d\delta_k/dt$ was computed for the postfault system (with operating values of the control variables). This ensures that $d\delta_k/dt$ is continuous when the switching takes place. Note that when the power flow is changed instantaneously (by neglecting UPFC dynamics), the angle $\delta_k$ also undergoes change. The exact computation of the frequency deviation across the line is not required, but only the instants when $d\delta_k/dt$ is zero, maximum, or minimum.

2. Another approach to determine the switching instants could be to utilize the difference in the open circuit voltages at the two ports of the individual UPFCs. However, this assumes that the system network (including load models) is linear. The transmission network, including the transformer, is generally linear but the loads are nonlinear. However, if the loads are modeled by fast dynamics as mentioned in Chapters 4 and 6, Thevenin equivalents can still be obtained with time-varying voltages.

3. The problem of determination of switching instants is less severe with TCSCs since $\delta_k$ is close to zero when $d\delta_k/dt$ is maximum or minimum. The insertion of capacitive reactance when $\delta_k = 0$ does not result in sudden changes in the power flow in the line.

## 9.5 Discussion and Directions for Further Research

As mentioned in Chapters 7 and 8, the control action (based on energy function analysis) suggested here is aimed at (a) maximizing the first swing stability and (b) fast damping of the nonlinear oscillations following a large disturbance to ensure that the system reaches a postfault stable equilibrium state as soon as possible. It is a discrete control action in the sense that the control action is enabled at the instant of detection (or prediction) of a large disturbance (not necessarily leading to loss of synchronism, as there is urgency to ensure system stability at the earliest). FACTS controllers should be located in sensitive lines (that are usually members of the critical cutsets), which can be determined by planning studies. The discrete control action is discontinued when the system reaches close to the postfault equilibrium state. Note that the control action of linear damping controllers (described in Chapter 7) will ensure that the postfault equilibrium state remains stable for small disturbances that are perennially present in a power system. It may be desirable that the linear damping control action is disabled when the discrete control action is active.

Kakimoto et al. (2004) describe the possibility of instability of interarea oscillation modes during large disturbances in Japan, even when the modes are stable under small disturbances. This phenomenon has been termed as "auto-parametric resonance" (Tamura and Yorino 1987). The main reason for this phenomenon is the interaction between various swing modes. See Chapter 4 for a brief discussion. This clearly shows that damping controllers may be ineffective during large disturbances. The damping action due to the switching control described in Chapter 8 is reliable and easy to implement although it may not be as fast as optimal control action using Pontryagin's principle. However, as mentioned in Chang and Chow (1998), the optimal control is not feasible in practice (nor it is required in power systems; achieving a decrement factor of 0.1 is considered to be adequate).

Although the emphasis in this book is on the application of SPEF for on-line monitoring and control of security related to synchronous (angle) stability, it should be possible to extend the applications to the detection and emergency control in the context of voltage and frequency stability problems. Whenever the balance between the generation and load is affected, frequency starts to decrease or increase (although slowly, as compared to a major fault), there is a need to drop load or generation in an optimal manner. Similarly, when the transmission congestion affects the transfer of power from generation and load, there is need to trip load or generation depending on the circumstance. The application of SPEF can help in these tasks. New energy storage technologies can substantially improve the system security and their locations and control strategies can be guided from the energy function concepts.

The WAM need not be limited to phasor measurements. The basic idea is the application of intelligent sensors and communication technology for simultaneous measurements of relevant quantities such as bus frequencies and power flows in lines in addition to phasors. There is a need to apply measurement and signal processing techniques that give accurate and real-time information about the state of the system to monitor system security and initiate automatic remedial action (emergency control) measures to enable the system to survive any emergency.

# References

Abed, E. and P. Varaiya. 1984. Nonlinear oscillations in power systems. *Elec. Power Energy Syst.*, 6(1): 37–43.

Adamiak, M.G. et al. 2006. Wide area protection—Technology and infrastructures. *IEEE Trans. Power Deliv.*, 21(2): 601–609.

Anderson, P.M. and A.A. Fouad. 1977. *Power System Control and Stability.* Ames: Iowa State University Press.

Andersson, G. et al. 2005. Causes of the 2003 major grid blackouts in North America and Europe, and recommended means to improve system dynamic performance. *IEEE Trans. Power Syst.*, 20(4): 1922–1928.

Arapostathis, A., S.S. Sastry, and P. Varaiya. 1982. Global analysis of swing dynamics. *IEEE Trans. Circuits Syst.*, 29(10): 673–679.

Arrillaga, J. 1998. *High Voltage Direct Current Transmission.* London: Peter Peregrinus.

Athans, M. and P.L. Falb. 1966. *Optimal Control: An Introduction to the Theory and Its Applications.* New York: McGraw-Hill.

Athay, T., R. Podmore, and S. Virmani. 1979. A practical method for direct analysis of transient stability. *IEEE Trans. Power Appar. Syst.*, 98(2): 573–584.

Aylett, P.D. 1958. The energy-integral criterion of transient stability limits of power systems. *IEE Proc. (London), Pt. C*, 105(8): 527–536.

Azbe, A. and R. Mihalic. 2008. The control strategy for an IPFC based on the energy function. *IEEE Trans. Power Syst.*, 23(4): 1662–1669.

Azbe, V. et al. 2005. The energy function of a general multimachine system with a unified power flow controller. *IEEE Trans. Power Syst.*, 20(3): 1478–1485.

Begovic, M. et al. 2005. Wide-area protection and emergency control. *IEEE Proc.*, 93(5): 876–891.

Behera, A.K. 1988. Transient stability analysis of multimachine power systems using detailed machine models. PhD thesis, University of Illinois at Urbana-Champaign.

Bergen, A.R. and D.J. Hill. 1981. Structure preserving model for power system stability analysis. *IEEE Trans. Power Appar. Syst.*, 100(1): 25–35.

Bergen, A.R., D.J. Hill, and C.L. deMarcot. 1986. Lyapunov functions for multimachine power systems with generator flux decay and voltage dependent loads. *Elec. Power Energy Syst.*, 8(1): 2–10.

Brereton, D.S. et al. 1957. Representation of induction motor loads during power system stability studies. *AIEE Trans. Pt. III*, 76: 451–460.

Bretas, N.G. and L.F.C. Alberto. 2003. Lyapunov function for power systems with transfer conductances: Extension of the invariance principle. *IEEE Trans. Power Syst.*, 18(2): 769–777.

Byerly, R.T., D.T. Poznaniak, and E.R. Taylor Jr. 1982. Static reactive compensation for power transmission systems. *IEEE Trans. PAS*, 101(10): 3997–4005.

Centeno, V. et al. 1997. An adaptive out of step relay. *IEEE Trans. Power Deliv.*, 12(1): 61–71.

Chandrashekhar, K.S. and D.J. Hill. 1986. Cutset stability criterion for power systems using a structure preserving model. *Elec. Power Energy Syst.*, 8(3): 146–157.

Chang, J. and J.H. Chow. 1998. Time-optimal control of power systems requiring multiple switchings of series capacitors. *IEEE Trans. Power Syst.*, 13(2): 367–373.

Chiang, H.D. 1989. Study of existence of energy functions for power systems with losses. *IEEE Trans. Circuits Syst.*, 36(11): 1423–1429.

Chiang, H.D., J. Tong, and K.N. Miu. 1993. Predicting unstable modes in power systems: Theory and computations. *IEEE Trans. Power Syst.*, 8(4): 1429–1437.

Chiang, H.D., F.F. Wu, and P. Varaiya. 1994. A BCU method for direct analysis of system transient stability. *IEEE Trans. Power Syst.*, 9(3): 1194–1208.

Chiang, H.D., F.F. Wu, and P.P. Varaiya. 1987. Foundations of direct methods for power system transient stability analysis. *IEEE Trans. Circuits Syst.*, 34(2): 160–173.

Chiang, H.D., F.F. Wu, and P.P. Varaiya. 1988. Foundations of the potential energy boundary surface method for power system transient stability analysis. *IEEE Trans. Circuits Syst.*, 35(6): 712–728.

Coleman, T., M.A. Branch, and A. Grace. 1999. *Optimization Toolbox—User's Guide, Version 2*. The MathWorks Inc.

Counan, C. et al. 1993. Major incidents on the French electric system: Potentiality and curative measures studies. *IEEE Trans. Power Syst.*, 8(3): 879–886.

Crary, S.B. 1947. *Power System Stability, Vol. II—Transient Stability.* New York: John Wiley.

Cresap, R.L., C.W. Taylor, and M.J. Kreipe. 1981. Transient stability enhancement by 120-degree phase rotation. *IEEE Trans. Power Appar. Syst.*, 100(2): 745–753.

Dandeno, P.L., R.L. Hauth, and R.P. Schulz. 1973. Effects of synchronous machine modeling in large scale system studies. *IEEE Trans. Power Appar. Syst.*, 92: 574–582.

Davy, R.J. and I.A. Hiskens. 1997. Lyapunov functions for multimachine power systems with dynamic loads. *IEEE Trans. Circuits Syst.–I: Fund. Theory Appl.*, 44(9): 796–812.

De La Ree, J. et al. 2005. Catastrophic failures in power systems: Causes, analyses, and countermeasures. *IEEE Proc.*, 93(5): 956–964.

Debs, A.S. and F. Dominguez. 1990. Sensitivity analysis of voltage dip compensation using the TEF method. *29th IEEE Conference on Decision and Control*, Honolulu, Hawaii.

DeMarco, C.L. and C.A. Canizares. 1992. A vector energy function approach for security analysis of AC/DC systems. *IEEE Trans. Power Syst.*, 7: 1000–1011.

DeMello, F.P. and C. Concordia. 1969. Concepts of synchronous machine stability as affected by excitation control. *IEEE Trans. Power Appar. Syst.*, 88(4): 316–329.

Deo, N. 1974. *Graph Theory with Applications to Engineering and Computer Science*. Englewood Cliffs, NJ: Prentice-Hall.

Dobson, I. and H.D. Chiang. 1989. Towards a theory of voltage collapse in electric power systems. *Syst. Control Lett.*, 13: 253–262.

Dommel, H.W. and N. Sato. 1972. Fast transient stability solutions. *IEEE Trans. Power Appar. Syst.*, 91: 1643–1650.

Donnelly, M.K. et al. 1993. Control of a dynamic brake to reduce turbine generator shaft transient torque. *IEEE Trans. Power Syst.*, 8(1): 67–73.

Dy Liacco, T.E. 1979. Security functions in power system control centers; the state-of-the-art in control center design. *IFAC Symposium on Computer Applications in Large Scale Power Systems*, New Delhi, pp. 16–19.

Edris, M.A. 1991. Enhancement of first swing stability using a high-speed phase shifter. *IEEE Trans. Power Syst.*, 6(3): 1113–1118.

El-Abiad, A.H. and K. Nagappan. 1966. Transient stability region of multi-machine power systems. *IEEE Trans. Power Appar. Syst.*, 85(2): 169–178.

El-Kady, M.A. et al. 1986. Dynamic security assessment utilizing the transient energy function method. *IEEE Trans. Power Syst.*, 1(3): 284–291.

Ellis, H.M. et al. 1966. Dynamic stability of Peace River transmission system. *IEEE Trans. Power Appar. Syst.*, 85: 586–600.

Findlay, J.A. et al. 1988. State of the art and key issues in power system security assessment. *Proceedings of the Workshop on Power System Security Assessment*, Ames, Iowa.

Fink, L.H. 1985. Emergency control practices (Report prepared by Task Force on Emergency Control). *IEEE Trans. Power Appar. Syst.*, 104(9): 2336–2341.

Fink, L.H. and K. Carlsen. 1978. Operating under stress and strain. *IEEE Spectr.*, 15: 43–58.

Fouad, A.A. and V. Vittal. 1988. The transient energy function method. *Int. J. Econ. Policy Stud.*, 10(4): 233–246.

Fouad, A.A and V. Vittal. 1992. *Power System Transient Stability Analysis Using the Transient Energy Function Method*. Englewood Cliffs, NJ: Prentice-Hall.

Fouad, A.A. 1988. Dynamic security assessment practices in N. America. *IEEE Trans. Power Syst.*, 3(3): 1310–1321.

Fouad, A.A. and V. Vittal. 1983. Power system response to a large disturbance energy associated with system separation. *IEEE Trans. Power Appar. Syst.*, 102: 3534–3540.

Fudeh, H. and C.M. Ong. 1981. A simple and efficient AC–DC load flow method for multiterminal DC systems. *IEEE Trans. Power Appar. Syst.*, 100: 4389–4396.

Gama, C. et al. 2000. Commissioning and operative experience of TCSC for damping power oscillations in the Brazilian North–South interconnection. *CIGRE Paper 14-104*.

Gandhari, M. et al. 2001. A control strategy for controllable series capacitor in electric power systems. *Automatica*, 37: 1575–1583.

Garver, L.L., P.R. Van Horne, and K.A. Wirgau. 1979. Load supplying capability of generation–transmission networks. *IEEE Trans. Power Appar. Syst.*, 98(3): 957–962.

Gless, G.E. 1966. Direct method of Lyapunov applied to transient power system stability. *IEEE Trans. Power Appar. Syst.*, 85(2): 164–179.

Glover, J.D. and M.S. Sarma. 2002. *Power System Analysis and Design*, 3rd Edition. ISBN 981-243-125X, Thomson (copyright is held by Wadsworth Group).

Goldstein, H. 1959. *Classical Mechanics*. Reading: Addison-Wesley.

Gronquist, J.F. et al. 1995. Power oscillation damping control strategies for FACTS devices using locally measurable quantities. *IEEE Trans. Power Syst.*, 10(3): 1598–1605.

Gyugyi, L., K.K. Sen, and C.D. Schauder. 1999. The interline power flow controller concept: A new approach to power flow management in transmission systems. *IEEE Trans. Power Deliv.*, 14(3): 1115–1123.

Henner, V.E. 1974. A multimachine power system Lyapunov function using generalized Popov criterion. *Int. J. Control*, 19(5): 969–976.

Hill, D.J. 1993. Nonlinear dynamic load models with recovery for voltage stability studies. *IEEE Trans. Power Syst.*, 8(1): 166–176.

Hingorani, N.G. and L. Gyugyi. 2000. *Understanding FACTS—Concepts and Technology of Flexible AC Transmission Systems*. New York: IEEE Press.

Hirsch, M.W., S. Smale, and R.L. Devaney. 2004. *Differential Equations, Dynamical Systems & an Introduction to Chaos*, 2nd Edition, San Diego: Academic Press.

Hiskens, I.A. and D.J. Hill. 1992. Incorporation of SVC into energy function methods. *IEEE Trans. Power Syst.*, 7(1): 133–140.

Hiskens, I.A. and M.A. Pai. 2000. Trajectory sensitivity analysis of hybrid systems. *IEEE Trans. Circuits Syst.–I: Fund. Theory Appl.*, 47(2): 204–220.

Horowitz, S.H. and A.G. Phadke. 1995. *Power System Relaying*. New York: Research Studies/Wiley.

IEEE Committee Report. 1969. Recommended phasor diagram for synchronous machines. *IEEE Trans. Power Appar. Syst.*, 88(11): 1593–1610.

IEEE Committee Report. 1978. A description of discrete supplementary controls for stability. *IEEE Trans. Power Appar. Syst.*, 97(1): 149–165.

IEEE Committee Report. 1981a. Dynamic performance characteristics of North American HVDC systems for transient and dynamic stability evaluations. *IEEE Trans. Power Appar. Syst.*, 100(7): 3356–3364.

IEEE Committee Report. 1981b. Excitation system models for power system stability studies. *IEEE Trans. Power Appar. Syst.*, 100(2): 494–509.

IEEE Committee Report. 1991. Dynamic models for fossil-fueled steam units in power system studies. *IEEE Trans. Power Syst.*, 6(2): 753–761.

IEEE Committee Report. 1992. Transient stability test systems for direct stability methods. *IEEE Trans.*, 7: 37–43.

IEEE PES Digital Excitation Task Force. 1996. Computer models for representation of digital-based excitation systems. *IEEE Trans. Energy Conv.*, 11(3): 607–615.

IEEE Standard. 1983. Test procedures for synchronous machines. No. 115.

IEEE SSR Working Group. 1985. Second benchmark model for computer simulation of subsynchronous resonance. *IEEE Trans. Power Appar. Syst.*, 104(5): 1057–1066.

IEEE Task Force. 1986. Current usage and suggested practices in power system stability simulations for synchronous machines. *IEEE Trans. Energy Convers.*, 1(1): 77–93.

IEEE Task Force. 1993. Load representation for dynamic performance analysis. *IEEE Trans. Power Syst.*, 8(2): 472–481.

IEEE Working Group Report. 1992. Hydraulic turbine and turbine control models for system dynamic studies. *IEEE Trans. Power Syst.*, 7(1): 167–179.

IEEE/CIGRE Joint Task Force. 2004. Definition and classification of power system stability. *IEEE Trans. Power Syst.*, 19(2): 1387–1401.

Ilic, M. and J. Zaborszky. 2000. *Dynamics and Control of Large Electric Power Systems*. New York: John Wiley.

Immanuel, V. 1993. Application of structure preserving energy functions for stability evaluation of power systems with static var compensators. PhD thesis. Indian Institute of Science, Bangalore.

Jing, C. et al. 1995. Incorporation of HVDC and SVC models in the Northern State Power Co. (NSP) network for on-line implementation of direct transient stability assessment. *IEEE Trans. Power Syst.*, 10(2): 898–906.

Kakimoto, N., A. Nakanishi, and K. Tomiyama. 2004. Instability of interarea oscillation mode by autoparametric resonance. *IEEE Trans. Power Syst.*, 19(4): 1961–1970.

Kakimoto, N., Y. Ohsawa, and M. Hayashi. 1978. Transient stability analysis of electric power system via Lure-type Liapunov functions, parts I and II. *Trans. IEE Japan*, 98: 63–79.

Kakimoto, N., Y. Ohsawa, and M. Hayashi. 1980. Transient stability analysis of multimachine power systems with field flux decays via Lyapunov's direct method. *IEEE Trans. Power Appar. Syst.*, 99(6): 1819–1827.

Kalman, R.E. 1963. Lyapunov functions for the problem of Lure in automatic control. *Proc. Nat. Acad. Sci. (USA)*, 49: 201–206.

Kamwa, I., R. Grondin, and G. Trudel. 2005. IEEE PSSS2B versus PSS4B: The limits of performance of modern power system stabilizers. *IEEE Trans. Power Syst.*, 20(2): 903–915.

Khorasani, K., M.A. Pai, and P.W. Sauer. 1986. Modal-based stability analysis of power systems using energy functions. *Elec. Power Energy Syst.*, 8(1): 11–16.

Kimbark, E.W. 1948. *Power System Stability, Vol. I: Elements of Stability Calculations.* New York: John Wiley.

Kimbark, E.W. 1956. *Power System Stability, Vol. III: Synchronous Machines.* New York: John Wiley.

Kimbark, E.W. 1966. Improvement of system stability by switched series capacitors. *IEEE Trans. Power Appar. Syst.*, 85: 180–188.

Klein, M., G.J. Rogers, and P. Kundur. 1991. A fundamental study of interarea oscillations in power systems. *IEEE Trans. Power Syst.*, 16(3): 914–921.

Kokotovic, P.V. and, R.S. Putman. 1965. Sensitivity of automatic control systems (Survey). *Autom. Remote Control.*, 26: 727–749.

Kosterev, D.N. and W.J. Kolodziej. 1995. Bang-bang series capacitor transient stability control. *IEEE Trans. Power Syst.*, 10(2): 915–923.

Krishna, S. 2003. Dynamic security assessment and control using unified power flow controller. PhD thesis. Indian Institute of Science, Bangalore.

Krishna, S. and K.R. Padiyar. 2000. Dynamic security analysis using structure preserving energy function. *11th National Power Systems Conference (NPSC)*, Bangalore.

Krishna, S. and K.R. Padiyar. 2005. Discrete control of unified power flow controller for stability improvement. *Elec. Power Syst. Res.*, 75: 178–189.

Krishna, S. and K.R. Padiyar. 2010. Online dynamic security assessment: Determination of critical transmission lines. *Elec. Power Comp. & Syst.*, 38(2): 152–165.

Kulkarni, A.M. and K.R. Padiyar. 1999. Damping of power swings using series FACTS controllers. *Elec. Power Energy Syst.*, 21: 475–495.

Kundur, P. 1994. *Power System Stability and Control.* New York: McGraw-Hill.

Landgren, G.L. and S.W. Anderson. 1973. Simultaneous power interchange capability analysis. *IEEE Trans. Power Appar. Syst.*, 92(6): 1973–1986.

Laufenberg, M.J. and M.A. Pai. 1998. A new approach to dynamic security assessment using trajectory sensitivities. *IEEE Trans. Power Syst.*, 13(3): 953–958.

Lefebvre, S. et al. 1991. Considerations for modeling MTDC systems in transient stability programs. *IEEE Trans. Power Deliv.*, 6(1): 397–404.

Liu, C.W. and J. Thorp. 1995. Application of synchronized phasor measurements to real-time transient stability prediction. *Proc. Inst. Elect. Eng. Gener. Transm. Distrib.*, 142(4): 355–360.

Liu, C.W. et al. 1999. Application of a novel fuzzy neural network to real-time transient stability swing prediction based on synchronized phasor measurements. *IEEE Trans. Power Syst.*, 14(2): 685–692.

Liu, S., X.P. Wang, and Q.Z. Yu. 1996. Hybrid transient stability analysis using structure preserving model. *Elec. Power Energy Syst.*, 18(6): 347–352.

Luenberger, D.G. 1984. *Linear and nonlinear programming, second edition.* Reading, MA: Addison Wesley.

Magnusson, P.C. 1947. Transient energy method of calculating stability. *AIEE Trans.*, 66: 747–755.

Maria, G.A., C. Tang, and J. Kim. 1990. Hybrid transient stability analysis. *IEEE Trans. Power Syst.*, 5(2): 384–391.

Marioka, Y. et al. 1993. System separation equipment to minimize power system instability using generator's angular velocity measurements. *IEEE Trans. Power Deliv.*, 8(3): 941–947.

Mathur, R.M. and R.K. Varma. 2002. *Thyristor-Based FACTS Controllers for Electrical Transmission Systems*. New York: IEEE Press and Wiley Interscience.

McLachlan, N.W. 1964. *Theory and Applications of Mathieu Functions*. New York: Dover Publications.

McNabb, D. et al. 2005. Transient and dynamic modeling of the new Langlois VFT asynchronous tie and validation with commissioning tools. *Paper presented at the International Conference on Power System Transients (IPST-05)*, Montreal, Canada.

Mihalic, R. and U. Gabrijel. 2004. A structure preserving energy function for a static series synchronous compensator. *IEEE Trans. Power Syst.*, 19(3): 1501–1507.

Mihalic, R., P. Zunko, and D. Povh. 1996. Improvement of transient stability using unified power flow controller. *IEEE Trans. Power Deliv.*, 11(1): 485–492.

Moore, J.B. and B.D.O. Anderson. 1968. A generalization of the Popov criterion. *J. Franklin Inst.*, 285(6): 488–492.

Narasimhamurthi, N. and M.T. Musavi. 1984. A generalized energy function for transient stability analysis of power systems. *IEEE Trans. Circuits Syst.*, 31(7): 637–645.

Ni, Y.X. and A.A. Fouad. 1987. A simplified two-terminal HVDC model and its use in direct transient stability assessment. *IEEE Trans. Power Syst.*, 2(2): 1006–1012.

Nishida, S. 1984. On-line identification of potential energy at the unstable equilibrium point and its application to adaptive emergency control of power system. *Electr. Eng. Japan*, 104(4): 497–504.

Ohura, Y. et al. 1990. A predictive out of step protection system based on observation of the phase-difference between substations. *IEEE Trans. Power Deliv.*, 5(4): 1695–1704.

Overbye, T.J. and C.L. DeMarco. 1991. Improved techniques for power system voltage stability assessment using energy methods. *IEEE Trans. Power Syst.*, 6(4): 1446–1452.

Overbye, T.J., M.A. Pai, and P.W. Sauer. 1992. Some aspects of the energy function approach to angle and voltage stability analysis in power systems. *Proceedings of the 31st Conference on Decision and Control*, Tucson, Arizona, pp. 2941–2946.

Padiyar, K.R. 1984. On the nature of power flow equations in security analysis. *Elec. Mach. Power Syst.*, 9: 297–306.

Padiyar, K.R. 1996. *Power System Dynamics: Stability and Control*. Singapore: John Wiley.

Padiyar, K.R. 1999. *Analysis of Subsynchronous Resonance in Power Systems*. Boston: Kluwer Academic Publishers.

Padiyar, K.R. 2002. *Power System Dynamics: Stability and Control*, 2nd Edition. Hyderabad: BS Publications.

Padiyar, K.R. 2007. *FACTS Controllers in Power Transmission and Distribution*. New Delhi: New Age Publishers.

Padiyar, K.R. 2010. *HVDC Power Transmission Systems*, 2nd Edition. New Delhi: New Age Publishers.

Padiyar, K.R. and K. Bhaskar. 2002. An integrated analysis of voltage and angle stability of a three node power system. *Elec. Power Energy Syst.*, 24: 489–501.

Padiyar, K.R., M.K. Geetha, and K.U. Rao. 1996. A novel power flow controller for controllable series compensation. Conference Publication, AC and DC Power Transmission, IEE, London.

Padiyar, K.R. and K.K. Ghosh. 1988. A new structure preserving energy function incorporating transmission line resistance. *Elec. Mach. Power Syst.*, 14(4): 324–340.

Padiyar, K.R. and K.K. Ghosh. 1989a. Direct stability evaluation of power systems with detailed generator models using structure preserving energy functions. *Elec. Power Energy Syst.*, 11(1): 47–56.

Padiyar, K.R. and K.K. Ghosh. 1989b. Dynamic security assessment of power systems using structure preserving energy functions. *Elec. Power Energy Syst.*, 11(1): 39–46.

Padiyar, K.R. and V. Immanuel. 1994. Modelling SVC for stability evaluation using structure preserving energy function. *Elec. Power Energy Syst.*, 16(5): 339–348.

Padiyar, K.R. and S. Krishna. 2006. Online detection of loss of synchronism using energy function criterion. *IEEE Trans. Power Delivery*, 21(1): 46–55.

Padiyar, K.R. and S.S. Rao. 1996. Dynamic analysis of small signal voltage instability decoupled from angle instability. *Elec. Power Energy Syst.*, 18(7): 445–452.

Padiyar, K.R. and H.S.Y. Sastry. 1986a. Fast evaluation of transient stability of power systems using a structure preserving energy function. *Elec. Mach. Power Syst.*, 11: 421–441.

Padiyar, K.R. and H.S.Y. Sastry. 1986b. Direct stability analysis of power systems with realistic generator models using topological energy function. *IFAC Symposium*, Beijing, China.

Padiyar, K.R. and H.S.Y. Sastry. 1984. Direct stability analysis of AC/DC power systems using a structure preserving energy function. *Proceedings of the International Conference on Computer Systems and Signal Processing*, Bangalore.

Padiyar, K.R. and H.S.Y. Sastry. 1987. Topological energy function analysis of stability of power systems. *Elec. Power Energy. Syst.*, 9(1): 9–16.

Padiyar, K.R. and H.S.Y. Sastry. 1993. A structure preserving energy function for stability analysis of AC/DC systems. *Sadhana*, 18(Pt. 5): 787–799.

Padiyar, K.R. and H.V. Sai Kumar. 2003. Modal inertia—A new concept for the location of PSS in multimachine systems. *Paper presented at the 27th National Systems Conference*, IIT, Kharagpur.

Padiyar, K.R. and H.V. Saikumar. 2005. Coordinated design and performance evaluation of UPFC supplementary modulation controllers. *Elec. Power Energy Syst.*, 27: 101–111.

Padiyar, K.R. and H.V. Sai Kumar. 2006. Investigations of strong resonance in multimachine power systems with STATCOM supplementary modulation controller. *IEEE Trans. Power Syst.*, 21(2): 754–762.

Padiyar, K.R. and V. Swayam Prakash. 2003. Tuning and performance evaluation of damping controller for a STATCOM. *Elec. Power Energy Syst.*, 25: 155–166.

Padiyar, K.R. and K. Uma Rao. 1997. Discrete control of series compensation for stability improvement in power systems. *Elec. Power Energy Syst.*, 19(5): 311–319.

Padiyar, K.R. and P.P. Varaiya. 1983. *A Network Analogy for Power System Stability Analysis*. Preprint, EECS Department, University of California, Berkeley.

Padiyar, K.R. and R.K. Varma. 1991. Damping torque analysis of static var system controllers. *IEEE Trans. Power Syst.*, 6(2): 458–465.

Pai, M.A. 1981. *Power System Stability*. New York: North Holland.

Pai, M.A. 1989. *Energy Function Analysis for Power System Stability*. Boston: Kluwer Academic Publishers.

Pai, M.A., M. Lafenberg, and P.W. Sauer. 1993. Evaluation of path dependent integrals in the energy function method. *North American Power Symposium*, Washington DC.

Pai, M.A. and P.G. Murthy. 1974. New Lyapunov functions for power systems based on minimal realizations. *Int. J. Control*, 19(2): 401–405.

Pai, M.A., K.R. Padiyar, and C. Radhakrishna. 1981. Transient stability of multimachine AC/DC power systems via energy function method. *IEEE Trans. Power Appar. Syst.*, 100: 5027–5035.

Pai, M.A. and V. Rai. 1974. Lyapunov–Popov stability analysis of synchronous machine with flux decay and voltage regulator. *Int. J. Control*, 20(2): 203–212.

Park, R.H. 1929. Two-reaction theory of synchronous machines—Part I; Generalized methods of analysis. *AIEE Trans.*, 48: 716–730.

Pavella, M., D. Ernst, and D. Ruiz-Vega. 2000. *Transient Stability of Power Systems—A Unified Approach to Assessment and Control.* Boston: Kluwer Academic Publishers.

Pavella, M. and P.G. Murthy. 1994. *Transient Stability of Power Systems—Theory and Practice.* New York: John Wiley.

Penfield, P., R. Spence, and S. Duinker. 1970. *Tellegen's Theorem and Electrical Networks.* Cambridge, MA: MIT Press.

Popov, V.M. 1962. Absolute stability of nonlinear systems of automatic control. *Autom. Remote Control*, 22: 857–887.

Rahimi, F.A. et al. 1993. Evaluation of the transient energy function method for on-line dynamic security analysis. *IEEE Trans. Power Syst.*, 8(2): 497–507.

Rama Rao, N. and D.K. Reitan. 1970. Improvement in power system transient stability using optimal control: Bang-bang control of reactance. *IEEE Trans. Power Appar. Syst.*, 89: 975–983.

Ribbens-Pavella, M. and F.J. Evans. 1985. Direct methods for studying of the dynamics of large scale electric power systems—A survey. *Automatica*, 21(1): 1–21.

Rovnyak, S. et al. 1994. Decision trees for real-time transient stability prediction. *IEEE Trans. Power Syst.*, 9(3): 1417–1426.

Rovnyak, S. et al. 1995. Predicting future behavior of transient events enough to evaluate remedial control options in real time. *IEEE Trans. Power Syst.*, 10(3): 1195–1203.

Sasaki, H. 1979. An approximate incorporation of field flux decay into transient stability analysis of multimachine power systems by second method of Lyapunov. *IEEE Trans. Power Appar. Syst.*, 98: 473–483.

Sastry, H.S.Y. 1984. Application of topological energy functions for the direct stability evaluation of power systems. PhD thesis, IIT Kanpur.

Sauer, P.W., K.D. Demaree, and M.A. Pai. 1983. Stability limited load supply and interchange capability. *IEEE Trans. Power Appar. Syst.*, 102(11): 3637–3643.

Sauer, P.W. and M.A. Pai. 1998. *Power System Dynamics and Stability.* Upper Saddle River: Prentice-Hall.

Sauer, P.W. et al. 1989. Trajectory approximation for direct methods that use sustained faults with detailed power system models. *IEEE Trans. Power Syst.*, 4(2): 499–506.

Schauder, C. and H. Mehta. 1993. Vector analysis and control of advanced static VAR compensators. *IEE Proc.-C*, 140(4): 299–306.

Shelton, M.L. et al. 1975. Bonneville Power Administration 1400 MW braking resistor. *IEEE Trans. Power Appar. Syst.*, 94: 602–611.

Shubhanga, K.N. and A.M. Kulkarni. 2002. Application of structure preserving energy margin sensitivity to determine the effectiveness of shunt and series FACTS devices. *IEEE Trans. Power Syst.*, 17(3): 730–738.

Shubhanga, K.N. and A.M. Kulkarni. 2004. Determination of effectiveness of transient stability controls using reduced number of trajectory sensitivity computations. *IEEE Trans. Power Syst.*, 19(1): 473–482.

Smith, O.J.M. 1969. Power system transient control by capacitor switching. *IEEE Trans. Power Appar. Syst.*, 88: 28–35.

Sood, V.K. 2004. *HVDC and FACTS Controllers-Application of Static Converters in Power Systems.* Boston: Kluwer Academic Publishers.

St. Clair, H.P. 1953. Practical concepts in capability and performance of transmission lines. *AIEE Trans.*, 72: 1152–1157.

Stanton, K.N. 1972. Dynamic energy balance studies for simulation of power-frequency transients. *IEEE Trans. Power Appar. Syst.*, 91: 110–117.

Stott, B. 1979. Power system dynamic response calculations. *IEEE Proc.*, 67(2): 219–241.

Susuki, Y., T. Hikihara, and H.D. Chiang. 2008. Discontinuous dynamics of electric power system with DC transmission: A study on DAE system. *IEEE Trans. Circuits Syst.–I: Regul. Pap.*, 55(2): 697–707.

Takahashi, M. et al. 1988. Fast generation shedding equipment based on the observation of swings of generators. *IEEE Trans. Power Syst.*, 3(2): 439–446.

Tamura, Y. and N. Yorino. 1987. Possibility of auto- & hetero-parametric resonances in power systems and their relationship with long-term dynamics. *IEEE Trans. Power Syst.*, 2(4): 890–896.

Tavora, C.J. and O.J.M. Smith. 1972a. Characterization of equilibrium and stability in power systems. *IEEE Trans. Power Appar. Syst.*, 91(3): 1127–1130.

Tavora C.J. and O.J.M. Smith. 1972b. Equilibrium analysis of power systems. *IEEE Trans. Power Appar. Syst.*, 91(3): 1131–1137.

Tavora, C.J. and O.J.M. Smith. 1972c. Stability analysis of power systems. *IEEE Trans. Power Appar. Syst.*, 91(3): 1138–1144.

Taylor, C.W. 1994. *Power System Voltage Stability.* New York: McGraw-Hill.

Taylor, C.W. et al. 1983. A new out-of-step relay with rate of change of apparent resistance augmentation. *IEEE Trans. Power Appar. Syst.*, 102(3): 631–639.

Tellegen, B.D.H. 1952. A general network theorem, with applications. *Phillips Res. Rept.*, 7: 259–269.

Thapar, J. et al. 1997. Application of the normal form of vector fields to predict inter-area separation in power systems. *IEEE Trans. Power Syst.*, 12(2): 844–850.

Tinney, W.F. and J.W. Walker. 1969. Direct solutions of sparse network equations by optimally ordered triangular factorization. *Proc. IEEE,* 55(11): 1801–1809.

Tomovic, R. 1963. *Sensitivity Analysis of Dynamical Systems.* New York: McGraw-Hill.

Tong, J., H.D. Chiang, and T.P. Conneen. 1993. A sensitivity-based BCU method for fast derivation of stability limits in electric power systems. *IEEE Trans. Power Syst.*, 8(4): 1418–1428.

Tsolas, N.A., A. Arapostathis, and P.P. Varaiya. 1985. A structure preserving energy function for power system transient stability analysis. *IEEE Trans. Circuits Syst.*, 32(10): 1041–1049.

Uma Rao, K. 1996. Power system stability evaluation and improvement using energy functions and FACTS controllers. PhD thesis, Indian Institute of Science, Bangalore.

Undrill, J.M. 1969. Structure in the computation of power system nonlinear dynamic response. *IEEE Trans. Power Appar. Syst.*, 88(1): 1–6.

Van Cutsem, T. and C. Vournas. 1998. *Voltage Stability of Electric Power Systems.* Boston: Kluwer Academic.

Varaiya, P.P., F.F. Wu, and R.L. Chen. 1985. Direct methods for transient stability analysis of power systems. *Proc. IEEE,* 73(12): 1703–1715.

Vittal, V. et al. 1988. Transient stability analysis of stressed power systems using the energy function method. *IEEE Trans. Power Syst.*, 3(1): 239–244.

Vittal, V. et al. 1989. Determination of stability limits using analytical sensitivity of the transient energy margin. *IEEE Trans. Power Syst.*, 4(4): 1363–172.

Vittal, V., N. Bhatia, and A.A. Fouad. 1991. Analysis of interarea phenomenon in power systems following large disturbances. *IEEE Trans. Power Syst.*, 6(4): 1515–1521.

Walve, K. 1986. Modelling of power system components at severe disturbances. *CIGRE Conference Paper 38-18*, Paris.

Wang, L. and A.A. Girgis. 1997. A new method for power system transient instability detection. *IEEE Trans. Power Deliv.*, 12(3): 1082–1089.

Willems, J.L. 1969. The computation of finite stability regions by means of open Lyapunov surfaces. *Int. J. Control*, 10: 537–544.

Willems, J.L. and J.C. Willems. 1970. The application of Lyapunov methods to the computation of transient stability regions for multimachine power systems. *IEEE Trans. Power Appar. Syst.*, 89: 795–801.

Xue, Y. and M. Pavella. 1993. Critical cluster identification in transient stability studies. *IEE Proc. (London): Pt. C*, 140(6): 481–489.

Xue, Y., Th. Van Cutsem, and M. Ribbens-Pavella. 1989. Extended equal area criterion: Justifications, generalizations and applications. *IEEE Trans. Power Syst.*, 4(1): 44–52.

Xue, Y. et al. 1992. Extended equal area criterion revisited. *IEEE Trans. Power Syst.*, 7(3): 1012–1022.

Yakubovitch, V.A. 1962. The solution of certain matrix inequalities in automatic control. *Dokl. Nauk. S.S.S.R.*, 143: 1304–1307.

Yorino, N. et al. 1989. A generalized analysis method of auto-parametric resonances in power systems. *IEEE Trans. Power Syst.*, 4(1): 1057–1084.

Zaborsky, J. et al. 1981. Stabilizing control in emergencies, Part I: Equilibrium point and state determination, Part II: Control by local feedback. *IEEE Trans. Power Appar. Syst.*, 100(5): 2374–2389.

Zhong, Z. et al. 2005. Power system frequency monitoring network (FNET) implementation. *IEEE Trans. Power Syst.*, 20(4): 1914–1921.

Zou, Y., M. Yin and H. Chiang. 2003. Theoretical foundation of the controlling UEP method for direct transient-stability analysis of network-preserving power system models. *IEEE Trans. Circuits Systems–I: Fund. Theory Appl.*, 50(10): 1324–1336.

# Appendix A: Synchronous Generator Model

## A.1 Synchronous Machine

The synchronous machine considered is shown in Figure A.1 (Padiyar 2002). It shows a three-phase armature windings (a, b, and c) on the stator and four windings on the rotor, including the filed winding $f$. The amortisseur (or damper) circuits in the salient pole machine or the eddy current effects in the rotor are represented by a set of coils with constant parameters. Three damper coils, $h$ in the $d$-axis and $g$ and $k$ in the $q$-axis, are shown in Figure A.1. In general, the number of damper coils represented can vary from zero (in the simplest model) to three in the most detailed model used in stability studies. The following assumptions are used in the derivation of the basic equations of the machine:

1. The MMF in the airgap is distributed sinusoidally and the harmonics are neglected.
2. Saliency is restricted to the rotor. The effect of slots in the stator is neglected.

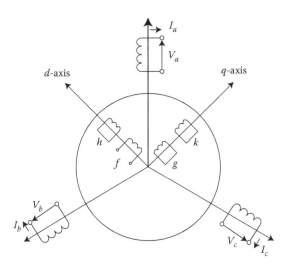

**FIGURE A.1**
Synchronous machine.

3. Magnetic saturation and hysteresis are ignored.

There is no loss of generality in assuming the machine has two poles, which implies that the mechanical speed of the rotor (in radians per second) is identical to the electrical frequency in steady state. The rotor angle $\theta$ (with respect to a stationary axis) can be expressed as

$$\theta = \omega_0 t + \delta$$

where $\omega_0$ is the operating (radian) frequency (normally assumed to be the rated frequency). $\delta$ is constant in steady state.

Based on the assumptions stated above, it is possible to derive the self- and mutual inductances of the stator and rotor coils. The self-inductances of the rotor are constants and the mutual inductances between the stator and rotor coils are sinusoidal functions of $\theta$. The self-inductances of the stator coils are constants for the round rotor machine and has a varying component (with $\theta$) for the salient pole machine.

## A.2 Park's Transformation

It would be advantageous if the time-varying machine equations can be transformed to a time invariant set. This would result in the simplification of the calculations both for steady-state and transient conditions. Park (1929) introduced the following transformation:

$$\begin{bmatrix} f_a \\ f_b \\ f_c \end{bmatrix} = [C_P] \begin{bmatrix} f_d \\ f_q \\ f_o \end{bmatrix} \tag{A.1}$$

where $f_\alpha$ can be stator voltage, current, or flux linkage of the stator winding $\alpha$ ($\alpha = a, b,$ or $c$). $[C_P]$ is defined as

$$[C_P] = \begin{bmatrix} k_d \cos\theta & k_q \sin\theta & k_o \\ k_d \cos\left(\theta - \dfrac{2\pi}{3}\right) & k_q \sin\left(\theta - \dfrac{2\pi}{3}\right) & k_o \\ k_d \cos\left(\theta + \dfrac{2\pi}{3}\right) & k_q \sin\left(\theta + \dfrac{2\pi}{3}\right) & k_o \end{bmatrix} \tag{A.2}$$

where $k_d$, $k_q$, and $k_o$ are constants appropriately chosen. In the original Park's transformation, $k_d = 1.0$, $k_q = -1.0$, and $k_o = 1$.

The inverse transformation is given by

$$\begin{bmatrix} f_d \\ f_q \\ f_o \end{bmatrix} = [C_P]^{-1} \begin{bmatrix} f_a \\ f_b \\ f_c \end{bmatrix} \tag{A.3}$$

where

$$[C_P]^{-1} = \begin{bmatrix} k_1 \cos\theta & k_1 \cos\left(\theta - \dfrac{2\pi}{3}\right) & k_1 \cos\left(\theta + \dfrac{2\pi}{3}\right) \\ k_2 \sin\theta & k_2 \sin\left(\theta - \dfrac{2\pi}{3}\right) & k_2 \sin\left(\theta + \dfrac{2\pi}{3}\right) \\ k_3 & k_3 & k_3 \end{bmatrix}$$

$$k_1 = \frac{2}{3k_d}, \quad k_2 = \frac{2}{3k_q}, \quad k_3 = \frac{2}{3k_o}$$

If we use a power invariant transformation by selecting $k_d = \sqrt{2/3} = k_1$, $k_q = \sqrt{2/3} = k_2$, $k_o = 1/\sqrt{3} = k_3$, we can observe that $[C_P]^{-1} = [C_P]^t$.

Note that the rotor quantities are not transformed. Park's transformation results in replacing stator coils ($a$, $b$, and $c$) by fictitious coils ($d$, $q$, and $o$). Out of these, $o$ coil has no coupling with rotor coils and may be neglected if the zero sequence current, $i_o = 0$. The transformed mutual inductances between $d$, $q$ coils and the rotor coils are constants. This can be interpreted that $d$ and $q$ coils rotate at the same speed as the rotor. It can also be assumed that the $d$ coil is aligned with the $d$-axis and the $q$ coil is aligned with the $q$-axis. We have assumed that the $d$-axis leads the $q$-axis (in the direction of rotation).

The positive value of $k_q$ indicates that the $q$-axis is lagging the direct axis, whereas in the original transformation by Park, the $q$-axis is assumed to lead the $d$-axis. Although an IEEE committee report (1969) recommended a revision of the old convention (of $q$ leading $d$), usage in power industry is often based on the old convention (Kundur 1994).

It should also be noted here that the use of generator convention in expressing the stator voltage equations is consistent with the choice of the $d$-axis leading the $q$-axis. Similarly, the earlier motor convention is consistent with the $q$-axis leading the $d$-axis.

Note that although the armature currents in the original convention are considered to flow out of the generator, the signs associated with these currents (in flux calculations) are negative (consistent with the motor convention).

## A.3 Per-Unit Quantities

It is common to express voltages, currents, and impedances in per unit by choosing appropriate base quantities. The advantages of an appropriate per-unit system are

1. The numerical values of currents and voltages are related to their rated values irrespective of the size of the machine.
2. The per-unit impedances on the machine base lie in a narrow range for a class of machines of similar design.
3. The number of parameters required is minimized.

It is to be noted that the base quantities for the stator and rotor circuits can be independently chosen with certain restrictions, which result in per-unit mutual reactances being reciprocal. If power invariant Park's transformation is used, the constraints imply selecting the same base power for all the circuits.

### A.3.1 Stator Base Quantities

The base quantities for the stator $d$–$q$ windings are chosen as follows:

Base power, $S_B$ = three-phase rated power

Base voltage, $V_B$ = rated line to line voltage (rms)

Base current, $I_B = \sqrt{3} \times$ rated line current

Base impedance, $Z_B = V_B/I_B$ = rated line to neutral voltage/rated line current

Base flux linkages, $\psi_B = V_B/\omega_B$

Base inductance, $L_B = \psi_B/I_B = Z_B/\omega_B$

$\omega_B$ is the base angular frequency in radians per second (this is also the rated angular speed for a two-pole machine).

## A.4 Synchronous Machine Model

One of the major assumptions in the analysis of dynamic performance involving low-frequency (<5 Hz) behavior of the system is to neglect the transients in the external network. This simplifies the analysis as the network is modeled by algebraic equations based on single-phase (positive sequence) representation. The network equations are conveniently expressed using voltage (and current) phasors with $D$–$Q$ components (expressed on a synchronously rotating or Kron's reference frame).

If network transients are to be neglected, it is logical to ignore the transients in the stator windings of the synchronous machine, which are connected to the external network. This implies that stator equations are also reduced to algebraic. The use of stator flux linkages or currents as state variables is not possible. Also, the degree of detail used in modeling a synchronous machine can vary depending on the requirement and the data available.

### A.4.1 Stator Equations

The stator equations in Park's reference frame (expressed in per unit) are

$$-\frac{1}{\omega_B}\frac{d\psi_d}{dt} - \frac{\omega}{\omega_B}\psi_q - R_a i_d = v_d \tag{A.4}$$

$$-\frac{1}{\omega_B}\frac{d\psi_q}{dt} + \frac{\omega}{\omega_B}\psi_d - R_a i_q = v_q \tag{A.5}$$

It is assumed that the zero sequence currents in the stator are absent. If stator transients are to be ignored, it is equivalent to ignoring the $p\psi_d$ and $p\psi_q$ terms in Equations A.4 and A.5 (note that $p$ is the differential operator $d/dt$). In addition, it is also advantageous to ignore the variations in the rotor speed $\omega$. This can be justified on the grounds that under disturbances considered, the variations in the speed are negligible. With these assumptions, Equations A.4 and A.5 can be expressed as

$$-(1 + S_{mo})\psi_q - R_a i_d = v_d \tag{A.6}$$

$$(1 + S_{mo})\psi_d - R_a i_q = v_q \tag{A.7}$$

where $S_{mo}$ is the initial operating slip defined as

$$S_{mo} = \frac{\omega_o - \omega_B}{\omega_B} \tag{A.8}$$

In most of the cases, it will be assumed that the initial operating slip is zero (the operating frequency is the rated (nominal) frequency).

### A.4.2 Rotor Equations

Since the stator Equations A.6 and A.7 are algebraic (neglecting stator transients), it is not possible to choose stator currents $i_d$ and $i_q$ as state variables (state variables have to be continuous functions of time whereas $i_d$ and $i_q$ can be discontinuous due to any sudden changes in the network). As rotor windings either remain closed (damper windings) or closed through finite voltage source (field winding), the flux linkages of these windings cannot change suddenly. This implies that if $i_d$ changes suddenly, the field and damper currents also change suddenly in order to maintain field and damper flux linkages continuous. The flux linkage immediately after a disturbance remains constant at the value just prior to the disturbance (this property is termed as the theorem of constant flux linkages in the literature—see Kimbark (1956)).

The previous discussion shows that rotor winding currents cannot be treated as state variables when stator transients are neglected. The obvious choice of state variables are rotor flux linkages or transformed variables, which are linearly dependent on the rotor flux linkages.

Depending on the degree of detail used, the number of rotor windings and corresponding state variables can vary from one to four. In a report by an IEEE Task Force (1986), the following models are suggested based on varying degrees of complexity:

1. Classical model (model 0.0)
2. Field circuit only (model 1.0)
3. Field circuit with one equivalent damper on the $q$-axis (model 1.1)
4. Field circuit with one equivalent damper on the $d$-axis
   a.  Model 2.1 (one damper on the $q$-axis)
   b.  Model 2.2 (two dampers on the $q$-axis)

It is to be noted that in the classification of the machine models, the first number indicates the number of rotor windings on the $d$-axis while the second number indicates the number of windings on the $q$-axis (alternately, the numbers represent the number of state variables considered in the $d$-axis and $q$-axis). Thus, the classical model that neglects damper circuits and field flux decay ignores all state variables for the rotor coils and is termed model 0.0.

In Figure A.1, it is assumed that the synchronous machine is represented by model 2.2. This is the most detailed model for which data are supplied by manufacturers of machines or obtained by tests described in IEEE Standard No. 115 (1983). It is to be noted that while higher-order models provide better results for special applications, they also require an exact determination of parameters. With constraints on data availability and for study of large systems, it may be adequate to use model 1.1 if the data are correctly determined (Dandeno et al. 1973).

In what follows, model 1.1 is assumed for the representation of synchronous machine.

## A.5 Application of Model 1.1

The stator and rotor flux linkages are given by

$$\psi_d = x_d i_d + x_{ad} i_f \tag{A.9}$$

$$\psi_f = x_{ad} i_d + x_f i_f \tag{A.10}$$

$$\psi_q = x_q i_q + x_{aq} i_g \tag{A.11}$$

$$\psi_g = x_{aq} i_q + x_g i_g \tag{A.12}$$

Solving Equations A.10 and A.12 for $i_f$ and $i_g$, and substituting in Equations A.9 and A.11, respectively, we get

$$\psi_d = x'_d i_d + E'_q \tag{A.13}$$

$$\psi_q = x'_q i_q - E'_d \tag{A.14}$$

where

$$x'_d = x_d - \frac{x_{ad}^2}{x_f}, \ x'_q = x_q - \frac{x_{aq}^2}{x_g}, \ E'_q = \frac{x_{ad}\psi_f}{x_f}, \ E'_d = -\frac{x_{aq}\psi_g}{x_g}$$

The voltage equations for the rotor windings are

$$\frac{1}{\omega_B}\frac{d\psi_f}{dt} = -R_f i_f + v_f \tag{A.15}$$

$$\frac{1}{\omega_B}\frac{d\psi_g}{dt} = -R_g i_g \tag{A.16}$$

Expressing $\psi_f$ and $\psi_g$ in terms of $E'_q$ and $E'_d$ and eliminating $i_f$ and $i_g$ from the above equations leads to

$$\frac{dE'_q}{dt} = \frac{1}{T'_{do}}[-E'_q + (x_d - x'_d)i_d + E_{fd}] \tag{A.17}$$

$$\frac{dE'_d}{dt} = \frac{1}{T'_{qo}}[-E'_d - (x_q - x'_q)i_g] \tag{A.18}$$

where

$$T'_{do} = \frac{x_f}{\omega_B R_f}, \quad T'_{qo} = \frac{x_g}{\omega_B R_g}, \quad E_{fd} = \frac{x_{ad}}{R_f} v_f$$

It is to be noted that in model 1.1, it is convenient to define the equivalent voltage sources $E'_d$ and $E'_q$, which are used as state variables instead of $\psi_f$ and $\psi_g$. The advantages of this will be self-evident when we consider the stator and torque equations.

### A.5.1 Stator Equations

Substituting Equations A.13 and A.14 in Equations A.6 and A.7 and letting $S_{m0} = 0$, we get

$$E'_q + x'_d i_d - R_a i_q = v_q \tag{A.19}$$

$$E'_d - x'_q i_q - R_a i_d = v_d \tag{A.20}$$

If transient saliency is neglected by letting

$$x'_d = x'_q = x' \tag{A.21}$$

we can combine Equations A.19 and A.20 into a single complex equation given by

$$(E'_q + jE'_d) - (R_a + jx')(i_q + ji_d) = v_q + jv_d \tag{A.22}$$

The above equation represents an equivalent circuit of the stator shown in Figure A.2a. This shows a voltage source $(E'_q + jE'_d)$ behind an equivalent impedance $(R_a + jx')$.

The variables $(D-Q)$ in Kron's frame of reference are related to the variables $(d-q)$ in Park's frame of reference by

$$(f_Q + jf_D) = (f_q + jf_a)e^{j\delta} \tag{A.23}$$

where $\delta = \theta - \omega_0 t$, $\omega_0$ is the operating frequency in radians per second, and $f$ can represent voltage or current. Applying Equation A.23 to Equation A.22, we get

$$(E'_Q + jE'_D) - (R_a + jx')(i_Q + ji_D) = v_Q + jv_D \tag{A.24}$$

Equation A.24 also represents an equivalent circuit of the stator shown in Figure A.2b.

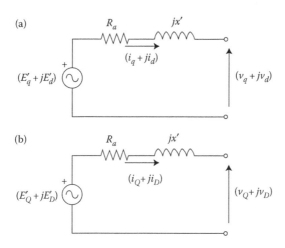

**FIGURE A.2**
Stator equivalent circuits. (a) Park's reference. (b) Kron's reference (synchronously rotating reference).

Unfortunately, no equivalent circuit of the stator exists when transient saliency is considered. This can pose a problem in the network calculations in multimachine systems. The methods of handling transient saliency will be discussed in Appendix C. For a single-machine system, however, saliency does not pose any problem.

### A.5.2 Rotor Mechanical Equations

The rotor mechanical equations in per unit can be expressed as

$$M\frac{d^2\delta}{dt^2} + D'\frac{d\delta}{dt} = T_m - T_e \qquad (A.25)$$

where $M = 2H/\omega_B$ and $T_e$ is electrical torque given by

$$T_e = \psi_d i_q - \psi_q i_d \qquad (A.26)$$

$D'$ is the damping term and $T_m$ is the mechanical torque acting on the rotor. Substituting Equations A.13 and A.14 in Equation A.26, we get

$$T_e = E'_d i_d + E'_q i_q + (x'_d - x'_q)i_d i_q \qquad (A.27)$$

If transient saliency is ignored ($x'_d = x'_q$), then the third term in the above expression is identically zero.

Equation A.25 can be expressed as two first-order equations as

$$\frac{d\delta}{dt} = \omega_B(S_m - S_{m0}) \tag{A.28}$$

$$2H\frac{dS_m}{dt} = -D(S_m - S_{m0}) + T_m - T_e \tag{A.29}$$

where the generator slip $S_m$ is defined below:

$$S_m = \frac{\omega - \omega_B}{\omega_B}$$

Normally, the operating speed is considered to be the same as the nominal or rated speed. In this case, $S_{m0} = 0$. $D$ is the per-unit damping given by

$$D = \omega_B D'$$

## A.6 Simpler Models

Model 1.0 can be handled by letting

$$x_q' = x_q, \quad T_{qo}' \neq 0, \tag{A.30}$$

Note that with $x_q' = x_q$, Equation A.34 reduces to

$$\frac{dE_d'}{dt} = \frac{-E_d'}{T_{qo}'}, \quad E_{d0}' = -(x_q - x_q')i_{qo} = 0$$

With the initial condition at zero, $E_d'$ remains at zero throughout the simulation as long as $T_{qo}' > 0$. The actual value of $T_{qo}'$ is unimportant and can be set at any arbitrary (convenient) value (say 1.0 s).

The models for excitation and prime mover (turbine-governor) controllers are described in IEEE Committee or Working Group Reports (1981b, 1991, 1992). A simplified model for a static excitation system is given in Chapter 4.

For the consideration of classical model 0.0, in addition to the constraints Equation A.30, it is necessary to set

$$T_{do}' = \text{large value (say 1000 s)} \tag{A.31}$$

If saliency is not to be considered, then it is necessary to set

$$x'_q = x_q = x'_d \tag{A.32}$$

With the constraints Equations A.30 through A.32, the model reduces to that of a voltage source $E'_q$ behind a transient reactance of $x'_d$; the large value of $T'_{do}$ ensures that $E'_q$ remains practically constant (neglecting flux decay).

Note that the constraint Equation A.32 can also be included in model 1.0 and has the effect of neglecting saliency while considering flux decay. The term "saliency" has been used rather loosely here. The normal definition of saliency applies when $x_d \neq x_q$. The saliency that we need to consider in dynamic analysis depends on the model used. The "dynamic saliency" has been defined (Undrill 1969) in this context to distinguish it from the usual definition of saliency. Dynamic saliency implies (i) $x_q \neq x'_d$ for model 1.0, (ii) $x'_d \neq x'_q$ for model 1.1, and (iii) $x''_d \neq x''_q$ for model 2.2.

# Appendix B: Boundary of Stability Region: Theoretical Results

Here, we summarize the results of Tsolas et al. (1985) and Chiang et al. (1987) in defining the boundary of the stability region for a nonlinear system under certain assumptions. Some results on gradient systems are also presented (Hirsch et al. 2004), which are of relevance in the analysis of the PEBS method.

## B.1 Stability Boundary

Consider a nonlinear autonomous dynamical system described by

$$\dot{x} = f(x) \tag{B.1}$$

where the vector field $f$ is of dimension $n$ and its derivatives are continuous. The solution curve, starting from $x_0$ at $t = 0$, is called a trajectory, denoted by $\Phi(x_0, t)$ with $\Phi(x_0, 0) = x_0$. For a SEP, $x_s$, there is a region in the state space from which trajectories converge to $x_s$. This is called the stability region of $x_s$ and is denoted by $A(x_s)$. The boundary of the stability region is denoted by $\partial A(x_s)$.

Let $x_i$ be a hyperbolic EP of Equation B.1; its stable manifold $W^s(x_i)$ and unstable manifold $W^u(x_i)$ are defined as

$$W^s(x_i) = \{x \mid \Phi(x,t) \to x_i, \text{ as } t \to \infty\} \tag{B.2}$$

$$W^u(x_i) = \{x \mid \Phi(x,t) \to x_i, \text{ as } t \to -\infty\} \tag{B.3}$$

## Assumptions

1. All the EPs on the stability boundary are hyperbolic.
2. The intersection of $W^s(x_i)$ and $W^u(x_i)$ satisfies the transversality conditions for all the EPs on the stability boundary.
3. There exists a $C^1$ function $V : R^n \to R$ such that
   a. $(dV/dt) [\Phi (x, t)] \leq 0$ for all $x \notin E$ ($E$ is a set of EPs)
   b. If $x$ is not an EP, then the set $\{t \in R; \dot{V}[\Phi(x,t)] = 0\}$ has a measure 0 in $R$.
   c. $V[\Phi(x, t)]$ is bounded implies that $\Phi(x, t)$ is bounded.

**Remarks**

1. The intersection of two manifolds A and B in $R^n$ satisfies the transversality condition if

   i.  At every point of the intersection $x \in (A \cap B)$, the tangent spaces of A and B at x satisfy

$$T_x(A) + T_x(B) = R^n \quad \text{for } x \in (A \cap B) \text{ or}$$

   ii. They do not intersect at all

2. The first two assumptions are generic properties of dynamic systems.

3. Conditions (a) and (b) in assumption 3 imply that every trajectory must either go to infinity or converge to one of the EPs. This ensures that limit cycles or chaotic motions do not exist for the system described by Equation B.1.

## B.1.1 Characterization of Stability Boundary

For the system described by Equation B.1 and satisfying the three assumptions stated earlier, let $x_i$, $i = 1,2,\ldots$ be the UEPs on the stability boundary $\partial A(x_s)$; then

$$\partial A(x_s) = \bigcup_{x_i \in E \cap \partial A} W^s(x_i) \tag{B.4}$$

## B.1.2 Some Interesting Results

1. If the system described by Equation B.1 satisfies assumption (3), then

$$V(x_i) = \min[V(x)] \quad \text{for } x \in W^s(x_i)$$

2. On the stability boundary, the point at which the Lyapunov function V achieves a minimum must be a type 1 UEP.

## B.1.3 Characterization of CUEP

Let the fault-on trajectory cross the stability boundary at the exit point $x_e$. If $x_e$ lies on the stable manifold of the UEP, $x_{cu}$, then we say that $x_{cu}$ is the CUEP. Finding the CUEP is a nontrivial problem.

## B.2 Gradient Systems

A gradient system on $R^n$ is a system described by

$$\dot{x} = -\text{grad}V(x) \tag{B.5}$$

where

$$\text{grad}V = \left[\frac{\partial V}{\partial x_1}, \dots, \frac{\partial V}{\partial x_n}\right]^T$$

### B.2.1 Properties of Gradient Systems

1. $V(x)$ is a Lyapunov function for the system described by Equation B.5. $\dot{V}(x) = 0$ if and only if $x$ is an EP.
2. If $c$ is a regular value of $V$, then the vector field is perpendicular to the level set $V^{-1}(c)$.
3. If a critical point is an isolated minimum of $V$, then this point is an asymptotically stable EP (Hirsch et al., 2004).

Note that a critical point is one where $\text{grad}V = 0$. $x \in V^{-1}(c)$ is a regular point if grad $V(x) \neq 0$. The level surfaces of the function $V$ are the subsets $V^{-1}(c)$ with $c \in R$. If all the points in $V^{-1}(c)$ are regular points, $c$ is said to be a regular value. For a regular value of $c$, the set $V^{-1}(c)$ is a surface of dimension $(n-1)$.

# Appendix C: Network Solution for Transient Stability Analysis

The power system model for transient stability analysis (including simulation) is nonlinear. The system equations are differential and algebraic (DAE). The algebraic equations describe the network, loads, and generator stator where the dynamics or transients are neglected. Even some FACTS controllers can be modeled by steady-state control characteristics represented by nonlinear algebraic equations (see Chapter 5).

In this appendix, we describe the network equations and their solution in the transient stability studies and how it can be interfaced with the solution of the dynamic subsystems. The dynamic subsystems are primarily the generator and its controllers—excitation and prime mover. The network controllers (HVDC and FACTS), if present, can also be included as components of the dynamic subsystem.

The AC electrical network consists of transmission lines, transformers, shunt reactors, shunt, and/or series capacitors. These are linear elements. The admittance matrix of the network is constant for a specified system configuration. This is augmented by including the equivalent circuits of the generators and voltage-dependent nonlinear loads. Only the diagonal elements of the admittance matrix are modified by the inclusion of the generator stator and load admittances.

## C.1 Inclusion of Generator Stator in the Network

A major assumption in the modeling of AC network is that it is symmetric. Hence, for steady-state analysis, the network can be represented on a single-phase basis using phasor quantities (for slowly varying sinusoidal voltages and currents in the network). Thus, the generator stator can also be represented on a single-phase basis.

The stator equations can be expressed as a single equation in phasor quantities if transient saliency is neglected, that is, $x_d' = x_q' = x'$. In this case (see Appendix A),

$$i_q + ji_d = \frac{1}{R_a + jx'}[(E_q' + jE_d') - (v_q + jv_d)] \tag{C.1}$$

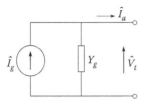

**FIGURE C.1**
Generator equivalent circuit.

The armature current phasor $\hat{I}_a$ can also be represented on a synchronously rotating reference frame as

$$\hat{I}_a = i_Q + ji_D = (i_q + ji_d)e^{j\delta} = \frac{1}{R_a + jx'}[\hat{E}' - \hat{V}_t] \tag{C.2}$$

where

$$\hat{E}' = E'_Q + jE'_D = (E'_q + E'_d)e^{j\delta}$$

$$\hat{V}_t = v_Q + jv_D = (v_q + jv_d)e^{j\delta} = V_t e^{j\theta}$$

Equation C.2 represents an equivalent circuit shown in Figure C.1 where

$$Y_g = \frac{1}{R_a + jx'} \tag{C.3}$$

$$\hat{I}_g = \hat{E}'Y_g = \frac{(E'_q + jE'_d)e^{j\delta}}{R_a + jx'} \tag{C.4}$$

It is to be noted that $\hat{I}_g$ is a function of state variables ($E'_q, E'_d$, and $\delta$) only. Hence, it does not change suddenly whenever there is a network switching. The equivalent circuit shown in Figure C.1 can be readily merged with the AC network (external to the generator).

## C.2 Treatment of Transient Saliency

If transient saliency is to be considered, the stator cannot be represented by a single-phase equivalent circuit shown in Figure C.1. Here, $i_D$ and $i_Q$ are solved from the stator equations given below.

$$\begin{bmatrix} i_q \\ i_d \end{bmatrix} = \frac{1}{R_a^2 + x_d' x_q'} \begin{bmatrix} R_a & x_d' \\ -x_q' & R_a \end{bmatrix} \begin{bmatrix} E_q' - v_q \\ E_d' - v_d \end{bmatrix} \tag{C.5}$$

Equation C.5 when transformed to a synchronously rotating reference frame results in

$$\begin{bmatrix} i_Q \\ i_D \end{bmatrix} = [Y_g^{DQ}(t)] \begin{bmatrix} E_Q' - v_Q \\ E_D' - v_D \end{bmatrix} \tag{C.6}$$

where

$$[Y_g^{DQ}(t)] = \frac{1}{R_a^2 + x_d' x_q'} \begin{bmatrix} \cos\delta & -\sin\delta \\ \sin\delta & \cos\delta \end{bmatrix} \begin{bmatrix} R_a & x_d' \\ -x_q' & R_a \end{bmatrix} \begin{bmatrix} \cos\delta & \sin\delta \\ -\sin\delta & \cos\delta \end{bmatrix} \tag{C.6}$$

$[Y_g^{DQ}]$ is a function of $\delta$, and as $\delta$ varies with time, it is a time-varying matrix. This poses problems as the overall network matrix, including $[Y_g^{DQ}]$ in real quantities (twice the size of the matrix in complex quantities), is not constant but time varying. This implies the network matrix has to be factored at every step, which increases computational complexity.

Hence, there is need for special techniques to handle transient saliency. There are two ways of doing this, namely

   i. Use of dependent source (application of compensation theorem)
   ii. Use of a dummy rotor coil

*i. Use of a dependent source*
The stator equations can be rewritten as

$$\left. \begin{array}{l} E_q' + x_d' i_d - R_a i_q \quad = v_q \\ E_d' + E_{dc}' - x_d' i_q - R_a i_d = v_d \end{array} \right\} \tag{C.7}$$

where

$$E_{dc}' = -(x_q' - x_d') i_q \tag{C.8}$$

The above equations can be expressed as a single equation in phasor quantities.

$$E_q' + j(E_d' + E_{dc}') - (R_a + jx_d')(i_q + ji_d) = v_q + jv_d \tag{C.9}$$

The equivalent circuit of Figure C.1 applies. Hence

$$\hat{I}_g = \frac{[E_q' + j(E_d' + E_{dc}')]}{R_a + jx_d'} e^{j\delta} = Y_g \hat{E}' + \hat{I}_{sal} \tag{C.10}$$

where

$$\hat{I}_{sal} = \frac{jE_{dc}'e^{j\delta}}{R_a + jx_d'}, \quad Y_g = \frac{1}{(R_a + jx_d')}$$

It is to be noted that the transient saliency is replaced by introducing a dependent current source $\hat{I}_{sal}$, which is a function of the terminal voltage $\hat{V}_t$ and $\delta$. It is not difficult to see that, in this approach, the choice of $Y_g$ and correspondingly $\hat{I}_{sal}$ is not unique. Dommel and Sato (1972) choose $Y_g$ as

$$Y_g = \frac{R_a - j(1/2)(x_d' + x_q')}{R_a^2 + x_d' x_q'} \tag{C.11}$$

$\hat{I}_{sal}$ can be derived as

$$\hat{I}_{sal} = \frac{-j(x_q' - x_d')}{2(R_a^2 + x_d' x_q')}(\hat{E}^* - \hat{V}^*)e^{j2\delta} \tag{C.12}$$

where $*$ denotes complex conjugate.

In the paper by Dommel and Sato (1972), the saliency is handled by iterative solution (improving on $\hat{I}_{sal}$ until convergence is obtained). It is claimed that the choice of $Y_g$ given in Equation C.11 speeds up convergence and two or three iterations are normally sufficient. As an initial guess for the terminal voltage, the value at the previous instant (say $t_k$) is used, except that the angle is changed by the same amount by which $\delta$ has changed from $t_k$ to $t_{k+1}$.

### ii. Use of a dummy rotor coil

The motivation for this approach is that if $E_{dc}'$ is a state variable (proportional to flux linkage of a dummy coil in the $q$-axis), then the problem of iterative solution for transient saliency is eliminated. In other words, the dependent source $\hat{I}_{sal}$ can be eliminated. This is an approximate treatment of transient saliency, but the degree of approximation can be directly controlled to obtain acceptable accuracy.

Consider a rotor dummy coil in the $q$-axis that is linked only with the $q$-axis coil in the armature but has no coupling with other coils. Just as $E_d'$ is a voltage source proportional to $-\psi_q$, $E_{dc}'$ can be considered as another voltage source proportional to the flux linkage $-\psi_c$ of the dummy coil. The differential equation for $E'_{dc}$ can be expressed as

$$\frac{dE'_{dc}}{dt} = \frac{1}{T_c}[-E'_{dc} - (x'_q - x'_d)i_q] \qquad \text{(C.13)}$$

where $T_c$ is the open circuit time constant of the dummy coil, which can be arbitrarily selected. In comparing Equation C.13 with Equation C.8, it is readily seen that the latter is a steady-state solution of the former. As $T_c$ tends to zero, the solution of Equation C.13 approaches that given by Equation C.8. From numerical experiments, it is observed that $T_c$ need not be smaller than 0.01 s for acceptable accuracy (Immanuel 1993). This is of similar order as the time constant of a high-resistance damper winding. Using implicit integration methods, the step size chosen need not be constrained by the value of $T_c$ chosen.

## C.3 Load Representation

Loads are represented as static voltage-dependent models given by

$$P_L = f_P(V_L) \qquad \text{(C.14)}$$

$$Q_L = f_Q(V_L) \qquad \text{(C.15)}$$

where $f_P$ and $f_Q$ are any general nonlinear functions. In many stability programs, $f_P$ and $f_Q$ are represented as

$$P_L = f_P(V_L) = a_0 + a_1 V_L + a_2 V_L^2 \qquad \text{(C.16)}$$

$$Q_L = f_Q(V_L) = b_0 + b_1 V_L + b_2 V_L^2 \qquad \text{(C.17)}$$

If $a_0 = a_1 = b_0 = b_1 = 0$, then we say that loads can be represented by constant impedances. The load at a bus can be represented by the equivalent circuit shown in Figure C.2, where $Y_l$ is given by

$$Y_l = \frac{P_{L0} - jQ_{L0}}{V_{L0}^2} \qquad \text{(C.18)}$$

where subscript 0 indicates operating values, and $Y_l$ is chosen such that $\hat{I}_l$ is zero at the operating point. During a transient, $\hat{I}_L$ is calculated from

$$\hat{I}_l = Y_l \hat{V}_l - \hat{I}_L, \hat{I}_L = \frac{P_L - jQ_L}{V_L^*} \qquad \text{(C.19)}$$

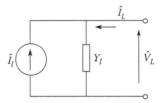

**FIGURE C.2**
Load equivalent circuit.

In general, this is a nonlinear function of $V_L$ and results in the overall network equations being nonlinear. For constant impedance load models, $I_l$ is identically equal to zero.

## C.4 AC Network Equations

Only quasi-steady-state response of the AC network is considered for stability evaluation, neglecting transients. Also, it is assumed that the network is symmetric. Hence, a single-phase representation (positive sequence network) is adequate. The network equations can be expressed using bus admittance matrix $Y_N$ as

$$[Y_N]\hat{V} = \hat{I}_N \tag{C.20}$$

where $\hat{V}$ is a vector of complex bus voltages and $\hat{I}_N$ is a vector of current injections. At a bus $j$, the component of $I_N$ is given by

$$\hat{I}_{Nj} = \hat{I}_{aj} - \hat{I}_{Lj} \tag{C.21}$$

where $I_{aj}$ is the generator armature current and $I_{Lj}$ is the load current at bus $j$.

## C.5 System Algebraic Equations

The generator and load equivalent circuits at all the buses can be integrated into the AC network and the overall system algebraic equations can be obtained as follows:

$$[Y]\hat{V} = \hat{I} \tag{C.22}$$

where [Y] is the complex admittance matrix that is obtained from augmenting $[Y_N]$ by inclusion of the shunt admittances $Y_g$ and $Y_l$. $\hat{I}$ is the vector of complex current sources the $j$th element of which can be expressed as

$$\hat{I}_j = \hat{I}_{gj} - \hat{I}_{lj} \tag{C.23}$$

It is to be noted that in general, $\hat{I}_j$ is a function of bus voltage $\hat{V}_j$. Hence, the solution of Equation C.22 has to be obtained by an iterative process. The initial estimate for $\hat{V}_k$ can be assumed as $\hat{V}_{k-1}$ (at the previous instant). Also, if transient saliency is neglected or represented by using a dummy coil, $\hat{I}_{gj}$ is only a function of the state variables $E'_{qj}, E'_{dj}, E'_{dcj}$, and $\delta_j$. In this case, the iterative solution of Equation C.22 is required only if loads are nonlinear (not represented by constant impedances).

It must be noted that transient saliency does not introduce nonlinearity. It only results in time-varying generator impedance. Nonlinearity of Equation C.22 is mainly due to nonlinear voltage-dependent loads that are modeled as constant admittances in parallel with nonlinear current sources.

Equation C.22 can also be expressed as real variables as

$$[Y^{DQ}]V^{DQ} = I^{DQ} \tag{C.24}$$

where the elements of $Y_{ij}^{DQ}$, $V_{ii}^{DQ}$, and $I_i^{DQ}$ can be expressed as

$$[Y_{ij}^{DQ}] = \begin{bmatrix} B_{ij} & G_{ij} \\ G_{ij} & -B_{ij} \end{bmatrix}, \quad [V_j^{DQ}] = \begin{bmatrix} V_{Qj} \\ V_{Dj} \end{bmatrix}, \quad [I_i^{DQ}] = \begin{bmatrix} I_{Di} \\ I_{Qi} \end{bmatrix}$$

The arrangement of the variables in Equation C.24 is deliberate such that $[Y^{DQ}]$ is a symmetric matrix if [Y] is symmetric. (Note that [Y] is symmetric if no phase shifting transformers are considered.)

## C.6 System Differential Equations

The differential equations for a generator are expressed as

$$\dot{x}_{gi} = f(x_{gi}, V_{gi}^{DQ}) \tag{C.25}$$

where $x_{gi}$ is a vector of state variables for generator $i$, including machine variables $E'_{qi}, E'_{di}, E'_{dci}$, and $\delta_i$, and other variables corresponding to the excitation system and turbine-governor. $V_{gi}^{DQ}$ are the D- and Q-axes components of the generator terminal voltage $V_{gi}$. It is interesting to note that the generator equations in a multimachine system are decoupled and the interconnection is provided through AC network variables.

The equations corresponding to all the generators in the system can be combined and solved either simultaneously (with multiple processors) or sequentially (using a single processor).

If an HVDC link or SVC or any controllable device is considered, the differential equations for these devices can also be expressed in the general form

$$\dot{x}_{cj} = f(x_{cj}, V_{cj}^{DQ}) \quad j = 1, 2, ..., n_c \tag{C.26}$$

where $n_c$ is the number of such devices in the system. The equations for dynamic loads can also be expressed in the above form.

## C.7  Solution of System Equations

The algorithm for numerical solution of the system equations has to tackle the problem of computing $x_k$ and $V_k$ at the end of step $k$, given the initial conditions $x_{k-1}$ and $V_{k-1}$.

### C.7.1  Partitioned Solution

Explicit or implicit methods are used to numerically integrate the differential equations of the type given in Equations C.25 and C.26. In doing this, values of $V_g$ and $V_c$ are required in the time interval $(t_{k-1}, t_k)$. In general, the bus voltage vector $V$ is required. One approach is to predict the values of $V$ by extrapolation. One extrapolation formula (Stott 1979) for $V_k$ based on two previous values is

$$V_k = \frac{V_{k-1}^2}{V_{k-2}} \tag{C.27}$$

Extrapolation is necessary to avoid the solution of the algebraic equations more than once in a step. This reduces the computations per step considerably as the network solution is a major component in the computational procedure. However, the interface error is not eliminated.

The solution of differential equations can be partitioned into subsets corresponding to the subsystems. This procedure not only helps in attaining a flexible program structure but can also improve the numerical performance, particularly when an implicit integration method (such as trapezoidal) is used (Immanuel 1993). There can be an optimal sequencing in the integration of the equations, which reduces error. For example, solving excitation system first results in the computation of $E_{fd}$, which can be used subsequently in the solution of rotor electrical equations. The solution for $E'_d$ and $E'_q$ can be used to compute electrical torque, which is the input to the mechanical system including turbine-governor.

After the solution of $x_k$, the network equations can be solved from

$$[Y]V_k = I(E_k, V_k) \tag{C.28}$$

Note that the current source vector $I$ is a function of $E_k$ (obtained from a subset of state variables $x_k$) and $V_k$. It is convenient to solve Equation C.28 by factorizing $[Y]$ and expressing Equation C.28 as

$$[L][U]V_k = I(E_k, V_k) \tag{C.29}$$

Sparsity-oriented triangular factorization (Tinney and Walker 1969) results in sparse triangular matrices $[L]$ and $[U]$, since $[Y]$ is sparse. The solution for $V_k$ is obtained from forward and backward substitutions as

$$[L]W_k = I_k, \quad [U]V_k = W_k \tag{C.30}$$

Whenever $I_k$ is a function of $V_k$ (in case of nonlinear loads), Equation C.28 is conveniently solved using Gauss–Jacobi method where the initial estimate of $V_k$ is used to compute $I_k$ and solve the equation using triangularized admittance matrix. This process is iterated until convergence is obtained in the solution of $V_k$. The number of iterations are the same as in the case when bus impedance matrix is used, typically 5–7. However, the use of triangularized admittance matrix instead of impedance matrix results in less memory and computations. Also, if $[Y]$ is a symmetric matrix, the lower triangular matrix $[L]$ is the transpose of the upper triangular matrix $[U]$ and need not be computed and stored.

Newton's method can also be used to solve Equation C.28. Here, the network equations have to be used in the real form (using $D–Q$ variables). At each iteration of Newton's method, the following linear equations have to be solved:

$$[J_k]\Delta V_k = F_k = I_k - [Y]V_k \tag{C.31}$$

where $[J_k]$ is the Jacobian matrix calculated at step $k$ and $F_k$ is the mismatch that should be zero at the solution. The Jacobian matrix differs from $[Y]$ (real admittance matrix) only in the self (diagonal) terms made of $2 \times 2$ blocks. However, updating the Jacobian matrix at every iteration is not practical. Even updating at every step is not recommended. The use of very dishonest

Newton (VDHN) method involves maintaining the Jacobian matrix constant for several time steps. The updating of the Jacobian can be based on the number of iterations required for convergence. An increase in the number of iterations required signals the need to update the Jacobian.

It is to be noted that the use of decoupled Jacobian is not recommended as in transient stability studies, the decoupling between $P - \theta$ and $Q - V$ is not valid during transient swings caused by large disturbances.

### C.7.2 Simultaneous Solution

The system equations can be expressed as

$$\dot{x} = f(x, V) \tag{C.32}$$

$$I(x, V) - [Y]V = 0 \tag{C.33}$$

The implicit method of integration is used to discretize Equation C.32. For example, using trapezoidal rule, we get

$$F_{xk} = x_k - x_{k-1} - \frac{h}{2}[f(x_k, V_k) + f(x_{k-1}, V_{k-1})] = 0 \tag{C.34}$$

Combining this with Equation C.33 and linearizing, we get

$$\begin{bmatrix} F_{xk} \\ F_{Vk} \end{bmatrix} = - \begin{bmatrix} A_x^k & B^k \\ C^k & J_N^k \end{bmatrix} \begin{bmatrix} \Delta x_k \\ \Delta V_k \end{bmatrix} = -[J^k] \begin{bmatrix} \Delta x_k \\ \Delta V_k \end{bmatrix} \tag{C.35}$$

where

$$F_{Vk} = I(x_k, V_k) - [Y]V_k$$

The VDHN method is used to solve for the vectors $\Delta x_k$ and $\Delta V_k$.

# Appendix D: Data for 10-Generator System

The data for the 39-bus, 10-generator system are given here. This is adapted from Behera (1988). The single-line diagram of the system is shown in Figure D.1. The machine data, line data, load flow, and transformer data are given in Tables D.1 through D.4, respectively. The AVR data are given in Table D.1.

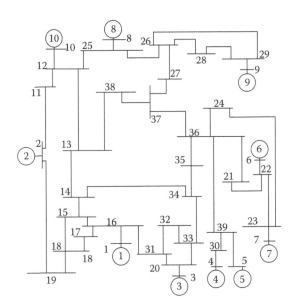

**FIGURE D.1**
Single-line diagram of the 10-generator system.

**TABLE D.1**

Machine Data

| Generator Number | $R_a$ | $X_d$ | $X'_d$ | $X_q$ | $X'_q$ | $H$ | $T'_{d0}$ | $T'_{q0}$ | $T_c$ | $D$ |
|---|---|---|---|---|---|---|---|---|---|---|
| 1 | 0.0 | 0.2950 | 0.0647 | 0.2820 | 0.0647 | 30.3 | 6.56 | 1.5 | 0.01 | 0.0 |
| 2 | 0.0 | 0.0200 | 0.0060 | 0.0190 | 0.0060 | 500.0 | 6.0 | 0.7 | 0.01 | 0.0 |
| 3 | 0.0 | 0.2495 | 0.0531 | 0.2370 | 0.0531 | 35.8 | 5.7 | 1.5 | 0.01 | 0.0 |
| 4 | 0.0 | 0.3300 | 0.0660 | 0.3100 | 0.0660 | 26.0 | 5.4 | 0.44 | 0.01 | 0.0 |
| 5 | 0.0 | 0.2620 | 0.0436 | 0.2580 | 0.0436 | 28.6 | 5.69 | 1.5 | 0.01 | 0.0 |
| 6 | 0.0 | 0.2540 | 0.0500 | 0.2410 | 0.0500 | 34.8 | 7.3 | 0.4 | 0.01 | 0.0 |
| 7 | 0.0 | 0.2950 | 0.0490 | 0.2920 | 0.0490 | 26.4 | 5.66 | 1.5 | 0.01 | 0.0 |
| 8 | 0.0 | 0.2900 | 0.0570 | 0.2800 | 0.0570 | 24.3 | 6.7 | 0.41 | 0.01 | 0.0 |
| 9 | 0.0 | 0.2106 | 0.0570 | 0.2050 | 0.0570 | 34.5 | 4.79 | 1.96 | 0.01 | 0.0 |
| 10 | 0.0 | 0.2000 | 0.0040 | 0.1900 | 0.0040 | 42.0 | 5.7 | 0.50 | 0.01 | 0.0 |

*Note:* $T_c$ is the time constant of the dummy coil considered to represent transient saliency.
AVR data: $K_A = 25$, $T_A = 0.025$ s, $V_{r\,max} = 10$, and $V_{r\,min} = -10$.

**TABLE D.2**

Line Data

| Bus Number | | | | |
|---|---|---|---|---|
| From | To | $R_L$ | $X_L$ | $B_c$ |
| 37 | 27 | 0.0013 | 0.0173 | 0.3216 |
| 37 | 38 | 0.0007 | 0.0082 | 0.1319 |
| 36 | 24 | 0.0003 | 0.0059 | 0.0680 |
| 36 | 21 | 0.0008 | 0.0135 | 0.2548 |
| 36 | 39 | 0.0016 | 0.0195 | 0.3040 |
| 36 | 37 | 0.0007 | 0.0089 | 0.1342 |
| 35 | 36 | 0.0009 | 0.0094 | 0.1710 |
| 34 | 35 | 0.0018 | 0.0217 | 0.3660 |
| 33 | 34 | 0.0009 | 0.0101 | 0.1723 |
| 28 | 29 | 0.0014 | 0.0151 | 0.2490 |
| 26 | 29 | 0.0057 | 0.0625 | 1.0290 |
| 26 | 28 | 0.0043 | 0.0474 | 0.7802 |
| 26 | 27 | 0.0014 | 0.0147 | 0.2396 |
| 25 | 26 | 0.0032 | 0.0323 | 0.5130 |
| 23 | 24 | 0.0022 | 0.0350 | 0.3610 |
| 22 | 23 | 0.0006 | 0.0096 | 0.1846 |
| 21 | 22 | 0.0008 | 0.0135 | 0.2548 |
| 20 | 33 | 0.0004 | 0.0043 | 0.0729 |
| 20 | 31 | 0.0004 | 0.0043 | 0.0728 |
| 19 | 2 | 0.0010 | 0.0250 | 1.2000 |

*continued*

**TABLE D.2    (continued)**

Line Data

| Bus Number | | | | |
| From | To | $R_L$ | $X_L$ | $B_c$ |
| --- | --- | --- | --- | --- |
| 18 | 19 | 0.0023 | 0.0363 | 0.3804 |
| 17 | 18 | 0.0004 | 0.0046 | 0.0780 |
| 16 | 31 | 0.0007 | 0.0082 | 0.1389 |
| 16 | 17 | 0.0006 | 0.0092 | 0.1130 |
| 15 | 18 | 0.0008 | 0.0112 | 0.1476 |
| 15 | 16 | 0.0002 | 0.0026 | 0.0434 |
| 14 | 34 | 0.0008 | 0.0129 | 0.1382 |
| 14 | 15 | 0.0008 | 0.0128 | 0.1342 |
| 13 | 38 | 0.0011 | 0.0133 | 0.2138 |
| 13 | 14 | 0.0013 | 0.0213 | 0.2214 |
| 12 | 25 | 0.0070 | 0.0086 | 0.1460 |
| 12 | 13 | 0.0013 | 0.0151 | 0.2572 |
| 11 | 12 | 0.0035 | 0.0411 | 0.6987 |
| 11 | 2 | 0.0010 | 0.0250 | 0.7500 |

**TABLE D.3**

Load Flow Data

| Bus Number | $V$ | $\phi$ | $P_G$ | $Q_G$ | $P_L$ | $Q_L$ |
| --- | --- | --- | --- | --- | --- | --- |
| 1 | 0.98200 | 0.00000 | 5.04509 | 1.36036 | 0.09200 | 0.04600 |
| 2 | 1.03000 | −9.55016 | 10.00000 | 1.95746 | 11.04000 | 2.50000 |
| 3 | 0.98310 | 3.20174 | 6.50000 | 1.59104 | 0.00000 | 0.00000 |
| 4 | 1.01230 | 4.61664 | 5.08000 | 1.58151 | 0.00000 | 0.00000 |
| 5 | 0.99720 | 5.57217 | 6.32000 | 0.95582 | 0.00000 | 0.00000 |
| 6 | 1.04930 | 6.62654 | 6.50000 | 2.76414 | 0.00000 | 0.00000 |
| 7 | 1.06350 | 9.46958 | 5.60000 | 2.35485 | 0.00000 | 0.00000 |
| 8 | 1.02780 | 3.16537 | 5.40000 | 0.63019 | 0.00000 | 0.00000 |
| 9 | 1.02650 | 9.04654 | 8.30000 | 0.84790 | 0.00000 | 0.00000 |
| 10 | 1.04750 | −2.47597 | 2.50000 | 1.46483 | 0.00000 | 0.00000 |
| 11 | 1.03829 | −7.79710 | 0.00000 | 0.00000 | 0.00000 | 0.00000 |
| 12 | 1.02310 | −4.89487 | 0.00000 | 0.00000 | 0.00000 | 0.00000 |
| 13 | 0.99576 | −8.07759 | 0.00000 | 0.00000 | 3.22000 | 0.02400 |
| 14 | 0.95894 | −9.35310 | 0.00000 | 0.00000 | 5.00000 | 1.84000 |
| 15 | 0.95660 | −8.29471 | 0.00000 | 0.00000 | 0.00000 | 0.00000 |
| 16 | 0.95688 | −7.56925 | 0.00000 | 0.00000 | 0.00000 | 0.00000 |
| 17 | 0.95140 | −9.97400 | 0.00000 | 0.00000 | 2.33800 | 0.84000 |
| 18 | 0.95276 | −10.5017 | 0.00000 | 0.00000 | 5.22000 | 1.76000 |
| 19 | 1.01028 | −9.92054 | 0.00000 | 0.00000 | 0.00000 | 0.00000 |

*continued*

**TABLE D.3** (continued)

Load Flow Data

| Bus Number | $V$ | $\phi$ | $P_G$ | $Q_G$ | $P_L$ | $Q_L$ |
|---|---|---|---|---|---|---|
| 20 | 0.95988 | -4.71314 | 0.00000 | 0.00000 | 0.00000 | 0.00000 |
| 21 | 0.99046 | -2.98024 | 0.00000 | 0.00000 | 2.74000 | 1.15000 |
| 22 | 1.01550 | 1.62430 | 0.00000 | 0.00000 | 0.00000 | 0.00000 |
| 23 | 1.01344 | 1.34841 | 0.00000 | 0.00000 | 2.74500 | 0.84660 |
| 24 | 0.98179 | -5.45955 | 0.00000 | 0.00000 | 3.08600 | 0.92200 |
| 25 | 1.02088 | -3.68918 | 0.00000 | 0.00000 | 2.24000 | 0.47200 |
| 26 | 1.01822 | -4.76321 | 0.00000 | 0.00000 | 1.39000 | 0.17000 |
| 27 | 1.00150 | -6.92554 | 0.00000 | 0.00000 | 2.81000 | 0.75500 |
| 28 | 1.02204 | -0.95906 | 0.00000 | 0.00000 | 2.06000 | 0.27600 |
| 29 | 1.02143 | 1.95588 | 0.00000 | 0.00000 | 2.83500 | 0.26900 |
| 30 | 0.98832 | -0.62515 | 0.00000 | 0.00000 | 6.28000 | 1.03000 |
| 31 | 0.95760 | -5.69316 | 0.00000 | 0.00000 | 0.00000 | 0.00000 |
| 32 | 0.93795 | -5.68713 | 0.00000 | 0.00000 | 0.07500 | 0.88000 |
| 33 | 0.95912 | -5.47342 | 0.00000 | 0.00000 | 0.00000 | 0.00000 |
| 34 | 0.96168 | -7.20767 | 0.00000 | 0.00000 | 0.00000 | 0.00000 |
| 35 | 0.96683 | -7.32475 | 0.00000 | 0.00000 | 3.20000 | 1.53000 |
| 36 | 0.98196 | -5.55956 | 0.00000 | 0.00000 | 3.29400 | 0.32300 |
| 37 | 0.99086 | -6.73437 | 0.00000 | 0.00000 | 0.00000 | 0.00000 |
| 38 | 0.99197 | -7.71437 | 0.00000 | 0.00000 | 1.58000 | 0.30000 |
| 39 | 0.98770 | 0.34648 | 0.00000 | 0.00000 | 0.00000 | 0.00000 |

**TABLE D.4**

Transformer Data

| Bus Number | | $R_T$ | $X_T$ | Tap |
|---|---|---|---|---|
| From | To | | | |
| 39 | 30 | 0.0007 | 0.0138 | 1.0 |
| 39 | 5 | 0.0007 | 0.0142 | 1.0 |
| 32 | 33 | 0.0016 | 0.0435 | 1.0 |
| 32 | 31 | 0.0016 | 0.0435 | 1.0 |
| 30 | 4 | 0.0009 | 0.0180 | 1.0 |
| 29 | 9 | 0.0008 | 0.0156 | 1.0 |
| 25 | 8 | 0.0006 | 0.0232 | 1.0 |
| 23 | 7 | 0.0005 | 0.0272 | 1.0 |
| 22 | 6 | 0.0000 | 0.0143 | 1.0 |
| 20 | 3 | 0.0000 | 0.0200 | 1.0 |
| 16 | 1 | 0.0000 | 0.0250 | 1.0 |
| 12 | 10 | 0.0000 | 0.0181 | 1.0 |

*Note:* The load flow results are given for the case when transmission losses are neglected.

# Index

**A**

AC, *see* Alternating current (AC)
Adaptive out-of-step relaying scheme, 196
Adaptive system protection, 196, 219
   controlled system separation, 221
   loss of synchronism, 220
   out-of-step relaying, 220
   swing curves, 222–225
$\alpha$ sensitivity, 235; *see also* Trajectory sensitivity
Alternating current (AC), 1
   electrical network, 331
   network equations, 156, 336
   system model, 155
   transmission lines, 1
Analogous network, 100; *see also* Structure preserving energy functions (SPEF)
   individual element energy in network, 103
   for lossless power system, 138
   RCFN, 248
   resistances of, 143
   in steady state, 138
ANN, *see* Artificial neural network (ANN)
Artificial neural network (ANN), 196
ASIL, *see* Asymptotically stable in the large (ASIL)
Asymptotically stable in the large (ASIL), 36, 38
Automatic voltage regulators (AVR), 2
Auto-parametric resonance, 302
AVR, *see* Automatic voltage regulators (AVR)

**B**

Back-to-back (BTB), 150
Bang-bang control, 257, 259
of TCSC, 184
   time-optimal control, 257
Basin of attraction, 28
   Lyapunov function, 38
BCU method, 51
   CUEP determination, 52
   in modified PEBS method, 120
$\beta$ sensitivity, 235–236; *see also* Trajectory sensit
BTB, *see* Back-to-back (BTB)

**C**

CA, *see* Constant angle (CA)
Capacitors, switched series, 257; *see also* FACTS controllers
   bang-bang control, 257
   control action, 257
   time-optimal control, 257–260
CC, *see* Constant current (CC)
CCT, *see* Critical clearing time (CCT)
CEA control, *see* Constant extinction angle control (CEA control)
Center of angle (COA), 41
Center of inertia (COI), 41
Closet UEP, 48; *see also* Unstable equilibrium point (UEP)
COA, *see* Center of angle (COA)
COI, *see* Center of inertia (COI)
Constant angle (CA), 182, 183
Constant current (CC), 162, 183
   load, 22, 62, 84
   source, 98
Constant extinction angle control (CEA control), 154
Constant reactance (CR), 189, 234
Constant voltage drop control (CVD control), 182, 185
Controlled series compensator (CSC), 185
   on bus voltages, 187

Controlled system separation, 283
  and load shedding, 283–284
  as protective measure, 215, 221
  simulation, 222
  swing curves, 224
Controlling unstable equilibrium point
      (CUEP), 18, 48
  characterization of, 328
  critical energy, 68–69, 76, 132
  determination of, 51–52
  generator rotor angles and, 55
  on stability boundary, 53
  TCSC capacitance and, 233
CR, *see* Constant reactance (CR)
Critical clearing time (CCT), 18
  actual, 55
  in classical model of generator, 267
  comparison of, 70, 97, 118, 19
  computation of, 49, 54
  and critical energy, 120
  for different generators, 97
  by digital simulation, 162
  voltage dependence of active power
      characteristics, 69
Critical cutset, 60–61, 103, 196, 197,
      286
  angles in, 198, 217, 218
  for generator models, 215
  identification algorithm, 207
  lines frequency, 220
  two-area system separated by, 286
Critical cutset prediction, 198, 204; *see
      also* Instability detection
  analysis, 198–202
  bus voltage vector, 200
  case study, 203
  energy margin computation, 231
  locus of voltage magnitudes, 201
  theorem, 200–201
Critical energy, 10, 17–18, 118, 119
  critical cutset and, 219
  CUEP and, 132
  determination of, 46–48, 50, 70
  stability region boundary, 44
  transient EM, 229
CSC, *see* Controlled series compensator
      (CSC)
CUEP, *see* Controlling unstable
      equilibrium point (CUEP)

CVD control, *see* Constant voltage drop
      control (CVD control)

**D**

DAE, *see* Differential and algebraic
      equations (DAE)
Damping controller; *see also* Energy
      function; Preventive control;
      FACTS controllers
  control signal synthesis, 242–243
  design, 237
  energy function-based design,
      236–237
  energy stored in inductor, 246
  equivalent circuit with controller,
      241
  example, 243–244
  interarea mode eigenvalues, 253
  linear inductance, 237
  linear IRCFN, 248
  linear network model, 247–249
  linearized system equations, 238
  network, 14
  NLF, 242, 249
  nonlinear inductor, 245
  one-port network, 238
  phase compensation, 253
  power swing, 285
  for preventive control, 227
  as preventive control, 255
  for series FACTS device, 242
  and shunt FACTS device, 245
  10-machine system case study,
      244–245
  Thevenin equivalent circuit, 240
  for UPFC, 251–254
DC, *see* Direct current (DC)
DEEAC, *see* Dynamic extended equal
      area criterion (DEEAC)
Differential and algebraic equations
      (DAE), 8, 331
  as implicit dynamic system, 9
  issues in solution of, 76
  singular surfaces, 9
  study in AC–DC systems, 150
Direct current (DC), 12
Discrete supplementary controls, 256,
      257

DSA, *see* Dynamic security assessment (DSA)
Dynamic extended equal area criterion (DEEAC), 54
Dynamic load, 105
  equations for, 338
  due to HVDC link, 150
Dynamic load models, 122–124; *see also* Structure preserving energy functions (SPEF)
  analogous network, 138
  dynamic load models, 135–136
  energy computation at UEP, 132–134
  energy function analysis, 128–130
  EPs characterization, 130
  equilibrium point computation, 130–132
  induction motor, 124–128
  load bus voltage vs. load, 131
  rotor angle vs. load, 131
  three-bus system, 129, 132
Dynamic security assessment (DSA), 7
  on-line techniques, 195
  purpose of, 256
  sensitivity analysis techniques for, 227

**E**

EAC, *see* Equal area criterion (EAC)
EEAC, *see* Extended equal area criterion (EEAC)
EHV, *see* Extra-high-voltage (EHV)
Electric power systems, 1, 8; *see also* Power system dynamics
  cost reduction, 1
  discrete controls, 256–257
  disturbances and postdisturbance system, 9
  emergency control and system protection, 7–8
  initial application, 1
  monitoring and enhancing, 6–7
  operating state of, 256
  reliability, 1
  swing equations, 23
  transient stability, 256
Electrical equipment, efficient, 1
Electrical torque, 323
  components, 21

developed in motor, 126
on generator rotor, 19
Electromotive forces (EMFs), 25
EM, *see* Energy margin (EM)
Emergency control measures, 283
  controlled system separation, 283
  generator tripping, 284
  load shedding, 283–284
  and system protection, 7
Emergency controller, 161, 162–163
  dynamic and algebraic equations, 162
  technological innovations, 8
EMFs, *see* Electromotive forces (EMFs)
Energy function, 9–10, 29; *see also* Lyapunov function; Transient energy function methods (TEF methods); Structure preserving energy functions (SPEF)
  analysis, 32–34
  analysis of synchronous and voltage stability, 128
  for STATCOM, 180
  center of inertia formulations, 40–43
  using COI formulation, 43–44
  damping controller design based on, 236
  damping controllers, 236
  detection criterion, 214
  for DSA and devise preventive control strategies, 18
  inclusion of transmission losses in, 85
  modified, 67
  for multimachine power systems, 39
  SMIB system analysis, 32–34
  SPM, 60
  for SVC, 169
  synchronism loss and instability mode detection, 14
  for system, 40
  terms of, 67–68
  time derivative of modified, 67
  as topological Lyapunov function, 60
  transfer conductance, 34, 44–46
  transformation of lossy network, 85–86
  transient, 227
  transient stability, 39–40, 150
  transmission line resistance incorporation, 87–90
  due to UPFC, 191, 192

Energy function application, 8, 17
    in conservative systems, 11–14
    issues in, 10
    to power systems, 17
Energy margin (EM), 18, 227, 228–229;
    *see also* Sensitivity analysis
    application to structure preserving
        model, 232–234
    computation based on critical
        cutsets, 231–232
    computation of, 229–231
    gradient system, 230
    PE, 233
    sensitivity, 232, 236
    transient, 229
EP, *see* Equilibrium point (EP)
EPC, *see* Equidistant pulse control (EPC)
Equal area criterion (EAC), 8, 17, 31
    extension of, 18, 221
    nomenclature of, 32
    swing equation, 31
    two-machine system and, 30
Equidistant pulse control (EPC), 152
Equilibrium point (EP), 26–27
    stability, 27
Exit point, 48, 70
Extended equal area criterion (EEAC),
    18
    based on proposition 1, 197
    critical cluster identification, 55
    faulted trajectory approximation, 54
    formulation, 53–54
    initial development of, 221
    machine equations, 53
Extra-high-voltage (EHV), 1

**F**

FACTS, *see* Flexible AC transmission
    system (FACTS)
FACTS controllers, 255; *see also* Damping
    controller
    application, 255
    basic concepts, 256–257
    COI, 286
    control algorithm, 288
    control strategy, 260–264, 278,
        285–289
    critical cutset, 286

    emergency control measures,
        282–284
    further research, 302–303
    line tripping and, 289–292
    MATLAB® optimization toolbox, 294
    multiple UPFC, 298–299
    objectives of, 255, 285
    potential energy in line with, 185–186
    potential energy with CC control,
        188–189
    power flow, 185
    practical implementation, 299–302
    single UPFC, 293–298
    SMIB system with, 264
    SPEF application, 303
    STATCOM discrete control, 269–272
    static synchronous series
        compensator, 184–185
    switched series compensation,
        257–260
    TCSC and SSSC comparative study,
        264–269
    TCSC application, 289
    thyristor-controlled series capacitor,
        182–184
    total system PE, 286
    transient stability improvement,
        280–282
    types of, 181
    UPFC application, 292
    UPFC discrete control, 272–280
Fast Fourier transform (FFT), 146
FC–TCR, *see* Fixed capacitor–thyristor-
    controlled reactor (FC–TCR)
FFT, *see* Fast Fourier transform (FFT)
Fifty-generator system, 74
Fixed capacitor–thyristor-controlled
    reactor (FC–TCR), 163
Flexible AC transmission system
    (FACTS), 8; *see also* FACTS
    controllers
    controllers, 255
    devices, 257
Flux linkage, 320
    current vs., 13
    expressions for stator, 19
    stator and rotor, 321
FNET, *see* Frequency monitoring
    network (FNET)

Frequency monitoring network (FNET), 204

**G**

Gate turn-off (GTO), 176
Generator; *see also* Structure preserving energy functions (SPEF); Synchronous generators; 10-generator system
energy variation and components, 121–122
modeling, 57
multimachine system with, 106
numerical examples, 114
simpler expression for, 112–113
slip, 324
SMIB system, 114–115
structure preserving energy function with, 109
ten-generator, 39-bus New England test system, 115–120
tripping, 284
Generator flux decay, 90; *see also* Structure preserving energy functions (SPEF)
example, 96–97
generator model, 91–92
load model, 92
power flow equations, 92–93
SPEF, 93–96
system model, 90
Global positioning satellite (GPS), 7
GPS, *see* Global positioning satellite (GPS)
Gradient system, 230, 329; *see also* Stability region boundary
PEBS as stability boundary of, 49
properties of, 329
GTO, *see* Gate turn-off (GTO)

**H**

HB, *see* Hopf bifurcation (HB)
HF, *see* Hidden failures (HF)
Hidden failures (HF), 8
High-power semiconductor devices, 257
High-voltage (HV), 1

High-voltage direct current (HVDC), 3; *see also* Power transmission links
controllers, 153
system model, 150
systems and energy functions, 149
Hopf bifurcation (HB), 2, 134
HV, *see* High-voltage (HV)
HVDC, *see* High-voltage direct current (HVDC)

**I**

IEEE test systems, 69; *see also* Structure preserving energy functions (SPEF)
fifty-generator system, 74–76
IEEE transient stability test systems, 70
PE component variations, 72, 73
seventeen-generator system, 70–74
IGBT, *see* Insulated gate bipolar transistor (IGBT)
Incremental RCFN (IRCFN), 248
Individual phase control (IPC), 153
Induction motor model, 124
electrical torque, 126
load dynamics, 23
motor slip, 125
simpler models of, 128
stator equivalent circuit of, 125
steady-state equivalent circuit of, 126
torque slip characteristics, 127
voltage instability, 126
Instability detection, 195, 210, 225–226
adaptive system protection, 219–225
basic concepts, 196–198
case studies, 210–215
critical cutset identification, 207–209
critical cutset prediction, 198–203
instability criterion, 204–207
instability prediction, 209–210
loss of synchronism, 220
by monitoring critical cutset, 203–204
practical system study, 215–219
seventeen-generator IEEE test system, 213–214
ten-generator New England test system, 210–213
transient instability, 215

Insulated gate bipolar transistor (IGBT),
    150
Interline power flow controller (IPFC),
    192
IPC, *see* Individual phase control (IPC)
IPFC, *see* Interline power flow controller
    (IPFC)
IRCFN, *see* Incremental RCFN (IRCFN)


**J**

Jacobian matrix, 9, 26


**K**

KE, *see* Kinetic energy (KE)
Kinetic energy (KE), 14, 33
    at clearing time, 231
    component, 88
Krasovskii's method, 35–36
Kyoto approach, 230


**L**

Lasalle's invariance principle, 29
LCC, *see* Line commutated converters
    (LCC)
Left-hand side (LHS), 81
LFC, *see* Load frequency control (LFC)
LHS, *see* Left-hand side (LHS)
Line commutated converters (LCC),
    150
Linear inductance, 237
Load buses, 61
    network transformation, 83
    voltage, 84, 134
Load dynamics, 22–23
Load frequency control (LFC), 3
Load shedding, 283–284
Load supply capability (LSC), 232
Lossless system, 136
    analogous network for, 138
    network equations, 83
    power angle curve for, 34
    power flow equations, 92
    Tellegen's theorem, 137
LSC, *see* Load supply capability (LSC)
Lyapunov function, 17, 28, 29; *see also*
    Energy function

construction of, 35, 38
for direct stability evaluation, 34
first integrals method, 35
Krasovskii's method, 35–36
for nonlinear pendulum, 29
Popov criterion methods, 36–37
for two-generator system, 46
Lyapunov-like energy functions, 57, 85
    for nonlinear system, 46
Lyapunov stability, 27
    advantages of, 28
    example, 28
    Lasalle's invariance principle, 29
    theorem, 27–28, 34


**M**

Magneto-motive Force (MMF), 42
Mathieu equation, 145
MMF, *see* Magneto-motive Force (MMF)
MOD, *see* Mode of disturbance (MOD)
Mode of disturbance (MOD), 51
Mode of instability (MOI), 44, 74, 197
Modern energy control centers, 6–7
MOI, *see* Mode of instability (MOI)
MTDC, *see* Multiterminal DC (MTDC)
Multimachine system, 48
    connected to nonlinear loads, 62
    connected to uncoupled loads, 79
    with detailed generator models, 106
    dynamic load models, 135–136
    loss of synchronism, 53
    *n*-bus, 62
    stability determination, 48
Multiterminal DC (MTDC), 150, 154


**N**

Network controllers, 57, 331
NLF, *see* Normalized location factor (NLF)
Noniterative solution of networks, 77;
        *see also* Structure preserving
        energy functions (SPEF)
    assumptions, 77–78
    generator dynamic equations, 78–79
    load characteristics transformation,
        84–85
    multimachine system, 79
    network transformation, 83–84

power flow equations, 79–81
quartic equation solutions, 82–83
special cases, 81
system equations, 78
Normalized location factor (NLF), 241, 249
computation, 242

**O**

OLTC, *see* On-load tap changers (OLTC)
OMIB systems, *see* One-machine, infinite-bus systems (OMIB systems)
One-machine, infinite-bus systems (OMIB systems), 17
On-load tap changers (OLTC), 3

**P**

Parametric resonances, 145
Park's transformation, 316; *see also* Synchronous generators
inverse transformation, 317
PE, *see* Potential energy (PE)
PEBS, *see* Potential energy boundary surface (PEBS)
Per-unit damping torque coefficient, 42
Per-unit system, 151, 318; *see also* Synchronous generators
advantages of, 318
rotor mechanical equations, 19, 323
stator base quantities, 318
stator equations, 18
Phasor measurement unit technology (PMU technology), 7
PMU technology, *see* Phasor measurement unit technology (PMU technology)
Potential energy (PE), 18
component, 68, 72, 93, 175
line contribution, 185
in lossless system network, 136
path-dependent term, 233
SEP, 48
for SIME, 33
UEPs, 48
Potential energy boundary surface (PEBS), 18, 48

advantage of, 49, 69
CCT computation, 49
critical clearing time determination, 49
crossing, 230
method, 69, 70
modified PEBS method, 120
theorem, 50
Power flow equations, 64, 108, 109
AC network, 156
at bus *i*, 64, 93
at load bus, 79
for lossless network, 61, 92
during transient, 79
Power swing, 220
angle and voltage instability effect, 134
instability, 133
Power swing damping controller (PSDC), 42, 183, 285
Power system dynamics, 25; *see also* Electric power systems
first-order differential equations, 26
Jacobian matrix, 26
operating states, 4–5
second-order vector differential equation, 25
Power system security, 3, 227, 255; *see also* Electric power systems; FACTS controllers
adaptive system protection, 196
blackout in US, 3–4
DSA, 256
emergency state, 5
in extremis state, 5–6
failures in, 8
loss of synchronism, 196
maintenance, 255
misoperation of, 7
power system operating states, 4–5
preventive control, 195; *see also* Instability detection
restorative state, 6
Power system stability, 2; *see also* Power system dynamics
constant-impedance-type loads, 69
load model effect, 69
region computation, 68

Power system stability *(Continued)*
  synchronous operation of
     generators, 2
  system frequency regulation, 3
  trajectory sensitivities, 236
  transient stability problem, 2
  voltage stability problem, 3
Power system stabilizers (PSS), 3, 42
Power transmission links, 149; *see also*
     Structure preserving energy
     functions (SPEF)
  auxiliary controller, 153, 161
  average DC voltage, 151
  case study and results, 162–163
  converter control model, 152–155
  converter model, 150–152
  current and extinction angle control,
    153
  DC network equations, 152
  emergency controller, 161, 162
  example, 160
  generator model, 155–156
  load model, 156
  Norton's equivalent of converter, 152
  power, 153
  reactive power, 151
  rectifier and inverter control, 155
  single-line diagram, 151, 160
  SPEF, 156–160
Preventive control, 195, 227; *see also*
     Damping controller; Sensitivity
     analysis
  control signal synthesis, 242–243
  damping controller design, 236–237,
    251
  linear network model, 247–249
  linearized system equations, 238–242
  series FACTS controllers, 237
  shunt FACTS controllers, 245–247
  10-machine system case study,
    244–245
Primary controllers, 257
PSDC, *see* Power swing damping
     controller (PSDC)
PSS, *see* Power system stabilizers (PSS)
Pulse width modulation (PWM), 177
  using IGBT devices, 179
PWM, *see* Pulse width modulation
    (PWM)

**R**

RCFN, *see* Reactive current flow
    network (RCFN)
Reactive current flow network (RCFN),
    143, 248
  incremental, 248
  power losses in, 144
Reduced bus admittance matrix, 24
  disadvantage of, 57
Reduced network models (RNM), 17
Remote terminal units (RTU), 6
RHS, *see* Right-hand side (RHS)
Right-hand side (RHS), 25, 139
RNM, *see* Reduced network models
    (RNM)
Rotor mechanical equations, 19, 323–324
RTU, *see* Remote terminal units (RTU)

**S**

Saddle-node bifurcation (SNB), 2, 133
Saliency, 325; *see also* Transient saliency
SCADA, *see* Supervisory control and
    data acquisition systems
    (SCADA)
SCR, *see* Short circuit ratios (SCR)
SE, *see* State estimation (SE)
Security, 255; *see also* Power system
    security
Sensitivity analysis, 227; *see also* Energy
    margin (EM); Preventive
    control
  basic concepts, 228
  DAE, 228
  dynamic security assessment, 229
  energy margin computation, 229–231,
    231–232
  energy margin sensitivity, 232–234
  path-dependent integral evaluation,
    231
  trajectory sensitivity, 235
  transient energy function methods,
    227
  transient energy margin, 229
  types of, 228
SEP, *see* Stable equilibrium point (SEP)
Sequential linear programming (SLP),
    244

Short circuit ratios (SCR), 163
Shunt FACTS devices, 123
   controller, 245
SIC, *see* Simultaneous interchange
      capability (SIC)
SIME, *see* Single-machine equivalent
      (SIME)
Simultaneous interchange capability
      (SIC), 232
Single machine infinite bus system
      (SMIB system), 13
   energy function analysis, 32
   energy function for, 114
   KE, 33
   modeling, 13–14
   PE, 33
   power angle curves for, 34
   with series FACTS controller, 264
   stability criterion, 33
   with STATCOM, 270
   TCSC in, 265
   trajectories in angle space for, 47
   with UPFC, 272
Single-machine equivalent (SIME), 18,
      197
   PE for, 33
SLP, *see* Sequential linear programming
      (SLP)
SM, *see* Swing mode (SM)
SMC, *see* Supplementary modulation
      controller (SMC)
SMIB system, *see* Single machine
      infinite bus system (SMIB
      system)
SNB, *see* Saddle-node bifurcation (SNB)
Sparsity-oriented triangular
      factorization, 339
SPEF, *see* Structure preserving energy
      functions (SPEF)
SPM, *see* Structure preserving model
      (SPM)
SPST, *see* Static phase-shifting
      transformer (SPST)
SSR, *see* Subsynchronous resonance
      (SSR)
SSSC, *see* Static synchronous series
      compensator (SSSC)
Stability region boundary, 50, 327
   assumptions, 327

BCU method, 51
characterization, 328
CUEP characterization, 328
gradient system, 329
nonlinear autonomous dynamical
      system, 327
results, 328
trajectory, 327
Stable equilibrium point (SEP), 9; *see also*
      Unstable equilibrium point
      (UEP)
STATCOM, *see* Static synchronous
      compensator (STATCOM)
State estimation (SE), 6–7
Static phase-shifting transformer
      (SPST), 257
   activation of, 282
   power angle curves, 281
Static synchronous compensator
      (STATCOM), 149, 175; *see also*
      Static var compensator (SVC);
      Structure preserving energy
      functions (SPEF)
   advantages, 176
   control characteristics of, 177
   controller, 178–180
   current controllers of, 179
   detailed model of, 270
   discrete control, 269
   equivalent circuit of, 178
   general, 175
   modeling of, 176–178, 270
   potential energy function for,
      180–181
   schematic of, 177
   simplified model, 270
   SMIB system with, 270
   swing curve, 271
   system trajectory, 271
   type 2 controller for, 272
Static synchronous series compensator
      (SSSC), 149, 181, 184, 263
   advantages, 185
   CCT, 267
   control strategy, 265
   controller performance, 264
   energy variations for, 266, 268
   injected reactive voltage, 187–188
   rotor angles, 266, 268

Static synchronous series compensator
(SSSC) *(Continued)*
SMIB system with, 264
system trajectories, 266, 268
and TCSC comparative study, 264
total energy derivative, 269
transient energy, 267
Static var compensator (SVC), 149, 163,
166; *see also* Static synchronous
compensator (STATCOM);
Structure preserving energy
functions (SPEF)
at bus 29, 292
bus voltage, 167
combined equivalent circuit of, 167
controller characteristics, 164–166
current, 167
equivalent circuit of, 166
example, 171–172
low-pass filter, 165
network solution, 166, 168–169
New England test system case study,
172–175
phase angle calculation, 167–168
potential energy function for, 169–170
TSC–TCR-type, 164
types of, 163
Stator base quantities, 318
Stator equations, 18–19, 319, 322
for generator *i*, 140
in Park's reference frame, 18
in phasor quantities, 331
Stator equivalent circuits, 21, 323
of induction motor, 125
Stator flux linkages, 19
Steady-state stability, 2
Strong resonance, 250
Structure preserving energy functions
(SPEF), 10, 14, 57, 64–68,
93–96, 105; *see also* Generator;
Dynamic load models; Energy
function; Generator flux decay;
IEEE test systems; Noniterative
solution of networks; Voltage-
dependent power loads
advantages of, 10
dynamic load role, 105
energy stored in machine reactances,
113

excitation system model, 107–108
generator model, 106–107
generator power output, 141
generator reactive output, 141
load model, 108
network analogy, 97–104
PE components, 93
potential energy, 140–144
potential energy fast fourier
transform, 146–148
power flow equations, 108–109
RCFN element, 143
reactive power loss, 141
results on, 136–140
static excitation system, 108
stator equations, 140
structure preserving model, 57–61
system equation solution, 76–77
system model, 106
systems with HVDC and FACTS
controllers, 149; *see also* Power
transmission links; FACTS
controllers; Static synchronous
compensator (STATCOM);
Static var compensator (SVC);
Unified power flow controller
(UPFC)
time derivative, 105
for transformed network, 88
transmission losses in energy
function, 85–90
unstable modes and parametric
resonance, 144–146
Structure preserving model (SPM), 8, 57;
*see also* Structure preserving
energy functions (SPEF)
application of, 62, 105
energy function, 60
power flow, 59
for stability analysis, 57
system equations, 58, 59
Subsynchronous resonance (SSR), 181
Supervisory control and data
acquisition systems (SCADA), 6
Supplementary modulation controller
(SMC), 42
SVC, *see* Static var compensator (SVC)
Swing mode (SM), 133
eigenvalues at operating point, 244

10-generator system, 245
  voltage across capacitor for, 240
Synchronized phasor measurements, 7
Synchronous generators, 18, 315; *see also*
      Synchronous machine
  classical model, 20
  connected to infinite bus, 21
  electrical torque, 19, 323
  equivalent circuit for generator
      stator, 21
  expressions for stator flux linkages,
      19–21
  field winding, 20
  generator slip, 324
  model 1. 1 application, 321–322
  Park's transformation, 316–317
  per-unit quantities, 318
  rotor angle, 19
  rotor mechanical equations, 19,
      323–324
  simpler models, 324–325
  stator and rotor flux linkages,
      321
  stator base quantities, 318
  stator equations, 18, 322–323
  stator equivalent circuits, 323
  two-axis model for, 140
  voltage equations, 321
Synchronous machine, 62, 315; *see also*
      Synchronous generators
  basic equations, 315–316
  flux linkage, 320
  initial operating slip, 319
  machine model classification, 320
  model, 318
  network transients, 319
  rotor angle, 316
  rotor equations, 319
  stator equations, 319
System algebraic equations, 336–337
System differential equations,
      337–338
System equation solution, 338
  extrapolation, 338
  partitioned solution, 338–340
  simultaneous solution, 340
  sparsity-oriented triangular
      factorization, 339
  during transient, 76–77

System equations; *see also* Lyapunov
      function
  DAE solution, 76
  linearized, 238
  for systems with nonlinear load
      models, 58
  during transient, 76–77
System security, *see* Power system
      security

**T**

TCBR, *see* Thyristor-controlled braking
      resistor (TCBR)
TCPAR, *see* Thyristor-controlled phase-
      angle regulators (TCPAR)
TCs, *see* Transfer conductances (TCs)
TCSC, *see* Thyristor-controlled series
      capacitor (TCSC)
TEF methods, *see* Transient energy
      function methods (TEF
      methods)
Tellegen's theorem, 101, 137
10-Generator system
  cases of, 117, 118, 119, 211, 212
  cases of faults, 116
  CCT and critical energy, 120
  cutset, 197, 214–215
  line data, 342–343
  load flow data, 343–344
  machine data, 342
  PE and FFT, 147, 148
  regulating slopes for SVCs, 173
  single-line diagram of, 341
  SVC effect on transient stability
      limit, 176
  swing curves, 121, 222
  swing modes, 245
  39-bus New England test system, 115,
      210–213
  transformer data, 344
  UPFC in multimachine power
      system, 299
Thyristor-controlled braking resistor
      (TCBR), 257
Thyristor-controlled phase-angle
      regulators (TCPAR), 257; *see
      also* Static phase-shifting
      transformer (SPST)

Thyristor-controlled series capacitor
        (TCSC), 149, 181, 182; *see also*
        FACTS controllers; Capacitors,
        switched series
    application of, 289–292
    CC or CA controller, 184
    CCT, 267
    control characteristics, 183
    control strategy, 265
    potential energy in line with, 186
    power scheduling control, 182
    power swing damping control, 183
    reactive voltage injected in line, 187
    rotor angles, 266, 268
    SMIB system with, 264
    SSSC comparative study, 264
    system trajectories, 266, 268
    total energy derivative, 269
    transient energy, 267
    transient stability, 183–184, 189
Thyristor-switched capacitor–thyristor-
        controlled reactor (TSC–TCR),
        163
    type SVC, 164
Time-optimal control, 257; *see also*
        Capacitors, switched series
    issues affecting robustness of, 260
    objective of, 258
    SMIB system, 258
    swing equations, 258
    system trajectory, 259
Trajectory sensitivity, 228, 235; *see also*
        Sensitivity analysis
    β sensitivity, 235–236
    differential equation for, 235
Transfer conductances (TCs), 25, 57
    in energy function, 44
    in two-machine system, 34
Transformer, 21, 178; *see also* Variable
        frequency transformer (VFT)
    data, 344
Transient EM, 229
Transient energy function methods
        (TEF methods), 7, 227
Transient saliency, 21, 332
    dependent source, 333–334
    dummy rotor coil, 334–335
    effects in one-axis models, 119
    methods of handling, 333

Transient stability, 256
    alternative measure of, 229
    control, 183, 282
    controllers, 257
    converter model, 150
    corrective actions, 256
    energy, 132
    generator limit, 172
    IEEE test systems, 70
    improvement, 280
    KE, 206
    limits, 189
    power angle curves, 281
    power flow, 276
    problem, 2
    SIME, 197
    SPST generator transformer, 281
    SVCs and, 176
    VFT, 282
Transient stability analysis, 331
    AC network equations, 336
    controller models, 154
    direct methods of, 14
    generator equivalent circuit, 332
    generator stator in network, 331–332
    load equivalent circuit, 336
    load representation, 335–336
    of multimachine AC/DC power
        systems, 150
    network equation solution, 331
    stator equations, 331
    system equations, 336–338
    transient saliency treatment, 332–335
Transient stability evaluations, 17
    BCU method, 51–52
    critical energy determination, 46
    direct, 62
    electrical power expressions, 23–25
    energy functions, 32–34, 39–44
    equal area criterion, 31–32
    equilibrium point, 26–27
    extended equal area criterion, 53–55
    load model, 22–23
    Lyapunov functions, 34
    Lyapunov stability, 27–29
    mathematical preliminaries, 25–26
    multimachine system, 48
    network equations, 21–22
    PE boundary surface, 48–50

single-machine system, 46–48
stability domain estimation, 44
synchronous generators, 18–21
system model, 18
transfer conductances incorporation, 44–46
two-machine system and EAC, 30–31
UEP method control, 50–51
Transient voltage stability, 123
Transmission network, 17, 57
  network equations, 173
  SVC, 163
TSC–TCR, *see* Thyristor-switched capacitor–thyristor-controlled reactor (TSC–TCR)
Two-axis generator model, 106
  electrical power output, 141
  reactive output, 141
  reactive power loss, 141
  stator equations, 140
Two-machine system, 30–31, 260
  control strategy, 260–264
  disturbance, 262
  into equivalent single machine, 260
  PE for SIME for, 33
  power angle curves, 263
  system PE, 261
  transfer conductance, 40

**U**

UEP, *see* Unstable equilibrium point (UEP)
Unified power flow controller (UPFC), 149, 189
  advantage, 192
  connected in line, 191
  constraints equation, 190
  control variables, 297
  controllers for, 251–254
  energy function, 191–192
  equivalent circuit of, 190
  multiple, 298
  phase compensation, 253
  power flow, 191
  single, 293
  in SMIB systems, 292
  as two-port network, 293
  VSC, 192

Unstable equilibrium point (UEP), 18
  classification, 27
  closest, 48
  control, 50
  energy computation at, 132
  in multimachine power system, 48
  PE at, 132
  in SMIB system, 46
  for system without TCSC/SSSC, 267
UPFC, *see* Unified power flow controller (UPFC)
UPFC discrete control, 272
  Kuhn–Tucker conditions, 272
  power angle curves, 276, 277
  power through DC link, 277
  series impedance, 274
  series reactive voltage, 279, 280
  series resistance variation, 274, 275
  shunt active current, 279, 280
  shunt admittance, 274
  shunt conductance variation, 275
  shunt reactive current, 279, 280
  shunt susceptance variation, 276
  SMIB system with, 272, 278
  special case, 273
  swing curve, 279, 280

**V**

Variable frequency transformer (VFT), 282
VDCOL, *see* Voltage-dependent current order limiter (VDCOL)
VDHN method, *see* Very dishonest Newton method (VDHN method)
Very dishonest Newton method (VDHN method), 339–340
VFT, *see* Variable frequency transformer (VFT)
Voltage-dependent current order limiter (VDCOL), 154
Voltage-dependent power loads, 61–62; *see also* Structure preserving energy functions (SPEF)
  energy function, 65
  generator equations, 62–63
  load model, 63–64
  modified energy function, 67

Voltage-dependent power
    loads (*Continued*)
    modified energy function time
        derivative, 67
    multimachine system, 62
    power flow equations, 64
    SPEF, 64–68
    stability region computation, 68–69
    system loadequations, 63
    voltage-dependent load models, 62
Voltage equations, 321
Voltage source converters (VSC), 150
    advanced SVC based on, 175
    configurations of, 192
    six-pulse, 177
    SSSC, 181
Voltage stability
    energy function analysis, 128
    problem, 3

    transient, 123
    UEP associated with, 132
VSC, *see* Voltage source converters
    (VSC)

**W**

WAM, *see* Wide area measurements
    (WAM)
WAMS, *see* Wide area measurement
    system (WAMS)
Weak resonance, 250
Wide area communication
    technologies, 7
Wide area measurements (WAM),
    203, 303
    control strategy deployment, 236
Wide area measurement system
    (WAMS), 195